LONDON MATHEMATICAL SOCIETY

D1353159

Managing Editor: Professor D. Benson,
Department of Mathematics, University of Aberdeen, UK

London Mathematical Society Student Texts 75

# An Introduction to the Theory of Graph Spectra

DRAGOŠ CVETKOVIĆ

*Mathematical Institute, Serbian Academy of Sciences
and Arts, Belgrade*

PETER ROWLINSON

*Department of Computing Science and Mathematics,
University of Stirling, Scotland*

SLOBODAN SIMIĆ

*Mathematical Institute, Serbian Academy of Sciences
and Arts, Belgrade*

CAMBRIDGE
UNIVERSITY PRESS

CAMBRIDGE UNIVERSITY PRESS
Cambridge, New York, Melbourne, Madrid, Cape Town, Singapore, São Paulo, Delhi

Cambridge University Press
The Edinburgh Building, Cambridge CB2 8RU, UK

Published in the United States of America by Cambridge University Press, New York

www.cambridge.org
Information on this title: www.cambridge.org/9780521118392

© D. Cvetković, P. Rowlinson and S. Simić 2010

This publication is in copyright. Subject to statutory exception
and to the provisions of relevant collective licensing agreements,
no reproduction of any part may take place without
the written permission of Cambridge University Press.

First published 2010

Printed in the United Kingdom at the University Press, Cambridge

*A catalogue record for this publication is available from the British Library*

ISBN 978-0-521-11839-2 Hardback
ISBN 978-0-521-13408-8 Paperback

Cambridge University Press has no responsibility for the persistence or
accuracy of URLs for external or third-party Internet websites referred to
in this publication, and does not guarantee that any content on such
websites is, or will remain, accurate or appropriate.

# Contents

# Preface

This book has been written primarily as an introductory text for graduate students interested in algebraic graph theory and related areas. It is also intended to be of use to mathematicians working in graph theory and combinatorics, to chemists who are interested in quantum chemistry, and in part to physicists, computer scientists and electrical engineers using the theory of graph spectra in their work. The book is almost entirely self-contained; only a little familiarity with graph theory and linear algebra is assumed.

In addition to more recent developments, the book includes an up-to-date treatment of most of the topics covered in *Spectra of Graphs* by D. Cvetković, M. Doob and H. Sachs [CvDSa], where spectral graph theory was characterized as follows:

> The theory of graph spectra can, in a way, be considered as an attempt to utilize linear algebra including, in particular, the well-developed theory of matrices, for the purposes of graph theory and its applications. However, that does not mean that the theory of graph spectra can be reduced to the theory of matrices; on the contrary, it has its own characteristic features and specific ways of reasoning fully justifying it to be treated as a theory in its own right.

*Spectra of Graphs* has been out of print for some years; it first appeared in 1980, with a second edition in 1982 and a Russian edition in 1984. The third English edition appeared in 1995, with new material presented in two Appendices and an additional Bibliography of over 300 items. The original edition summarized almost all results related to the theory of graph spectra published before 1978, with a bibliography of 564 items. A review of results in spectral graph theory which appeared mostly between 1978 and 1984 can be found in *Recent Results in the Theory of Graph Spectra* by D. Cvetković, M. Doob, I. Gutman and A. Torgašev [CvDGT]. This second monograph, published in 1988, contains over 700 further references, reflecting the rapid

growth of interest in graph spectra. Today we are witnessing an explosion of
the literature on the topic: there exist several thousand papers in mathematics,
chemistry, physics, computer science and other scientific areas that develop
or use some parts of the theory of graph spectra. Consequently a truly comprehensive text with a complete bibliography is no longer practicable, and we
have concentrated on what we see as the central concepts and the most useful
techniques.

The monograph [CvDSa] has been used for many years both as an introductory text book and as a reference book. Since it is no longer available, we
decided to write a new book which would nowadays be more suitable for both
purposes. In this sense, the book is a replacement for [CvDSa]; but it is not a
substitute because *Spectra of Graphs* will continue to serve as a reference for
more advanced topics not covered here. The content has been influenced by
our previous books from the same publisher, namely *Eigenspaces of Graphs*
[CvRS2] and *Spectral Generalizations of Line Graphs: on Graphs with Least
Eigenvalue* $-2$ [CvRS7]. Nevertheless, very few sections of the present text
are taken from these more specialized sources. For further reading we recom-
mend not only the books mentioned above but also [BroCN], [Big2], [Chu2]
and [GoRo].

The spectra considered here are those of the adjacency matrix, the Laplacian, the normalized Laplacian, the signless Laplacian and the Seidel matrix
of a finite simple graph. In Chapters 2–6, the emphasis is on the adjacency
matrix. In Chapter 1, we introduce the various matrices associated with a
graph, together with the notation and terminology used throughout the book.
We include proofs of the necessary results in matrix theory usually omitted
from a first course on linear algebra, but we assume familiarity with the funda-
mental concepts of graph theory, and with basic results such as the orthogonal
diagonalizability of symmetric matrices with real entries. Chapter 2 is concerned with the effects of constructing new graphs from old, and graph angles
are used in place of walk generating functions to provide streamlined proofs
of some classical results. Chapter 3 deals with the relations between the spectrum and structure of a graph, while Chapter 4 discusses the extent to which
the spectrum can characterize a graph. Chapter 5 explores the relation between
structure and just one eigenvalue, a relation made precise by the relatively
recent notion of a star complement. Chapter 6 is concerned with spectral
techniques used to prove graph-theoretical results which themselves make no
reference to eigenvalues. Chapter 7 is devoted to the Laplacian, the normalized
Laplacian and the signless Laplacian; here the emphasis is on the Laplacian
because the normalized Laplacian is the subject of the monograph *Spectral
Graph Theory* by F. R. K. Chung [Chu2], while the theory of the signless

Laplacian is still in its infancy. In Chapter 8 we discuss sundry topics that did not fit readily into earlier sections of the book, and in Chapter 9 we provide a small selection of applications, mostly outwith mathematics.

The tables in the Appendix provide lists of the various spectra, characteristic polynomials and angles of all connected graphs with up to 5 vertices, together with relevant data for connected graphs with 6 vertices, trees with up to 9 vertices, and cubic graphs with up to 12 vertices. We are indebted to M. Lepović for creating the graph catalogues for Tables A1, A3, A4 and A5, and for computing the data. We are grateful to D. Stevanović for the graph diagrams that appear with these tables: they were produced using *Graphviz* (open source graph visualization software developed by AT&T, www.graphviz.org/), in particular, the programs 'circo' (Tables A1,A3,A5) and 'neato' (Table A4). Table A2 is taken from *Eigenspaces of Graphs*.

Chapters 2, 4 and 9 were drafted by D. Cvetković, Chapters 1, 5 and 6 by P. Rowlinson, and Chapters 3, 7 and 8 by S. Simić. However, each of the authors added contributions to all of the chapters, which were then re-written in an effort to refine the text and unify the material. Hence all three authors are collectively responsible for the book. We have endeavoured to find a style that is concise enough to enable the extensive material to be treated in a book of limited size, yet intuitive enough to make the book readily accessible to the intended readership. The choice of consistent notation was a challenge because of conflicts in the 'standard' notation for several of the topics covered; accordingly we hope that readers will understand if their preferred notation has not been used. The proofs of some straightforward results in the text are relegated to the exercises. These appear at the end of the relevant chapter, along with notes which serve as a guide to a bibliography of over 500 selected items.

D. CVETKOVIĆ
P. ROWLINSON
S. SIMIĆ

# 1

# Introduction

In Section 1.1 we define various types of graph spectra, and in Section 1.2 we introduce graph-theoretic notation and terminology which will be used throughout the book. In Section 1.3 we establish the results from matrix theory that will be required.

## 1.1 Graph spectra

Let $G$ be a finite undirected graph without loops or multiple edges, and suppose that its vertices are labelled $1, 2, \ldots, n$. If vertices $i$ and $j$ are joined by an edge, we say that $i$ and $j$ are *adjacent* and write $i \sim j$. We consider first the spectrum of the $(0, 1)$-adjacency matrix $A$ of $G$ defined as follows: $A = A(G) = (a_{ij})$ where

$$a_{ij} = \begin{cases} 1 \text{ if } i \sim j \\ 0 \text{ otherwise.} \end{cases}$$

Thus $A$ is a symmetric matrix with zero diagonal; its entries may be taken as 0 and 1 in any field, but throughout this book the entries are treated as real numbers. An example of a graph and its adjacency matrix is given in Fig. 1.1.

The eigenvalues of $A$ are the $n$ roots of the characteristic polynomial $\det(xI - A)$, and so they are algebraic integers. They are independent of the labelling of the vertices of $G$ because similar matrices have the same characteristic polynomial: if the labels are permuted we obtain a $(0, 1)$-adjacency matrix $A' = P^{-1}AP$ where $P$ is a permutation matrix. Accordingly we speak of the *characteristic polynomial of* $G$, denoted by $P_G(x)$, and the *spectrum of* $G$, which consists of the $n$ eigenvalues of $G$. Since $A$ is a symmetric matrix with real entries, these eigenvalues are real. We usually denote them by $\lambda_1, \lambda_2, \ldots, \lambda_n$, and unless we indicate otherwise, we shall assume that

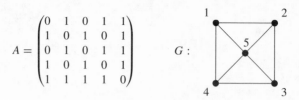

$$A = \begin{pmatrix} 0 & 1 & 0 & 1 & 1 \\ 1 & 0 & 1 & 0 & 1 \\ 0 & 1 & 0 & 1 & 1 \\ 1 & 0 & 1 & 0 & 1 \\ 1 & 1 & 1 & 1 & 0 \end{pmatrix} \qquad G:$$

Figure 1.1  A labelled graph $G$ and its adjacency matrix $A$.

$\lambda_1 \geq \lambda_2 \geq \cdots \geq \lambda_n$. Where necessary, we use the notation $\lambda_i = \lambda_i(G)$ ($i = 1, 2, \ldots, n$). The largest eigenvalue $\lambda_1(G)$ is called the *index* of $G$. For an integer $k \geq 0$, the $k$-th *spectral moment* of $G$ is $\sum_{i=1}^n \lambda_i^k$, denoted by $s_k$. Note that $s_k$ is the trace of $A^k$, and that the first $n$ spectral moments determine the spectrum of $G$.

The eigenvalues of $A$ are the real numbers $\lambda$ satisfying $A\mathbf{x} = \lambda\mathbf{x}$ for some non-zero vector $\mathbf{x} \in \mathbb{R}^n$. Each such vector $\mathbf{x}$ is called an *eigenvector* of the matrix $A$ (or of the labelled graph $G$) corresponding to the eigenvalue $\lambda$. The relation $A\mathbf{x} = \lambda\mathbf{x}$ can be interpreted in the following way: if $\mathbf{x} = (x_1, x_2, \ldots, x_n)^\top$ then

$$\lambda x_u = \sum_{v \sim u} x_v \quad (u = 1, 2, \ldots, n), \tag{1.1}$$

where the summation is over all neighbours $v$ of the vertex $u$. We note two straightforward consequences of these equations, which are called the *eigenvalue equations* for $G$.

**Proposition 1.1.1.** *If the graph $G$ has maximum degree $\Delta(G)$ then $|\lambda| \leq \Delta(G)$ for every eigenvalue $\lambda$ of $G$.*

**Proof.** With the notation above, let $u$ be a vertex for which $|x_u|$ is maximal. Using Equation (1.1), we have:

$$|\lambda||x_u| \leq \sum_{v \sim u} |x_v| \leq |\Delta(G)||x_u|.$$

Since $x_u \neq 0$, the result follows.                                                   □

The second observation is left as an exercise for the reader.

**Proposition 1.1.2.** *The graph $G$ is regular (of degree $r$) if and only if the all-1 vector is an eigenvector of $G$ (with corresponding eigenvalue $r$).*

If $\lambda$ is an eigenvalue of $A$ then the set $\{\mathbf{x} \in \mathbb{R}^n : A\mathbf{x} = \lambda\mathbf{x}\}$ is a subspace of $\mathbb{R}^n$, called the *eigenspace* of $\lambda$ and denoted by $\mathcal{E}(\lambda)$ or $\mathcal{E}_A(\lambda)$. Such eigenspaces are called *eigenspaces of $G$*. Of course, relabelling the vertices of

$G$ will result in a permutation of coordinates in eigenvectors (and eigenspaces). Since $A$ is symmetric with real entries, it can be diagonalized by an orthogonal matrix. Hence the eigenspaces are pairwise orthogonal; and by stringing together orthonormal bases of the eigenspaces we obtain an orthonormal basis of $\mathbb{R}^n$ consisting of eigenvectors (cf. Section 1.3). Moreover, the dimension of $\mathcal{E}_A(\lambda)$ is equal to the multiplicity of $\lambda$ as a root of $P_G(x)$. In other words, the geometric multiplicity of $\lambda$ is the same as the algebraic multiplicity of $\lambda$; accordingly we refer only to the *multiplicity* of $\lambda$. A *simple* eigenvalue is an eigenvalue of multiplicity 1. If $G$ has distinct eigenvalues $\mu_1, \mu_2, \ldots, \mu_m$ with multiplicities $k_1, k_2, \ldots, k_m$ respectively, we shall write $\mu_1^{k_1}, \mu_2^{k_2}, \ldots, \mu_m^{k_m}$ for the spectrum of $G$. (We often omit those $K_i$ equal to 1.)

**Example 1.1.3.** For the graph $G$ in Fig. 1.1 we have

$$P_G(x) = \begin{vmatrix} x & -1 & 0 & -1 & -1 \\ -1 & x & -1 & 0 & -1 \\ 0 & -1 & x & -1 & -1 \\ -1 & 0 & -1 & x & -1 \\ -1 & -1 & -1 & -1 & x \end{vmatrix}$$

$$= x^5 - 8x^3 - 8x^2 = x^2(x+2)(x^2 - 2x - 4).$$

The eigenvalues in non-increasing order are $\lambda_1 = 1 + \sqrt{5}$, $\lambda_2 = 0$, $\lambda_3 = 0$, $\lambda_4 = 1 - \sqrt{5}$, $\lambda_5 = -2$, with linearly independent eigenvectors $\mathbf{x}_1$, $\mathbf{x}_2$, $\mathbf{x}_3$, $\mathbf{x}_4$ and $\mathbf{x}_5$, where $\mathbf{x}_1 = (1, 1, 1, 1, -1 + \sqrt{5})^\top$, $\mathbf{x}_2 = (0, 1, 0, -1, 0)^\top$, $\mathbf{x}_3 = (1, 0, -1, 0, 0)^\top$, $\mathbf{x}_4 = (1, 1, 1, 1, -1 - \sqrt{5})^\top$ and $\mathbf{x}_5 = (1, -1, 1, -1, 0)^\top$.

We have $\mathcal{E}(1 + \sqrt{5}) = \langle \mathbf{x}_1 \rangle$, $\mathcal{E}(0) = \langle \mathbf{x}_2, \mathbf{x}_3 \rangle$, $\mathcal{E}(1 - \sqrt{5}) = \langle \mathbf{x}_4 \rangle$ and $\mathcal{E}(-2) = \langle \mathbf{x}_5 \rangle$, where angle brackets denote the subspace spanned by the enclosed vectors. □

**Example 1.1.4.** The eigenvalues of an $n$-cycle are $2\cos\frac{2\pi j}{n}$ ($j = 0, 1, \ldots, n - 1$). One way to see this is to observe that an adjacency matrix has the form $A = P + P^{-1}$ where $P$ is the permutation matrix determined by a cyclic permutation of length $n$. If $\omega$ is an $n$-th root of unity then $(1, \omega, \omega^2, \ldots, \omega^{n-1})^\top$ is an eigenvector of $P$ with corresponding eigenvalue $\omega$. Hence the eigenvalues of $A$ are the numbers $\omega + \omega^{-1}$, where $\omega^n = 1$. Thus the largest eigenvalue is 2 (with multiplicity 1) and the second largest is $2\cos\frac{2\pi}{n}$ (with multiplicity 2). The least eigenvalue is $-2$ (with multiplicity 1) if $n$ is even, and $2\cos\frac{(n-1)\pi}{n}$ (with multiplicity 2) if $n$ is odd. □

**Example 1.1.5.** The well-known Petersen graph (Fig. 1.2) has spectrum $3^1, 1^5, (-2)^4$. □

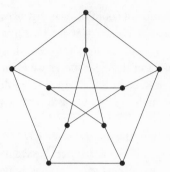

Figure 1.2 The Petersen graph.

Figure 1.3 Two pairs of non-isomorphic cospectral graphs.

We say that two graphs are *cospectral* if they have the same spectrum; clearly, isomorphic graphs are cospectral (in other words, the spectrum is a graph invariant). However, cospectral graphs are not necessarily isomorphic: the non-isomorphic graphs shown in Fig. 1.3(a) share the spectrum $2^1, 0^3, (-2)^1$. This is an example with fewest vertices. Fig. 1.3(b) shows non-isomorphic cospectral *connected* graphs with fewest vertices: their common characteristic polynomial is $(x-1)(x+1)^2(x^3 - x^2 - 5x + 1)$. Various graphs which *are* characterized by their spectrum, or by their spectrum together with related algebraic invariants, are discussed in Chapter 4.

Symmetric matrices other than the $(0, 1)$-adjacency matrix $A$ can be used to specify a graph, and we mention next the spectra of those that feature in this book. For a graph $G$ with vertex set $\{1, \ldots, n\}$, let $D$ be the diagonal matrix $\text{diag}(d_1, \ldots, d_n)$, where $d_i$ denotes the degree of vertex $i$ ($i = 1, \ldots, n$). The *Laplacian matrix* of a graph $G$ is the matrix $D - A$, and the *signless Laplacian* is the matrix $D + A$; their spectra are discussed in Chapter 7. The *Seidel matrix* of $G$ is the matrix $S = J - I - 2A$, where $J$ denotes the all-1 matrix (of size $n \times n$); thus the $(i, j)$-entry of $S$ is 0 if $i = j$, $-1$ if $i \sim j$, and 1 otherwise. As far as regular graphs are concerned, there is little to choose between these matrices from the spectral point of view, for suppose that $G$ is regular of degree $r$, and that $A$ has eigenvalues $\lambda_1, \lambda_2, \ldots, \lambda_n$ in non-increasing order. By Propositions 1.1.1 and 1.1.2, $\lambda_1 = r$ and the all-1 vector may be extended to an orthogonal basis of $I\!R^n$ consisting of eigenvectors common to the matrices $A, rI \pm A$ and $J - I - 2A$. Then we find that $D \pm A$ has eigenvalues

$$r \pm r, \ r \pm \lambda_2, \ \ldots, \ r \pm \lambda_n,$$

while $S$ has eigenvalues

$$n - 1 - 2r, \ -1 - 2\lambda_2, \ \ldots, \ -1 - 2\lambda_n.$$

Similar remarks apply to the generalized adjacency matrix $yJ - A$ discussed in [DamHK]. For non-regular graphs, there is no simple relation between the various spectra; Theorem 1.3.15 will provide some inequalities, but meanwhile we give an explicit example.

**Example 1.1.6.** For the graph in Fig. 1.1, the eigenvalues of the Laplacian are $5, 5, 3, 3, 0$; the eigenvalues of the signless Laplacian are $\frac{1}{2}(9 + \sqrt{17}), 3, 3, \frac{1}{2}(9 - \sqrt{17}), 1$; and the Seidel eigenvalues are $3, \frac{1}{2}(-1 + \sqrt{17}), -1, -1, \frac{1}{2}(-1 - \sqrt{17})$. □

The Seidel matrix is of particular relevance to *graph switching* (often called *Seidel switching*): given a subset $U$ of vertices of the graph $G$, the graph $G_U$ obtained from $G$ by switching with respect to $U$ differs from $G$ as follows. For $u \in U, v \notin U$ the vertices $u, v$ are adjacent in $G_U$ if and only if they are non-adjacent in $G$. Suppose that $G$ has adjacency matrix $A(G) = \begin{pmatrix} A_U & B^\top \\ B & C \end{pmatrix}$, where $A_U$ is the adjacency matrix of the subgraph induced by $U$, and $B^\top$ denotes the transpose of $B$. Then $G_U$ has adjacency matrix $A(G_U) = \begin{pmatrix} A_U & \overline{B}^\top \\ \overline{B} & C \end{pmatrix}$, where $\overline{B}$ is obtained from $B$ by interchanging 0 and 1. When $G$ is regular, this formulation makes it straightforward (Exercise 1.3) to find a necessary and sufficient condition on $U$ for $G_U$ to be regular of the same degree:

**Proposition 1.1.7.** *Suppose that $G$ is regular with $n$ vertices and degree $r$. Then $G_U$ is regular of degree $r$ if and only if $U$ induces a regular subgraph of degree $k$, where $|U| = n - 2(r - k)$.*

Note that switching with respect to the subset $U$ of the vertex-set is the same as switching with respect to its complement. Switching is described easily in terms of the Seidel matrix $S$ of $G$: the Seidel matrix of $G_U$ is $T^{-1}ST$ where $T$ is the (involutory) diagonal matrix whose $i$-th diagonal entry is 1 if $i \in U, -1$ if $i \notin U$. Now it is easy to see that switching with respect to $U$ and then with respect to $V$ is the same as switching with respect to $(U \setminus V) \dot{\cup} (V \setminus U)$; it follows that switching determines an equivalence relation on graphs. Note that switching-equivalent graphs have similar Seidel matrices and hence the same Seidel spectrum. In view of the relation between spectrum and Seidel spectrum for regular graphs, we have the following consequence:

**Proposition 1.1.8.** *If $G$ and $G_U$ are regular of the same degree, then $G$ and $G_U$ are cospectral.*

## 1.2 Some more graph-theoretic notions

As usual, $K_n, C_n$ and $P_n$ denote respectively the *complete graph*, the *cycle* and the *path* on $n$ vertices. A connected graph with $n$ vertices is said to be *unicyclic* if it has $n$ edges, for then it contains a unique cycle. If this cycle has odd length, then the graph is said to be *odd-unicyclic*. A connected graph with $n$ vertices and $n + 1$ edges is called a *bicyclic* graph. The *girth* of a graph $G$ is the length of a shortest cycle in $G$. A complete subgraph of $G$ is called a *clique* of $G$, while a *coclique* is an induced subgraph without edges. The *complete bipartite* graph with parts of size $m$ and $n$ is denoted by $K_{m,n}$. A graph of the form $K_{1,n}$ is called an *n-claw* or a *star*. (The term 'star' is used in different contexts in Sections 3.4 and 5.1.) More generally, $K_{n_1,n_2,\ldots,n_k}$ denotes the *complete k-partite graph* with parts (colour classes) of size $n_1, n_2, \ldots, n_k$. The *m*-dimensional *hypercube* is denoted by $Q_m$; its vertices are the $2^m$ *m*-tuples of 0s and 1s, and two such *m*-tuples are adjacent if and only if they differ in just one place.

Vertices, or edges, are said to be *independent* if they are pairwise non-adjacent. In the literature, a set of independent vertices is often referred to as a *stable* set. Any set of independent edges in a graph $G$ is called a *matching* of $G$. A matching of $G$ is *perfect* if each vertex of $G$ is the endvertex of an edge from the matching; perfect matchings are also called 1-*factors*. The *cocktail party graph* $CP(n)$ is the unique regular graph with $2n$ vertices of degree $2n - 2$; it is obtained from $K_{2n}$ by deleting a perfect matching. The degree of a vertex $v$ is denoted by $\deg(v)$ or $d_v$. The least degree in $G$ is denoted by $\delta(G)$, the largest by $\Delta(G)$. An edge that contains a vertex of degree 1 is called a *pendant* edge.

A regular graph of degree $r$ is said to be *r-regular*, and a 3-regular graph is called a *cubic* graph. A *strongly regular* graph, with parameters $(n, r, e, f)$, is an $r$-regular graph with $n$ vertices $(0 < r < n - 1)$ such that any two adjacent vertices have $e$ common neighbours and any two non-adjacent vertices have $f$ common neighbours. For example, the Petersen graph (Fig. 1.2) is strongly regular with parameters $(10, 3, 0, 1)$. The restriction $0 < r < n - 1$ simply excludes the complete graphs and their complements.

A graph is called *semi-regular bipartite*, with parameters $(n_1, n_2, r_1, r_2)$, if it is bipartite (i.e. 2-colourable) and vertices in the same colour class have the same degree ($n_1$ vertices of degree $r_1$ and $n_2$ vertices of degree $r_2$, where $n_1 r_1 = n_2 r_2$).

If $\mathcal{B}$ is a collection of subsets of the set $S$ then the *incidence graph* determined by $\mathcal{B}$ and $S$ is the bipartite graph $G_\mathcal{B}$ with vertex set $\mathcal{B} \mathbin{\dot\cup} S$, and with an edge between $x \in S$ and $B \in \mathcal{B}$ whenever $x \in B$. Thus if $\mathcal{B}$ is a design with $v$ points and $b$ blocks, in which each block has $k$ points and each point lies in $r$ blocks, then $G_\mathcal{B}$ is a semi-regular bipartite graph with parameters $(v, b, r, k)$. In this case, we call $G_\mathcal{B}$ the graph of the design. Recall that in a *t*-design with parameters $(v, k, \lambda)$, any $t$ points lie in exactly $\lambda$ blocks; and a *symmetric* design is a 2-design for which $b = v > k$ (equivalently, $r = k < v$).

The *complement* of a graph $G$ is denoted by $\overline{G}$, while $mG$ denotes the graph consisting of $m$ disjoint copies of $G$. The *subdivision graph* $S(G)$ is obtained from $G$ by inserting a vertex of degree 2 in each edge of $G$.

We write $V(G)$ for the vertex set of $G$, and $E(G)$ for the edge set of $G$. We say that $G$ is *empty* if $V(G) = \emptyset$, *trivial* if $|V(G)| = 1$, and *null* if $E(G) = \emptyset$. A subgraph $H$ with $V(H) = V(G)$ is called a *spanning* subgraph of $G$. A spanning cycle is called a *Hamiltonian cycle*, and a graph with such a cycle is said to be *Hamiltonian*.

An *automorphism* of $G$ is a permutation $\pi$ of $V(G)$ such that $u \sim v$ if and only if $\pi(u) \sim \pi(v)$. Clearly, the automorphisms of $G$ form a group (with respect to composition of functions). We say that $G$ is *vertex-transitive* if, for any $u, v \in V(G)$, there exists an automorphism $\pi$ of $G$ such that $\pi(u) = v$.

The *union* of disjoint copies of the graphs $G$ and $H$ is denoted by $G \mathbin{\dot\cup} H$. The *join* $G \triangledown H$ of (disjoint) graphs $G$ and $H$ is the graph obtained from $G \mathbin{\dot\cup} H$ by joining each vertex of $G$ to each vertex of $H$. The graph $K_1 \triangledown H$ is called the *cone* over $H$, while $K_2 \triangledown H$ ($= K_1 \triangledown (K_1 \triangledown H)$) is called the *double cone* over $H$. The graph $K_1 \triangledown C_n$ ($n \geq 3$) is the *wheel* $W_{n+1}$ with $n + 1$ vertices; thus the graph of Example 1.1.3 is the wheel $W_5$.

If $uv$ is an edge of $G$ we write $G - uv$ for the graph obtained from $G$ by deleting $uv$. More generally, if $E$ is a set of edges of $G$ we write $G - E$ for the graph obtained from $G$ by deleting the edges in $E$. For $v \in V(G)$, $G - v$ denotes the graph obtained from $G$ by deleting the vertex $v$ and all edges incident with $v$. For $U \subseteq V(G)$, $G - U$ denotes the subgraph of $G$ induced by $V(G) \setminus U$. If each vertex of $G - U$ is adjacent to a vertex of $U$ then $U$ is called a *dominating set* in $G$.

If $u, v$ are vertices of a connected graph $G$ then the *distance* between $u$ and $v$, denoted by $d(u, v)$, is the length of a shortest $u$-$v$ path in $G$.

**Definition 1.2.1.** The line graph $L(H)$ of a graph $H$ is the graph whose vertices are the edges of $H$, with two vertices in $L(H)$ adjacent whenever the corresponding edges in $H$ have exactly one vertex in common.

If $G = L(H)$ for some graph $H$, then $H$ is called a *root graph* of $G$. If $E(H) = \emptyset$ then $G$ is the empty graph. Accordingly, we take a line graph to mean a graph of the form $L(H)$, where $E(H)$ is non-empty; note that we may assume if necessary that $H$ has no isolated vertices. If $H$ is connected, then the same is true of $L(H)$. If $H$ is disconnected, then each non-trivial component of $H$ gives rise to a connected component of $L(H)$.

We mention a simple, but useful, observation (Exercise 1.10):

**Proposition 1.2.2.** *If $H$ is a connected graph and $L(H)$ is regular, then $H$ is either regular or semi-regular bipartite.*

The *incidence matrix* of the graph $H$ is a matrix $B$ whose rows and columns are indexed by the vertices and edges of $H$, respectively. The $(v, e)$-entry of $B$ is

$$b_{ve} = \begin{cases} 0 & \text{if } v \text{ is not incident with } e, \\ 1 & \text{if } v \text{ is incident with } e. \end{cases}$$

Thus the columns of $B$ are the characteristic vectors of the edges of $H$ as subsets of $V(H)$. Now we find easily that

$$B^\top B = A(L(H)) + 2I. \tag{1.2}$$

If $A(L(H))\mathbf{x} = \lambda\mathbf{x}$ then $(\lambda + 2)\mathbf{x}^\top\mathbf{x} = \mathbf{x}^\top B^\top B\mathbf{x} \geq 0$. Thus every eigenvalue of $L(H)$ is greater than or equal to $-2$; this is a notable spectral property of line graphs.

The class of graphs with spectrum in the interval $[-2, \infty)$ also contains the *generalized line graphs*, defined as follows. First we say that a *petal* is added to a graph when we add a pendant edge and then duplicate this edge to form a pendant 2-cycle. A *blossom* $B_k$ consists of $k$ petals ($k \geq 0$) attached at a single vertex; thus $B_0$ is just the trivial graph. A graph with blossoms (possibly empty) at each vertex is called a *B-graph*. Now we extend Definition 1.2.1 to the line graph of a *B*-graph $\hat{H}$: vertices in $L(\hat{H})$ are adjacent if and only if the corresponding edges in $\hat{H}$ have exactly one vertex in common. In particular, duplicate edges between two vertices of $\hat{H}$ are non-adjacent in $L(\hat{H})$; thus $L(B_k) = CP(k)$. If $G = L(\hat{H})$ then we call the multigraph $\hat{H}$ a *root graph* of $G$.

**Definition 1.2.3.** Let $H$ be a graph with vertex set $\{v_1, \ldots, v_n\}$, and let $a_1, \ldots, a_n$ be non-negative integers. The generalized line graph $G =$

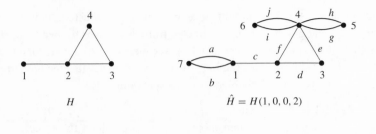

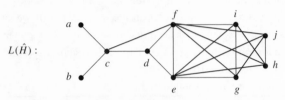

Figure 1.4 Construction of a generalized line graph.

$L(H; a_1, \ldots, a_n)$ is the graph $L(\hat{H})$, where $\hat{H}$ is the $B$-graph $H(a_1, \ldots, a_n)$ obtained from $H$ by adding $a_i$ petals at vertex $v_i$ ($i = 1, \ldots, n$). If not all $a_i$ are zero, $G$ is called a proper generalized line graph.

This construction of a generalized line graph is illustrated in Fig. 1.4.

An incidence matrix $C = (c_{ve})$ of $\hat{H} = H(a_1, \ldots, a_n)$ is defined as for $H$ with the following exception: if $e$ and $f$ are the edges between $v$ and $w$ in a petal at $v$ then $\{c_{we}, c_{wf}\} = \{-1, 1\}$. (Note that all other entries in row $w$ are zero.) For example, an incidence matrix of the multigraph $\hat{H}$ from Fig. 1.4 is:

$$\begin{pmatrix} 1 & 1 & 1 & 0 & 0 & 0 & 0 & 0 & 0 & 0 \\ 0 & 0 & 1 & 1 & 0 & 1 & 0 & 0 & 0 & 0 \\ 0 & 0 & 0 & 1 & 1 & 0 & 0 & 0 & 0 & 0 \\ 0 & 0 & 0 & 0 & 1 & 1 & 1 & 1 & 1 & 1 \\ 0 & 0 & 0 & 0 & 0 & 0 & -1 & 1 & 0 & 0 \\ 0 & 0 & 0 & 0 & 0 & 0 & 0 & 0 & -1 & 1 \\ -1 & 1 & 0 & 0 & 0 & 0 & 0 & 0 & 0 & 0 \end{pmatrix}.$$

Here the rows are indexed by $1, 2, \ldots, 7$ and the columns are indexed by $a, b, \ldots, j$.

With the incidence matrix $C$ defined above, we have $A(L(\hat{H})) = C^\top C - 2I$ and so $\lambda(L(\hat{H})) \geq -2$. Note that the least eigenvalue is strictly greater than $-2$ if and only if the rank of the matrix $C$ is $|V(\hat{H})|$. Not all connected graphs $G$ with $\lambda(G) \geq -2$ are generalized line graphs; however there are only finitely many exceptions, and they are discussed in Section 3.4.

We conclude this section with several examples to illustrate how various strongly regular graphs can be constructed from line graphs by switching. The relation between the eigenvalues and the parameters of a strongly regular graph will be discussed in Section 3.6. In particular, we shall see that the property of strong regularity can be identified from the spectrum.

**Examples 1.2.4.** If we switch the graph $L(K_{4,4})$ with respect to four independent vertices, then we obtain another 6-regular graph on 16 vertices, called the *Shrikhande* graph; it is strongly regular with parameters $(16, 6, 2, 2)$. By Proposition 1.1.8, this graph is cospectral with $L(K_{4,4})$. If we switch $L(K_{4,4})$ with respect to the vertices of an induced subgraph $L(K_{4,2})$ then we obtain a 10-regular graph with 16 vertices, called the *Clebsch* graph; it is strongly regular with parameters $(16, 10, 6, 6)$.

These graphs are represented in Fig. 1.5. In Fig. 1.5(a), the vertices of $L(K_{4,4})$ are shown as the points of intersection of four horizontal and four vertical lines, two vertices being adjacent in $L(K_{4,4})$ if and only if the corresponding points are collinear. In Figs. 1.5(b) and 1.5(c), the white vertices are those in switching sets which yield the Shrikhande and Clebsch graphs, respectively.                                                                          □

**Example 1.2.5.** If we switch a graph $G$ with respect to the set of neighbours of a vertex $v$, we obtain a graph $H$ in which $v$ is an isolated vertex. If $G = L(K_8)$ then $H - v$ is a 16-regular graph on 27 vertices which is called the *Schläfli* graph $Sch_{16}$; it is strongly regular with parameters $(27, 16, 10, 8)$.          □

**Example 1.2.6.** Let $S_1, S_2, S_3$ be sets of vertices of $L(K_8)$ which induce subgraphs isomorphic to $4K_1$, $C_5 \dot\cup C_3$ and $C_8$, respectively. The graphs $Ch_1, Ch_2, Ch_3$ obtained from $L(K_8)$ by switching with respect to $S_1, S_2, S_3$ respectively are called the *Chang graphs*. The graphs $L(K_8), Ch_1, Ch_2, Ch_3$ are regular of degree 12, and hence cospectral by Proposition 1.1.8. They are pairwise non-isomorphic, and strongly regular with parameters $(28, 12, 6, 4)$.                                                                          □

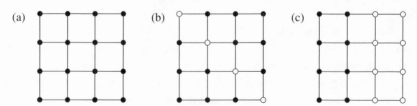

Figure 1.5 Construction of the graphs in Example 1.2.4.

## 1.3 Some results from linear algebra

First we note that a graph is determined by eigenvalues and corresponding eigenvectors in the following way. Let $A$ be the adjacency matrix of a graph $G$ with vertices $1, 2, \ldots, n$ and eigenvalues $\lambda_1 \geq \lambda_2 \geq \cdots \geq \lambda_n$. If $\mathbf{x}_1, \mathbf{x}_2, \ldots, \mathbf{x}_n$ are linearly independent eigenvectors of $A$ corresponding to $\lambda_1, \lambda_2, \ldots, \lambda_n$ respectively, if $X = (\mathbf{x}_1 | \mathbf{x}_2 | \cdots | \mathbf{x}_n)$ and if $E = \mathrm{diag}(\lambda_1, \lambda_2, \ldots, \lambda_n)$, then $AX = XE$ and so

$$A = XEX^{-1}.$$

Since $G$ is determined by $A$, we have the following elementary result:

**Theorem 1.3.1.** *Any graph is determined by its eigenvalues and a basis of corresponding eigenvectors.*

Since $A$ is a symmetric matrix with real entries there exists an orthogonal matrix $U$ such that $U^\top A U = E$. Here the columns of $U$ are eigenvectors which form an orthonormal basis of $\mathbb{R}^n$. If this basis is constructed by stringing together orthonormal bases of the eigenspaces of $A$ then $E = \mu_1 E_1 + \cdots + \mu_m E_m$, where $\mu_1, \ldots, \mu_m$ are the distinct eigenvalues of $A$ and each $E_i$ has block diagonal form $\mathrm{diag}(O, \ldots, O, I, O, \ldots O)$ $(i = 1, \ldots, m)$. Then $A$ has the *spectral decomposition*

$$A = \mu_1 P_1 + \cdots + \mu_m P_m \tag{1.3}$$

where $P_i = U E_i U^\top$ $(i = 1, \ldots, m)$. For fixed $i$, if $\mathcal{E}(\mu_i)$ has $\{\mathbf{x}_1, \ldots, \mathbf{x}_d\}$ as an orthonormal basis then

$$P_i = \mathbf{x}_1 \mathbf{x}_1^\top + \cdots + \mathbf{x}_d \mathbf{x}_d^\top \tag{1.4}$$

and $P_i$ represents the orthogonal projection of $\mathbb{R}^n$ onto $\mathcal{E}(\mu_i)$ with respect to the standard orthonormal basis $\{\mathbf{e}_1, \ldots, \mathbf{e}_n\}$ of $\mathbb{R}^n$. Moreover, $\sum_{i=1}^m P_i = I$, $P_i^2 = P_i = P_i^\top$ $(i = 1, \ldots, m)$ and $P_i P_j = O$ $(i \neq j)$. We shall also need the observation that for any polynomial $f$, we have

$$f(A) = f(\mu_1) P_1 + \cdots + f(\mu_m) P_m.$$

In particular, $P_i$ is a polynomial in $A$ for each $i$; explicitly, $P_i = f_i(A)$ where

$$f_i(x) = \frac{\prod_{s \neq i} (x - \mu_s)}{\prod_{s \neq i} (\mu_i - \mu_s)}. \tag{1.5}$$

Next we mention an eigenvector technique which is often employed to find the graphs with maximal or minimal index in a given class of graphs. A *Rayleigh quotient* for $A$ is a scalar of the form $\mathbf{y}^\top A \mathbf{y} / \mathbf{y}^\top \mathbf{y}$ where $\mathbf{y}$ is a non-zero vector in $I\!R^n$. The supremum of the set of such scalars is the largest eigenvalue $\lambda_1$ of $A$, equivalently

$$\lambda_1 = \sup\{\mathbf{x}^\top A\mathbf{x} : \mathbf{x} \in I\!R^n, \ \|\mathbf{x}\| = 1\}. \tag{1.6}$$

This well-known fact follows immediately from the observation that if $\{\mathbf{x}_1, \ldots, \mathbf{x}_n\}$ is an orthonormal basis of eigenvectors of $A$ and if $\mathbf{x} = \alpha_1\mathbf{x}_1 + \cdots + \alpha_n\mathbf{x}_n$ then $\alpha_1^2 + \cdots + \alpha_n^2 = 1$, while

$$\mathbf{x}^\top A\mathbf{x} = \lambda_1\alpha_1^2 + \cdots + \lambda_n\alpha_n^2, \tag{1.7}$$

where $A\mathbf{x}_i = \lambda_i\mathbf{x}_i$ $(i = 1, \ldots, n)$.

Note that for $\mathbf{y} \neq \mathbf{0}$, we have $\mathbf{y}^\top A\mathbf{y}/\mathbf{y}^\top \mathbf{y} \leq \lambda_1$, with equality if and only if $A\mathbf{y} = \lambda_1\mathbf{y}$. More generally, *Rayleigh's Principle* may be stated as follows:

$$\text{if } \mathbf{0} \neq \mathbf{y} \in \langle \mathbf{x}_i, \ldots, \mathbf{x}_n \rangle \text{ then } \lambda_i \geq \mathbf{y}^\top A\mathbf{y}/\mathbf{y}^\top \mathbf{y},$$

with equality if and only if $A\mathbf{y} = \lambda_i\mathbf{y}$; and

$$\text{if } \mathbf{0} \neq \mathbf{y} \in \langle \mathbf{x}_1, \ldots, \mathbf{x}_i \rangle \text{ then } \lambda_i \leq \mathbf{y}^\top A\mathbf{y}/\mathbf{y}^\top \mathbf{y},$$

with equality if and only if $A\mathbf{y} = \lambda_i\mathbf{y}$.

Moreover, each eigenvalue $\lambda_i$ $(i = 1, \ldots, n)$ can be characterized in terms of subspaces of $I\!R^n$ as follows. Let $U$ be an $(n - i + 1)$-dimensional subspace of $I\!R^n$, so that $\langle \mathbf{x}_1, \ldots, \mathbf{x}_i \rangle \cap U \neq \{\mathbf{0}\}$. If $\mathbf{x}$ is a unit vector in this intersection of subspaces then $\alpha_{i+1} = \cdots = \alpha_n = 0$ and so $\mathbf{x}^\top A\mathbf{x} \geq \lambda_i$ by (1.7). It follows that $\sup\{\mathbf{x}^\top A\mathbf{x} : \mathbf{x} \in U, \ \|\mathbf{x}\| = 1\} \geq \lambda_i$. On the other hand, by (1.7) again, this lower bound is attained when $U = \langle \mathbf{x}_i, \ldots, \mathbf{x}_n \rangle$ because in this case $\alpha_1 = \cdots = \alpha_{i-1} = 0$ for every vector in $U$. Hence for each $i \in \{1, \ldots, n\}$ we have

$$\lambda_i = \inf\{\sup\{\mathbf{x}^\top A\mathbf{x} : \mathbf{x} \in U, \ \|\mathbf{x}\| = 1\} : U \in \mathcal{U}_{n-i+1}\}, \tag{1.8}$$

where $\mathcal{U}_{n-i+1}$ denotes the set of all $(n - i + 1)$-dimensional subspaces of $I\!R^n$.

An $n \times n$ symmetric matrix $M$ (with real entries) is said to be *positive semi-definite* if all its eigenvalues are non-negative, equivalently $\mathbf{x}^\top M\mathbf{x} \geq 0$ for all $\mathbf{x} \in I\!R^n$.

**Theorem 1.3.2.** *Let $M$ be a positive semi-definite matrix with eigenvalues $\lambda_1 \geq \lambda_2 \geq \cdots \geq \lambda_n$. Then*

$$\lambda_1 + \lambda_2 + \cdots + \lambda_r = \sup\{\mathbf{u}_1^\top M\mathbf{u}_1 + \mathbf{u}_2^\top M\mathbf{u}_2 + \cdots + \mathbf{u}_r^\top M\mathbf{u}_r\} \ (r = 1, 2, \ldots, n),$$

*where the supremum is taken over all orthonormal vectors* $\mathbf{u}_1, \mathbf{u}_2, \ldots, \mathbf{u}_r$. *In particular,* $\lambda_1 + \lambda_2 + \cdots + \lambda_r$ *is bounded below by the sum of the r largest diagonal entries of M.*

**Proof.** Let $M\mathbf{x}_i = \lambda_i \mathbf{x}_i$ $(i = 1, 2, \ldots, n)$, where $\mathbf{x}_1, \mathbf{x}_2, \ldots, \mathbf{x}_n$ are orthonormal. Let $U = (\mathbf{u}_1 | \mathbf{u}_2 | \cdots | \mathbf{u}_r)$, $X = (\mathbf{x}_1 | \mathbf{x}_2 | \cdots | \mathbf{x}_n)$ and $\mathbf{u}_j = \sum_{i=1}^{n} c_{ij} \mathbf{x}_i$ $(j = 1, 2, \ldots, r)$. Then $U = XC$, where $C = (c_{ij})$; moreover, $I = U^\top U = C^\top C$. Using Equation (1.7), we have

$$\sum_{j=1}^{r} \mathbf{u}_j^\top M \mathbf{u}_j = \sum_{j=1}^{r} \sum_{i=1}^{n} c_{ij}^2 \lambda_i = \sum_{l=1}^{n} \left( \sum_{j=1}^{r} c_{ij}^2 \right) \lambda_i.$$

Note that $\sum_{j=1}^{r} c_{ij}^2 = b_i$, where $b_i$ is the $i$-th diagonal entry of $CC^\top$. Now $CC^\top$ and $C^\top C$ have the same non-zero eigenvalues and so the spectrum of $CC^\top$ is $1^r, 0^{n-r}$. By (1.7) again, $b_i = \mathbf{e}_i^\top CC^\top \mathbf{e}_i \leq 1$ $(i = 1, 2, \ldots, n)$. Now we have:

$$\sum_{j=1}^{r} \mathbf{u}_j^\top M \mathbf{u}_j = \sum_{i=1}^{n} b_i \lambda_i, \ 0 \leq b_i \leq 1, \ \sum_{i=1}^{n} b_i = \mathrm{tr}(CC^\top) = r,$$

and it follows that $\sum_{j=1}^{r} \mathbf{u}_j^\top M \mathbf{u}_j \leq \sum_{j=1}^{r} \lambda_j$. Equality holds when $\mathbf{u}_i = \mathbf{x}_i$ $(i = 1, 2, \ldots, r)$, and so the first statement of the theorem is proved. For the second statement, we may suppose without loss of generality that the $r$ largest diagonal entries of $M$ are the first $r$ diagonal entries; the assertion follows by taking $\mathbf{u}_i = \mathbf{e}_i$ $(i = 1, 2, \ldots, r)$. $\square$

If $M$ is a positive semi-definite matrix of rank $r$ then there exists an orthogonal matrix $U$ such that

$$U^\top M U = \begin{pmatrix} \theta_1 & & & & & \\ & \ddots & & & & \\ & & \theta_r & & & \\ & & & 0 & & \\ & & & & \ddots & \\ & & & & & 0 \end{pmatrix},$$

where $\theta_1 \geq \cdots \geq \theta_r > 0$. Now this matrix can be written as $X^\top X$, where

$$X = \begin{pmatrix} \sqrt{\theta_1} & \cdots & 0 & 0 & \cdots & 0 \\ 0 & \ddots & 0 & 0 & \cdots & 0 \\ 0 & \cdots & \sqrt{\theta_r} & 0 & \cdots & 0 \end{pmatrix},$$

of size $r \times n$. Thus $M = Q^\top Q$, where $Q = XU^\top$. If $Q = (\mathbf{q}_1| \cdots |\mathbf{q}_n)$ then each column $\mathbf{q}_i$ lies in $\mathbb{R}^r$, and the $(i, j)$-entry of $M$ is the scalar product $\mathbf{q}_i^\top \mathbf{q}_j$. The matrix $Q^\top Q$ is called the *Gram matrix* of the vectors $\mathbf{q}_1, \ldots, \mathbf{q}_n$. We shall often make use of Gram matrices in the case that $M = A - \lambda I$ and $\lambda$ is the least eigenvalue of $G$; in this situation, the multiplicity of $\lambda$ is $n - r$.

Since in general a graph is not determined by its eigenvalues, it is natural to seek further algebraic invariants which might serve to distinguish non-isomorphic cospectral graphs. For our first such definition, recall that $\{\mathbf{e}_1, \ldots, \mathbf{e}_n\}$ is the standard orthonormal basis of $\mathbb{R}^n$. The $mn$ numbers $\alpha_{ij} = \|P_i \mathbf{e}_j\|$ are called the *angles* of $G$; they are the cosines of the (acute) angles between axes and eigenspaces. We shall assume that $\mu_1 > \cdots > \mu_m$. If also we order the columns of the matrix $(\alpha_{ij})$ lexicographically then this matrix is a graph invariant, called the *angle matrix* of $G$. We shall see in the next chapter that the spectrum of the vertex-deleted subgraph $G - j$ is determined by the spectrum of $G$ and the angles $\alpha_{1j}, \ldots, \alpha_{mj}$. The basic relations between angles are the following:

**Proposition 1.3.3.** *The angles $\alpha_{ij}$ of a graph satisfy the equalities*

$$\sum_{j=1}^{n} \alpha_{ij}^2 = \dim \mathcal{E}(\mu_i), \qquad \sum_{i=1}^{m} \alpha_{ij}^2 = 1. \tag{1.9}$$

**Proof.** We have $\alpha_{ij}^2 = \|P_i \mathbf{e}_j\|^2 = \mathbf{e}_j^\top P_i \mathbf{e}_j$, and so the numbers $\alpha_{i1}^2$, $\alpha_{i2}^2, \ldots, \alpha_{in}^2$ appear on the diagonal of $P_i$. Now $\sum_{j=1}^{n} \alpha_{ij}^2 = \mathrm{tr}(P_i) = \mathrm{tr}(E_i) = \dim \mathcal{E}(\mu_i)$, and $\sum_{i=1}^{m} \alpha_{ij}^2 = 1$ because $\sum_{i=1}^{m} P_i = I$. $\qquad\square$

Next we discuss the relation between eigenvalues, angles and walks in a graph. By a *walk of length $k$* in a graph we mean any sequence of (not necessarily different) vertices $v_0, v_1, \ldots, v_k$ such that for each $i = 1, 2, \ldots, k$ there is an edge from $v_{i-1}$ to $v_i$. The walk is *closed* if $v_k = v_0$. The following result has a straightforward proof by induction on $k$.

**Proposition 1.3.4.** *If $A$ is the adjacency matrix of a graph, then the $(i, j)$-entry $a_{ij}^{(k)}$ of the matrix $A^k$ is equal to the number of walks of length $k$ that start at vertex $i$ and end at vertex $j$.*

It follows from Proposition 1.3.4 that the number of closed walks of length $k$ is equal to the $k$-th spectral moment, since $\sum_{j=1}^{n} a_{jj}^{(k)} = \mathrm{tr}(A^k) = \sum_{j=1}^{n} \lambda_j^k$. From the spectral decomposition of $A$ we have

$$A^k = \mu_1^k P_1 + \mu_2^k P_2 + \cdots + \mu_m^k P_m \tag{1.10}$$

and so $a_{jj}^{(k)} = \sum_{i=1}^{m} \mu_i^k \alpha_{ij}^2$, where the $\alpha_{ij}$ are the angles of $G$. In particular, the vertex degrees $a_{jj}^{(2)}$ are determined by the spectrum and angles.

We write $\mathbf{j}$ (or $\mathbf{j}_n$) for the all-1 vector in $I\!R^n$, and $\mathbf{j}^{\perp}$ for the subspace of vectors orthogonal to $\mathbf{j}$. It follows from (1.10) that the number $N_k$ of all walks of length $k$ in $G$ is given by

$$N_k = \sum_{u,v} a_{uv}^{(k)} = \mathbf{j}^{\top} A^k \mathbf{j} = \sum_{i=1}^{n} \mu_i^k \|P_i \mathbf{j}\|^2, \qquad (1.11)$$

The numbers $\beta_i = \|P_i \mathbf{j}\|/\sqrt{n}$ $(i = 1, \ldots, m)$ are called the *main angles* of $G$; they are the cosines of the (acute) angles between eigenspaces and $\mathbf{j}$. Note that $\sum_{i=1}^{m} \beta_i^2 = 1$ because $\mathbf{j} = \sum_{i=1}^{m} P_i \mathbf{j}$. The eigenvalue $\mu_i$ is said to be a *main* eigenvalue if $\mathcal{E}(\mu_i) \not\subseteq \mathbf{j}^{\perp}$, equivalently $P_i \mathbf{j} \neq \mathbf{0}$. In view of (1.11) we have the following result.

**Theorem 1.3.5.** *The total number $N_k$ of walks of length $k$ in a graph $G$ is given by*

$$N_k = n \Sigma' \mu_i^k \beta_i^2, \qquad (1.12)$$

*where the sum $\Sigma'$ is taken over all main eigenvalues $\mu_i$.*

We shall see in Chapter 2 that the spectrum of the complement $\overline{G}$, the spectrum of the cone $K_1 \nabla G$ and the Seidel spectrum of $G$ are all determined by the spectrum and main angles of $G$. A means of calculating main angles is described in Section 6.7.

Now we turn to some more general results from matrix theory that have implications for the spectra of graphs.

A symmetric matrix $M$ is *reducible* if there exists a permutation matrix $P$ such that $P^{-1}MP$ is of the form $\begin{pmatrix} X & O \\ O & Y \end{pmatrix}$, where $X$ and $Y$ are square matrices. Otherwise, $M$ is called *irreducible*. If $M = (m_{ij})$, of size $n \times n$, then we define the graph $G^M$ as follows. The vertices of $G^M$ are $1, \ldots, n$, and distinct vertices $i$, $j$ are adjacent if and only if $m_{ij} \neq 0$. Thus $G^M$ is connected if and only if $M$ is irreducible.

**Theorem 1.3.6.** *Let $M$ be an irreducible symmetric matrix with non-negative entries. Then the largest eigenvalue $\lambda_1$ of $M$ is simple, with a corresponding eigenvector whose entries are all positive. Moreover, $|\lambda| \leq \lambda_1$ for all eigenvalues $\lambda$ of $M$.*

**Proof.** Let $\mathbf{x} = (x_1, \ldots, x_n)^{\top}$ be a unit eigenvector corresponding to $\lambda_1$. Let $\mathbf{y} = (y_1, \ldots, y_n)^{\top}$, where $y_i = |x_i|$ $(i = 1, \ldots, n)$. Then $\mathbf{y}^{\top}\mathbf{y} = 1$ and $\mathbf{y}^{\top}M\mathbf{y} \geq \mathbf{x}^{\top}M\mathbf{x} = \lambda_1$. Hence $\mathbf{y}$ is also an eigenvector corresponding to $\lambda_1$.

We show that no $y_i$ (and hence no $x_i$) is zero by considering adjacencies in $G^M$. The eigenvalue equations may be written:

$$\lambda_1 y_i = m_{ii} y_i + \sum_{j \sim i} m_{ij} y_j \quad (i = 1, \ldots, n). \tag{1.13}$$

If $y_i = 0$ then by (1.10), $y_j = 0$ for all $j \sim i$. Since $G^M$ is connected, $y_j = 0$ for all $j$, a contradiction. Now $\lambda_1$ is a simple eigenvalue, for if $\dim \mathcal{E}(\lambda_1) > 1$ then there exists an eigenvector with a zero entry in any chosen position. In particular, $\mathcal{E}(\lambda_1)$ is spanned by $\mathbf{y}$ (and $\mathbf{x} = \pm \mathbf{y}$). Finally, if $M\mathbf{z} = \lambda \mathbf{z}$ where $\mathbf{z}^\top \mathbf{z} = 1$ and $\mathbf{z} = (z_1, \ldots, z_n)^\top$ then

$$|\lambda| = |\mathbf{z}^\top M \mathbf{z}| = |\sum_{i,j} z_i m_{ij} z_j| \leq \sum_{i,j} |z_i| \, m_{ij} \, |z_j| \leq \lambda_1.$$

□

We say that a vector $\mathbf{x} = (x_1, \ldots, x_n)^\top$ is non-negative (positive) if each $x_i$ is non-negative (positive); we write $\mathbf{x} \geq \mathbf{0}$, $\mathbf{x} > \mathbf{0}$ respectively. In the situation of Theorem 1.3.6, $M$ has a unique positive unit eigenvector corresponding to $\lambda_1$, and this is called the *principal* eigenvector of $M$. In the case that $M$ is the adjacency matrix of a (labelled) connected graph $G$, we refer to this vector as the *principal eigenvector* of $G$.

**Corollary 1.3.7.** *Let $M$ be an irreducible symmetric $n \times n$ matrix with non-negative entries $m_{ij}$, and let $\lambda_1$ be the largest eigenvalue of $M$. For any positive vector $\mathbf{y} = (y_1, y_2, \ldots, y_n)^\top$, we have*

$$\min_{1 \leq i \leq n} \sum_{j=1}^{n} \frac{m_{ij} y_j}{y_i} \leq \lambda_1 \leq \max_{1 \leq i \leq n} \sum_{j=1}^{n} \frac{m_{ij} y_j}{y_i}. \tag{1.14}$$

*Either equality holds if and only if $\mathbf{y}$ is an eigenvector of $M$ corresponding to $\lambda_1$.*

**Proof.** Let $\mathbf{x} = (x_1, x_2, \ldots, x_n)^\top$ be the principal eigenvector of $M$. Then

$$\lambda_1 \sum_{i=1}^{n} x_i y_i = \mathbf{y}^T M \mathbf{x} = \mathbf{x}^T M \mathbf{y} = \sum_{i=1}^{n} x_i y_i \left( \frac{\sum_{j=1}^{n} m_{ij} y_j}{y_i} \right). \tag{1.15}$$

The inequalities follow, since $\sum_{i=1}^{n} x_i y_i > 0$. Let $z_i = \lambda_1 y_i - \sum_{i=1}^{n} m_{ij} y_j$ $(i = 1, \ldots, n)$. If an equality holds in (1.14) then either all $z_i$ are non-negative or all $z_i$ are non-positive. From (1.15), we have $\sum_{i=1}^{n} x_i z_i = 0$, and so all $z_i$ are zero. In this situation, $\mathbf{y}$ is an eigenvector of $M$ corresponding to $\lambda_1$, as required. □

If we apply Theorem 1.3.6 to the adjacency matrix of a graph, we obtain:

**Corollary 1.3.8.** *A graph is connected if and only if its index is a simple eigenvalue with a positive eigenvector.*

We can also use Theorem 1.3.6 to prove:

**Proposition 1.3.9.** *For any vertex u of a connected graph G, we have $\lambda_1(G - u) < \lambda_1(G)$.*

**Proof.** Let $A = \begin{pmatrix} A' & \mathbf{r} \\ \mathbf{r}^\top & 0 \end{pmatrix}$, where $A' = A(G - u)$, and let $\mathbf{x}$ be a unit eigenvector of $A'$ corresponding to $\lambda_1(G - u)$. If $\mathbf{y} = \begin{pmatrix} \mathbf{x} \\ 0 \end{pmatrix}$ then $\mathbf{y}^\top \mathbf{y} = 1$ and $\lambda_1(G - u) = \mathbf{y}^\top A \mathbf{y} \leq \lambda_1(G)$. If equality holds then $\mathbf{y}$ is an eigenvector of $A$ corresponding to $\lambda_1(G)$; but this is a contradiction because $\mathbf{y}$ has a zero entry. $\square$

If we apply Corollary 1.3.8 to each component of an arbitrary graph $G$ which has index $\lambda_1(G)$, we can see that there is a non-negative eigenvector corresponding to $\lambda_1(G)$. This vector may also be used in Rayleigh quotients to obtain bounds for the index of modified graphs, as for example in the following:

**Proposition 1.3.10.** *If $G - uv$ is the graph obtained from a connected graph $G$ by deleting the edge $uv$, then $\lambda_1(G - uv) < \lambda_1(G)$.*

**Proof.** Let $\mathbf{x} = (x_1, \ldots, x_n)^\top$ be a non-negative unit eigenvector of $G - uv$ corresponding to $\lambda_1(G - uv)$. Then

$$\lambda_1(G - uv) = \mathbf{x}^\top A(G - uv)\mathbf{x} \leq \mathbf{x}^\top A(G)\mathbf{x} \leq \lambda_1(G).$$

If $\lambda_1(G - uv) = \lambda_1(G)$ then $\mathbf{x}$ is the principal eigenvector of $G$ and hence has no zero entries. Now $\mathbf{x}^\top A(G - uv)\mathbf{x} = \mathbf{x}^\top A(G)\mathbf{x} - 2x_u x_v < \lambda_1(G - uv)$, a contradiction. $\square$

Next we consider interlacing of eigenvalues.

**Theorem 1.3.11.** *Let $Q$ be a real $n \times m$ matrix such that $Q^\top Q = I$, and let $A$ be an $n \times n$ real symmetric matrix with eigenvalues $\lambda_1 \geq \cdots \geq \lambda_n$. If the eigenvalues of $Q^\top A Q$ are $\mu_1 \geq \cdots \geq \mu_m$ then*

$$\lambda_{n-m+i} \leq \mu_i \leq \lambda_i \ (i = 1, \ldots, m). \tag{1.16}$$

**Proof.** Let $\mathbf{x}_1, \ldots, \mathbf{x}_n$ be orthonormal eigenvectors of $A$, and let $\mathbf{y}_1, \ldots, \mathbf{y}_m$ be orthonormal eigenvectors of $Q^\top A Q$, taken in order. For each $i \in \{1, \ldots, m\}$, let $\mathbf{z}_i$ be a non-zero vector in the subspace

$$\langle \mathbf{y}_1, \ldots, \mathbf{y}_i \rangle \cap \langle Q^\top \mathbf{x}_1, \ldots, Q^\top \mathbf{x}_{i-1} \rangle^\perp.$$

Then $Q\mathbf{z}_i \in \langle \mathbf{x}_1, \ldots, \mathbf{x}_{i-1} \rangle^{\perp}$, and so (by Rayleigh's Principle)

$$\lambda_i \geq \frac{(Q\mathbf{z}_i)^{\top} A (Q\mathbf{z}_i)}{(Q\mathbf{z}_i)^{\top}(Q\mathbf{z}_i)} = \frac{\mathbf{z}_i^{\top} Q^{\top} A Q \mathbf{z}_i}{\mathbf{z}_i^{\top} \mathbf{z}_i} \geq \mu_i.$$

The second inequality in (1.16) is obtained by applying the above argument to $-A$ and $-Q^{\top}AQ$. ☐

When the inequalities (1.16) are satisfied, we say that the eigenvalues $\mu_i$ *interlace* the eigenvalues $\lambda_j$.

**Corollary 1.3.12.** *Let $G$ be a graph with $n$ vertices and eigenvalues $\lambda_1 \geq \lambda_2 \geq \cdots \geq \lambda_n$, and let $H$ be an induced subgraph of $G$ with $m$ vertices. If the eigenvalues of $H$ are $\mu_1 \geq \mu_2 \geq \cdots \geq \mu_m$ then $\lambda_{n-m+i} \leq \mu_i \leq \lambda_i$ $(i = 1, \ldots, m)$.*

**Proof.** Let $V(G) = \{1, \ldots, n\}$ and $V(H) = \{1, \ldots, m\}$. Then $A(H) = Q^{\top}A(G)Q$, where $Q^{\top}$ has the form $(I \mid O)$, and so the result follows from Theorem 1.3.11. ☐

The inequalities in Corollary 1.3.12 are known as *Cauchy's inequalities* and this result is generally known as the *Interlacing Theorem*. It is used frequently as a spectral technique in graph theory. In particular, when $H$ is a vertex-deleted subgraph we have $m = n - 1$ and:

$$\lambda_n \leq \mu_{n-1} \leq \lambda_{n-1} \leq \cdots \leq \lambda_2 \leq \mu_1 \leq \lambda_1.$$

The next result is a further consequence of Theorem 1.3.11.

**Corollary 1.3.13.** *Let $A$ be a real symmetric matrix with eigenvalues $\lambda_1 \geq \lambda_2 \geq \cdots \geq \lambda_n$. Given a partition $\{1, 2, \ldots, n\} = \Delta_1 \dot\cup \Delta_2 \dot\cup \cdots \dot\cup \Delta_m$ with $|\Delta_i| = n_i > 0$, consider the corrresponding blocking $A = (A_{ij})$, where $A_{ij}$ is an $n_i \times n_j$ block. Let $e_{ij}$ be the sum of the entries in $A_{ij}$ and set $B = (e_{ij}/n_i)$ (Note that $e_{ij}/n_i$ is the average row sum in $A_{ij}$.) Then the eigenvalues of $B$ interlace those of $A$.*

**Proof.** Suppose that the vertex-block incidence matrix has columns $\mathbf{c}_1, \ldots, \mathbf{c}_m$, and let $Q$ be the matrix with columns $\frac{1}{\sqrt{n_1}}\mathbf{c}_1, \ldots, \frac{1}{\sqrt{n_m}}\mathbf{c}_m$. Then $Q^{\top}Q = I$, $Q^{\top}AQ = B$ and the result follows from Theorem 1.3.11. ☐

If we assume that in each block $A_{ij}$ from Corollary 1.3.13 all row sums are equal then we can say more:

**Theorem 1.3.14.** *Let $A$ be any matrix partitioned into blocks as in Corollary 1.3.13. Suppose that the block $A_{ij}$ has constant row sums $b_{ij}$, and* *let $B = (b_{ij})$. Then the spectrum of $B$ is contained in the spectrum of $A$ (taking into account the multiplicities of the eigenvalues).*

**Proof.** It is straightforward to check that if $(x_1, \ldots, x_m)^\top$ is an eigenvector of $B$ then $\begin{pmatrix} x_1\mathbf{j}_{n_1} \\ \vdots \\ x_m\mathbf{j}_{n_m} \end{pmatrix}$ is an eigenvector of $A$ corresponding to the same eigenvalue. $\qquad\square$

Theorem 1.3.12 will be used in Section 3.9 to provide a link between spectral and structural properties of a graph. Next we establish the Courant–Weyl inequalities, embodied in the following result; as usual, the eigenvalues here are in non-increasing order.

**Theorem 1.3.15.** *Let $A$ and $B$ be $n \times n$ Hermitian matrices. Then*

$$\lambda_i(A + B) \leq \lambda_j(A) + \lambda_{i-j+1}(B) \quad (n \geq i \geq j \geq 1),$$

$$\lambda_i(A + B) \geq \lambda_j(A) + \lambda_{i-j+n}(B) \quad (1 \leq i \leq j \leq n).$$

**Proof.** Let $\{\mathbf{x}_1, \ldots, \mathbf{x}_n\}$, $\{\mathbf{y}_1, \ldots, \mathbf{y}_n\}$, $\{\mathbf{z}_1, \ldots, \mathbf{z}_n\}$ be orthonormal bases of eigenvectors for $A, B, A + B$ respectively. Suppose first that $i \geq j$, and consider the subspaces

$$V_1 = \langle \mathbf{x}_j, \ldots, \mathbf{x}_n \rangle, \quad V_2 = \langle \mathbf{y}_{i-j+1}, \ldots, \mathbf{y}_n \rangle, \quad V_3 = \langle \mathbf{z}_1, \ldots, \mathbf{z}_i \rangle.$$

Since $\dim(V_1 \cap V_2) \geq \dim V_1 + \dim V_2 - n$, we have

$$\dim((V_1 \cap V_2) \cap V_3) \geq \dim V_1 + \dim V_2 + \dim V_3 - 2n = 1,$$

and so $V_1 \cap V_2 \cap V_3$ contains a unit vector $\mathbf{x}$. Applying Rayleigh's Principle, we have:

$$\lambda_j(A) + \lambda_{i-j+1}(B) \geq \mathbf{x}^\top A\mathbf{x} + \mathbf{x}^\top B\mathbf{x} = \mathbf{x}^\top(A + B)\mathbf{x} \geq \lambda_i(A + B).$$

When $i \leq j$, we obtain the second inequality of the theorem by applying the first inequality to $-A$ and $-B$. $\qquad\square$

Theorem 1.3.15 applies to a graph on $n$ vertices specified as the edge-disjoint union of two spanning subgraphs. For example, if $A$ and $B$ are the adjacency matrices of $G$ and $\overline{G}$ then $A + B = J - I$ and so (for $n \geq 2$) $\lambda_2(G) + \lambda_{n-1}(\overline{G}) \geq \lambda_n(K_n) = -1$, $\lambda_2(G) + \lambda_n(\overline{G}) \leq \lambda_2(K_n) = -1$. We can also use Theorem 1.3.15 to obtain inequalities that relate the spectrum of an adjacency matrix $A$ to the spectra of the Laplacian $D - A$, the signless Laplacian $D + A$ and the Seidel matrix $J - I - 2A$: we apply the theorem to $A$ and $D - A$, to $-A$ and $D + A$, and to $2A$ and $J - I - 2A$ respectively. For example, $\lambda_k(D \pm A) \geq \lambda_n(A) \pm \lambda_{n-k+1}(A)$ and $\lambda_k(J - I - 2A) \geq -2\lambda_{n-k+1}(A) - 1$.

**Proposition 1.3.16.** *Let $M$ be a symmetric $n \times n$ matrix with real entries. If*

$$M = \begin{bmatrix} P & Q \\ Q^\top & R \end{bmatrix},$$

*then*

$$\lambda_1(M) + \lambda_n(M) \leq \lambda_1(P) + \lambda_1(R).$$

**Proof.** Let $\lambda = \lambda_n(M)$. Then we have $M - \lambda I = S + T$, where

$$S = \begin{pmatrix} P - \lambda I & O \\ Q^\top & O \end{pmatrix}, \quad T = \begin{pmatrix} O & Q \\ O & R - \lambda I \end{pmatrix}.$$

Any non-zero eigenvalue of $S$ is an eigenvalue of $P - \lambda I$, and so the eigenvalues of $S$ are real. Similarly, the eigenvalues of $T$ are real. Using Theorem 1.3.15, we have

$$\lambda_1(M) - \lambda = \lambda_1(S + T) \leq \lambda_1(S) + \lambda_1(T) =$$
$$\lambda_1(P - \lambda I) + \lambda_1(R - \lambda I) = \lambda_1(P) - \lambda + \lambda_1(R) - \lambda,$$

and the result follows.                                                    $\square$

Using an induction argument, we obtain the following:

**Corollary 1.3.17.** *Let $M$ be a symmetric $n \times n$ matrix with real entries. If $M$ is partitioned into $k^2$ blocks $M_{ij}$ (of size $n_i \times n_j$) then*

$$\lambda_1(M) + (k - 1)\lambda_n(M) \leq \sum_{i=1}^{k} \lambda_1(M_{ii}).$$

Finally we prove a result on determinants required in Chapter 7. For an $n \times m$ matrix $R$ ($n \leq m$), we write $R_{k_1,\ldots,k_n}$ for the matrix consisting of rows $k_1, \ldots, k_n$ of $R$; and for an $m \times n$ matrix $S$ ($n \leq m$) we write $S^{k_1,\ldots,k_n}$ for the matrix consisting of columns $k_1, \ldots, k_n$ of $S$. (Here, $k_1, \ldots, k_n$ are not necessarily distinct.) If $F$ is an $n$-element subset of $\{1, \ldots, m\}$, say $F = \{k_1, \ldots, k_n\}$ where $k_1 < k_2 < \cdots < k_n$, then we write $R_F = R_{k_1,\ldots,k_n}$ and $S^F = S^{k_1,\ldots,k_n}$.

**Theorem 1.3.18** (The Binet–Cauchy Theorem). *If $R$ is an $n \times m$ matrix and $S$ is an $m \times n$ matrix ($n \leq m$), then*

$$\det(RS) = \sum_{|F|=n} \det(R_F) \det(S^F).$$

**Proof.** Let $R = (r_{ij})$ and $S = (s_{ij})$. We have

$$\det(RS) = \sum_\sigma \text{sgn}(\sigma) \prod_{i=1}^{n} \left( \sum_{k=1}^{n} r_{ik} s_{k\sigma(i)} \right)$$

$$= \sum_\sigma \text{sgn}(\sigma) \left( \sum_{k_1=1}^{m} r_{1k_1} s_{k_1\sigma(1)} \right) \left( \sum_{k_2=1}^{m} r_{2k_2} s_{k_2\sigma(2)} \right) \cdots \left( \sum_{k_n=1}^{m} r_{nk_n} s_{k_n\sigma(n)} \right)$$

$$= \sum_{k_1=1}^{m} \sum_{k_2=1}^{m} \cdots \sum_{k_n=1}^{m} r_{1k_1} r_{2k_2} \cdots r_{nk_n} \sum_\sigma \text{sgn}(\sigma) s_{k_1\sigma(1)} s_{k_2\sigma(2)} \cdots s_{k_n\sigma(n)}$$

$$= \sum_{k_1=1}^{m} \sum_{k_2=1}^{m} \cdots \sum_{k_n=1}^{m} r_{1k_1} r_{2k_2} \cdots r_{nk_n} \det(S^{\{k_1,\dots,k_n\}}).$$

Now $\det(S^{\{k_1,\dots,k_n\}}) = 0$ when $k_1, \dots, k_n$ are not distinct, and so we may take the sum over $n$-element subsets $\{k_1, \dots, k_n\}$ of $\{1, \dots, m\}$. Then $\det(S^{\{\tau(k_1),\dots,\tau(k_n)\}}) = \text{sgn}(\tau) \det(S^{\{k_1,\dots,k_n\}})$ for any permutation $\tau$ of $k_1, \dots, k_n$, and so

$$\sum_{k_1=1}^{m} \sum_{k_2=1}^{m} \cdots \sum_{k_n=1}^{m} r_{1k_1} r_{2k_2} \cdots r_{nk_n} \det(S^{\{k_1,\dots,k_n\}})$$

$$= \sum_\tau \sum_{k_1<k_2<\cdots<k_n} \text{sgn}(\tau) r_{1\tau(1)} r_{2\tau(2)} \cdots r_{n\tau(n)} \det(S^{\{k_1,\dots,k_n\}})$$

$$= \sum_{|F|=n} \det(R_F) \det(S^F).$$

$\square$

## Exercises

**1.1** Prove Proposition 1.1.2.

**1.2** By considering the nullspace of an all-1 matrix, or otherwise, show that $K_n$ ($n > 1$) has spectrum $(n-1)^1, (-1)^{n-1}$.

**1.3** Prove Proposition 1.1.7.

**1.4** Show that $L(K_{4,4})$ has spectrum $6^1, 2^6, (-2)^9$.

**1.5** Let $G$ be a graph with $n$ vertices. Show that $\lambda_1(G) \le n - 1$, with equality if and only if $G = K_n$.

**1.6** Let $G$ be a bipartite graph, with each edge joining a vertex in $\{1, \dots, k\}$ to a vertex in $\{k+1, \dots, n\}$. Show that if $(x_1, \dots, x_n)^\top$ is an eigenvector of $G$ corresponding to $\lambda$, then $(x_1, \dots, x_k, -x_{k+1}, \dots, -x_n)^\top$ is an eigenvector of $G$ corresponding to $-\lambda$. Deduce that the spectrum of a bipartite graph is symmetric about 0.

**1.7** Let $G$ be a graph with $p$ vertices of odd degree and $q$ vertices of even degree, where $p$ and $q$ have the same parity. Show that if $G'$ is switching equivalent to $G$ then either $G'$ has $p$ vertices of odd degree and $q$ vertices of even degree, or $G'$ has $q$ vertices of odd degree and $p$ vertices of even degree [Sei2].

**1.8** Show that for any graph $G$ and any vertex $v$ of $G$ there exists a unique switching-equivalent graph $G'$ which has $v$ as an isolated vertex [Sei3].

**1.9** Let $I(G)$ be the collection of graphs obtained by isolating in turn the vertices of the graph $G$. Show that the graphs $G_1$ and $G_2$ are switching equivalent if and only if $I(G_1) = I(G_2)$ [BuCS1].

**1.10** Prove Proposition 1.2.2.

**1.11** Show that a regular connected generalized line graph is either a line graph or a cocktail party graph.

**1.12** Prove Proposition 1.3.4.

**1.13** Suppose that $G$, $\overline{G}$ have adjacency matrices $A$, $\overline{A}$. Show that if $\mu$ is a non-main eigenvalue of $G$ then $\mathcal{E}_A(\mu) \subseteq \mathcal{E}_{\overline{A}}(-\mu - 1)$. Provide an example of proper inclusion.

**1.14** Let $G$ be a graph with adjacency matrix $A$ and vertex degrees $d_1, \ldots, d_n$. Let $\mathbf{d} = (d_1, \ldots . d_n)$. Then $G$ is said to be *harmonic* if $\mathbf{d}$ is an eigenvector of $A$. Show that both $G$ and $\overline{G}$ are harmonic if and only if $G$ is regular.

**1.15** With the notation of Section 1.1, show that the vector $(d_1, \ldots, d_n)^\top$ is orthogonal to (i) $\mathcal{E}(0)$, and (ii) $\mathcal{E}(\lambda)$ for every non-main eigenvalue $\lambda$.

**1.16** Show that no line graph has $-2$ as a main eigenvalue.

**1.17** Show that if $G$ is a strongly regular graph then each vertex-deleted subgraph $G - v$ ($v \in V(G)$) has exactly two main eigenvalues.

**1.18** Show that in a connected graph $G$, the minimum degree of a vertex is bounded above by the index of $G$.

**1.19** Show that if $(\alpha_{ij})$ is the angle matrix of the connected graph $G$ then $(\alpha_{11}, \ldots, \alpha_{1n})^\top$ is the principal eigenvector of $G$.

**1.20** Show that if the graphs $G$, $G'$ differ in only one edge then $|\lambda_1(G) - \lambda_1(G')| \leq 1$.

**1.21** Use Theorem 1.3.15 to show that if the adjacency matrix of $G$ has eigenvalues $\lambda_1 \geq \cdots \geq \lambda_n$ and the Laplacian of $G$ has eigenvalues $\nu_1 \geq \cdots \geq \nu_n$ then

$$\delta(G) - \lambda_i \leq \nu_{n-i+1} \leq \Delta(G) - \lambda_i \quad (i = 1, \ldots, n).$$

State and prove an analogous result relating the eigenvalues of the signless Laplacian to $\lambda_1, \ldots, \lambda_n$.

**1.22** Show that if $A$ is a symmetric matrix with eigenvalues $\lambda_1 \geq \cdots \geq \lambda_n$ then

$$\lambda_1 - \lambda_n = \sup\{\mathbf{u}^\top A\mathbf{u} - \mathbf{v}^\top A\mathbf{v}\},$$

where the supremum is taken over all pairs of orthonormal vectors $\mathbf{u}, \mathbf{v}$ [Mir].

# Notes

For a background in graph theory and linear algebra, the reader is referred to the monographs [Mer5] and [Str] respectively; earlier texts are [Har2] and [Hal]. Most undergraduate texts on linear algebra discuss the orthogonal diagonalization of a matrix with real entries; a more advanced text is [Pra]. For results on matrices (not necessarily symmetric) with non-negative entries, [Gan, Vol. 2] is a standard reference. The interlacing property of the eigenvalues arising in Theorem 1.3.11 is taken from [Hae2]; Corollary 1.3.13 appears in the earlier paper [Hae1]. Theorem 1.3.14 appears in [Hay] and [PeSa1]. The proofs of Theorems 1.3.15 and 1.3.18 are taken from [Pra].

Line graphs are characterized by a collection of 9 forbidden induced subgraphs; see [Har2, Chapter 8] or the original proof by L. W. Beineke [Bei]. The concept of a strongly regular graph was introduced in 1963 by R. C. Bose [Bos], and there is now an extensive literature on graphs of this type; see, for example, [BroLi]. Generalized line graphs were introduced by A. J. Hoffman [Hof5] in 1970, and studied extensively by D. Cvetković, M. Doob and S. Simić [CvDS1, CvDS2] in 1980. They were characterized by a collection of 31 forbidden induced subgraphs in [CvDS1, CvDS2], and independently by S. B. Rao, N. M. Singhi and K. S. Vijayan in [RaoSV]; a recent proof appears in [CvRS8] and the monograph [CvRS7]. A survey of results concerning main eigenvalues, together with an explanation of their relation to harmonic graphs (Exercise 1.14), can be found in [Row16].

The modifications $G - u$, $G - uv$ may be regarded as perturbations of $G$; other perturbations are considered in Section 8.1.

# 2

# Graph operations and modifications

In this chapter we describe some procedures for determining characteristic polynomials of graphs derived from simpler graphs by certain operations or modifications. Typically, we define an $n$-ary operation on graphs $G_1, G_2, \ldots, G_n$ $(n = 1, 2, \ldots)$ to obtain a graph $G$, and then describe relations between the spectra of $G_1, G_2, \ldots, G_n$ and the spectrum of $G$. In some important cases, the spectrum of $G$ is determined by the spectra of $G_1, G_2, \ldots, G_n$; in other cases, additional invariants of $G_1, G_2, \ldots, G_n$ are required in the form of graph angles or walk generating functions. The modifications considered include the deletion and addition of a vertex.

Naturally, several proofs rely simply on determinantal expansions, but others require an interpretation of the coefficients in a characteristic polynomial, and this is presented in Section 2.4. At the end of the chapter, in Section 2.6, we use the theory we have developed to derive the spectra, or characteristic polynomials, of several special classes of graphs.

## 2.1 Complement, union and join of graphs

The operations of complement, union and join are connected by the relation

$$\overline{G \bigtriangledown H} = \overline{G} \;\dot{\cup}\; \overline{H}.$$

First we consider the (disjoint) union of graphs. If $G$ has adjacency matrix $A$ and $H$ has adjacency matrix $B$, then the adjacency matrix of $G \dot{\cup} H$ is the direct sum

$$A \dotplus B = \begin{pmatrix} A & O \\ O & B \end{pmatrix}.$$

Consideration of determinants leads immediately to the following result.

**Theorem 2.1.1.** *The characteristic polynomial of the disjoint union of two graphs is given by:*

$$P_{G \dot\cup H}(x) = P_G(x) P_H(x).$$

It follows that if $G_1, G_2, \ldots, G_s$ are the components of the graph $G$, then we have

$$P_G(x) = P_{G_1}(x) P_{G_2}(x) \cdots P_{G_s}(x).$$

If $G$ is a regular graph, then the characteristic polynomial $P_{\overline{G}}(x)$ of the complement $\overline{G}$ of $G$ can be expressed by means of $P_G(x)$ (and vice versa). The relation is given by the following theorem.

**Theorem 2.1.2.** *If $G$ is a regular graph of degree $r$ with $n$ vertices, then*

$$P_{\overline{G}}(x) = (-1)^n \frac{x - n + r + 1}{x + r + 1} P_G(-x - 1), \qquad (2.1)$$

*i.e., if the eigenvalues of $G$ are $\lambda_1 = r, \lambda_2, \ldots, \lambda_n$, then the eigenvalues of $\overline{G}$ are $n - 1 - r, -\lambda_2 - 1, \ldots, -\lambda_n - 1$.*

**Proof.** If $G$ has adjacency matrix $A$ then $\overline{G}$ has adjacency matrix $J - I - A$. Let $\mathbf{x}_1, \mathbf{x}_2, \ldots, \mathbf{x}_n$ be an orthogonal basis of $\mathbb{R}^n$ consisting of eigenvectors of $A$, with $\mathbf{x}_1 = \mathbf{j}$. Then we have $A\mathbf{x}_1 = r\mathbf{x}_1, (J - I - A)\mathbf{x}_1 = (n - 1 - r)\mathbf{x}_1$ and $(J - I - A)\mathbf{x}_i = (-1 - r)\mathbf{x}_i$ $(i = 2, \ldots, n)$. $\qquad \square$

In the general case, the spectrum of $G$ does not determine the spectrum of $\overline{G}$; for example the complements of the cospectral graphs $C_4 \dot\cup K_1, K_{1,4}$ are not cospectral. However the spectrum of $\overline{G}$ *is* determined by the spectrum and main angles of $G$:

**Proposition 2.1.3.** *For any graph $G$ with $n$ vertices, the complement $\overline{G}$ of $G$ has characteristic polynomial*

$$P_{\overline{G}}(x) = (-1)^n P_G(-x - 1) \left( 1 - n \sum_{i=1}^{m} \frac{\beta_i^2}{x + 1 + \mu_i} \right). \qquad (2.2)$$

**Proof.** We use a multilinear determinantal expansion in conjunction with the spectral decomposition of $A$ (Equation (1.3)). The characteristic polynomial of $\overline{G}$ is given by:

$$\begin{aligned}
P_{\overline{G}}(x) &= \det((x + 1)I + A - J) \\
&= \det((x + 1)I + A) - \mathbf{j}^\top \mathrm{adj}((x + 1)I + A)\mathbf{j}
\end{aligned}$$

$$= (-1)^n P_G(-x - 1)(1 - \mathbf{j}^\top((x+1)I + A)^{-1}\mathbf{j})$$

$$= (-1)^n P_G(-x - 1)\left(1 - n\sum_{i=1}^{m}\frac{\beta_i^2}{x + 1 + \mu_i}\right). \tag{2.3}$$

□

We may apply exactly the same argument to $J - 2A - I$ to obtain:

**Proposition 2.1.4.** *For any graph $G$ with $n$ vertices, the characteristic polynomial $S_G(x)$ of the Seidel adjacency matrix of $G$ is given by*

$$S_G(x) = (-2)^n P_G\left(-\frac{1}{2}(x+1)\right)\left(1 - n\sum_{i=1}^{m}\frac{\beta_i^2}{x + 1 + 2\mu_i}\right). \tag{2.4}$$

We may also apply the argument to $G \mathbin{\dot\cup} H$. By Proposition 2.1.1, the eigenvalues of $G \mathbin{\dot\cup} H$ are the eigenvalues of $G$ or $H$ (or both). We suppose that $G$ has $n_1$ vertices and $H$ has $n_2$ vertices. The adjacency matrix of $G \mathbin{\dot\cup} H$ has spectral decomposition

$$\begin{pmatrix} A & O \\ O & B \end{pmatrix} = \xi_1 \begin{pmatrix} P_1 & O \\ O & Q_1 \end{pmatrix} + \cdots + \xi_s \begin{pmatrix} P_s & O \\ O & Q_s \end{pmatrix},$$

where $P_i$ represents the orthogonal projection $\mathbb{R}^{n_1} \to \mathcal{E}_A(\xi_i)$ and $Q_i$ represents the orthogonal projection $\mathbb{R}^{n_2} \to \mathcal{E}_B(\xi_i)$ $(i = 1, \ldots, s)$. Here, $\mathcal{E}_A(\xi_i) = \{0\}$ if $\xi_i$ is not an eigenvalue of $G$, and $\mathcal{E}_B(\xi_i) = \{0\}$ if $\xi_i$ is not an eigenvalue of $H$. As in Proposition 2.1.3 we have

$$P_{\overline{G \mathbin{\dot\cup} H}}(x) = (-1)^{n_1 + n_2} P_{G \mathbin{\dot\cup} H}(-x - 1)$$

$$\times \left(1 - n_1\sum_{i=1}^{s}\frac{\beta_i^2}{x + 1 + \xi_i} - n_2\sum_{i=1}^{s}\frac{\gamma_i^2}{x + 1 + \xi_i}\right), \tag{2.5}$$

where the non-zero $\beta_i$ are precisely the non-zero main angles of $G$ and the non-zero $\gamma_i$ are precisely the non-zero main angles of $H$. The arguments here extend to the disjoint union of arbitrarily many graphs. We note in passing that the main angles $\delta_i$ of $G \mathbin{\dot\cup} H$ are given by:

$$(n_1 + n_2)\delta_i^2 = n_1\beta_i^2 + n_2\gamma_i^2 \quad (i = 1, \ldots, s).$$

This relation follows from the definition or from a comparison of Equations (2.2) and (2.5).

We can rewrite Equation (2.5) using Propositions 2.1.1 and 2.1.3 to obtain:

$$P_{\overline{G \mathbin{\dot\cup} H}}(x) = (-1)^{n_2} P_{\overline{G}}(x)P_H(-x - 1) + (-1)^{n_1} P_G(-x - 1)P_{\overline{H}}(x)$$

$$- (-1)^{n_1 + n_2} P_G(-x - 1)P_H(-x - 1). \tag{2.6}$$

Replacing $G$ with $\overline{G}$, and $H$ with $\overline{H}$, we obtain the following:

**Theorem 2.1.5.** *Let $G, H$ be graphs with $n_1, n_2$ vertices respectively. The characteristic polynomial of the join $G \bigtriangledown H$ is given by the relation*

$$P_{G\bigtriangledown H}(x) = (-1)^{n_2} P_G(x) P_{\overline{H}}(-x-1) + (-1)^{n_1} P_H(x) P_{\overline{G}}(-x-1)$$
$$- (-1)^{n_1+n_2} P_{\overline{G}}(-x-1) P_{\overline{H}}(-x-1). \tag{2.7}$$

**Corollary 2.1.6.** *Let $G, H$ be graphs with $n_1, n_2$ vertices respectively. Then*

$$P_{G\bigtriangledown H}(x) = P_G(x) P_H(x) \left( 1 - n_1 n_2 \sum_{i=1}^{m} \sum_{k=1}^{p} \frac{\beta_i^2 \gamma_k^2}{(x - \mu_i)(x - \nu_k)} \right)$$

*where $G$ has distinct eigenvalues $\mu_1, \ldots, \mu_m$ with corresponding main angles $\beta_1, \ldots, \beta_m$, and $H$ has distinct eigenvalues $\nu_1, \ldots, \nu_p$ with corresponding main angles $\gamma_1, \ldots, \gamma_p$.*

**Proof.** The result follows from Theorem 2.1.5 and Proposition 2.1.3. $\square$

From Proposition 2.1.3 and Theorem 2.1.5 we can also find an expression for the characteristic polynomial of the cone over a graph $G$ (i.e. the graph obtained from $G$ by adding a vertex adjacent to every vertex of $G$):

**Proposition 2.1.7.** *The cone over $G$ has characteristic polynomial*

$$P_{K_1 \bigtriangledown G}(x) = P_G(x) \left( x - \sum_{i=1}^{m} \frac{n\beta_i^2}{x - \mu_i} \right). \tag{2.8}$$

Next we discuss the join of regular graphs. First we can deduce the following from Proposition 2.1.2 and Theorem 2.1.5.

**Theorem 2.1.8.** *If $G_1$ is $r_1$-regular with $n_1$ vertices, and $G_2$ is $r_2$-regular with $n_2$ vertices, then the characteristic polynomial of the join $G_1 \bigtriangledown G_2$ is given by:*

$$P_{G_1 \bigtriangledown G_2}(x) = \frac{P_{G_1}(x) P_{G_2}(x)}{(x - r_1)(x - r_2)} ((x - r_1)(x - r_2) - n_1 n_2). \tag{2.9}$$

Note that if $G_1 \bigtriangledown G_2$ is a regular graph, then both $G_1$ and $G_2$ are regular. On the other hand, if $G_1$ is $r_1$-regular with $n_1$ vertices, and $G_2$ is $r_2$-regular with $n_2$ vertices, then $G_1 \bigtriangledown G_2$ is a regular graph if and only if $r_1 + n_2 = r_2 + n_1$. In this situation, $G_1 \bigtriangledown G_2$ has $n^{(1)} = n_1 + n_2$ vertices and is regular of degree

$r^{(1)} = r_1 + n_2 = r_2 + n_1$. Hence, the relations $n_1 - r_1 = n_2 - r_2 = n^{(1)} - r^{(1)}$ hold, and from (2.9) we have

$$P_{G_1 \triangledown G_2}(x) = (x - r^{(1)})(x + n^{(1)} - r^{(1)}) \frac{P_{G_1}(x) P_{G_2}(x)}{(x - r_1)(x - r_2)}. \qquad (2.10)$$

This equation can now be used to determine $P_{(G_1 \triangledown G_2) \triangledown G_3}(x)$ from $P_{G_i}(x)$ ($i = 1, 2, 3$). The necessary condition for $(G_1 \triangledown G_2) \triangledown G_3$ to be regular (of degree $r^{(2)}$ and with $n^{(2)}$ vertices) is that $n^{(1)} - r^{(1)} = n_3 - r_3 = n^{(2)} - r^{(2)}$; in this case, from (2.9) and (2.10) we have

$$P_{(G_1 \triangledown G_2) \triangledown G_3}(x) = (x - r^{(2)})(x + n^{(1)} - r^{(1)})(x + n^{(2)} - r^{(2)})$$
$$\times \frac{P_{G_1}(x) P_{G_2}(x) P_{G_3}(x)}{(x - r_1)(x - r_2)(x - r_3)}.$$

Continuing this reasoning, we arrive at the following result (where associativity of the join operation allows us to omit parentheses in $G_1 \triangledown G_2 \triangledown \cdots \triangledown G_k$).

**Theorem 2.1.9** [FiGr]. *Let $G_1, G_2, \ldots, G_k$ be regular graphs; let $G_i$ have degree $r_i$ and $n_i$ vertices ($i = 1, 2, \ldots, k$), where the relations $n_1 - r_1 = n_2 - r_2 = \cdots = n_k - r_k = s$ hold. Then the graph $G = G_1 \triangledown G_2 \triangledown \cdots \triangledown G_k$ has $n = n_1 + n_2 + \cdots + n_k$ vertices and is regular of degree $r = n - s$, so that we have*

$$P_G(x) = (x - r)(x + n - r)^{k-1} \prod_{i=1}^{k} \frac{P_{G_i}(x)}{x - r_i}. \qquad (2.11)$$

We conclude this section with some remarks on main angles and walk generating functions. From Propositions 2.1.3, 2.1.4 and 2.1.7, we see that, given the eigenvalues of $G$, knowledge of the main angles of $G$ is equivalent to knowledge of the spectrum of $\overline{G}$, or the spectrum of the Seidel matrix of $G$, or the spectrum of the cone over $G$. On the other hand, given the eigenvalues of $G$, knowledge of the main angles of $G$ is equivalent to knowledge of the *walk generating function*

$$H_G(t) = \sum_{k=0}^{\infty} N_k t^k, \qquad (2.12)$$

where $N_k$ is the number of walks of length $k$ in $G$. For by Theorem 1.3.5 we have

$$H_G(t) = n \sum_{p=1}^{m} \beta_p^2 / (1 - t \mu_p). \qquad (2.13)$$

Accordingly we may write formulae (2.2) and (2.4) in the form:

$$P_{\overline{G}}(x) = (-1)^n P_G(-1 - x)(1 - (x + 1)^{-1} H_G(-1/(1 + x))),$$

$$S_G(x) = (-1)^n 2^n P_G(-(x-1)/2)(1 - (x+1)^{-1} H_G(-2/(x+1))).$$

We can use the first equation to express the walk generating function in terms of characteristic polynomials:

$$H_G(t) = \frac{1}{t} \left\{ (-1)^n \frac{P_{\overline{G}}\left(-\frac{t+1}{t}\right)}{P_G\left(\frac{1}{t}\right)} - 1 \right\}. \tag{2.14}$$

This enables us to express $H_{\overline{G}}$ in terms of $H_G$, and $H_{G_1 \triangledown G_2}$ in terms of $H_{G_1}$ and $H_{G_2}$:

**Theorem 2.1.10.** (i) $H_{\overline{G}}(t) = \dfrac{H_G\left(\frac{-t}{t+1}\right)}{t + 1 - t H_G\left(\frac{-t}{t+1}\right)};$

(ii) $H_{G_1 \dot\cup G_2}(t) = H_{G_1}(t) + H_{G_2}(t);$

(iii) $H_{G_1 \triangledown G_2}(t) = \dfrac{H_{G_1}(t) + H_{G_2}(t) + 2t H_{G_1}(t) H_{G_2}(t)}{1 - t^2 H_{G_1}(t) H_{G_2}(t)}.$

**Proof.** From Equation (2.14), we have

$$H_{\overline{G}}(t) = \frac{1}{t} \left\{ (-1)^n \frac{P_G\left(-\frac{t+1}{t}\right)}{P_{\overline{G}}\left(\frac{1}{t}\right)} - 1 \right\} \tag{2.15}$$

and

$$H_G\left(\frac{-t}{t+1}\right) = -\frac{t+1}{t} \left\{ (-1)^n \frac{P_{\overline{G}}\left(\frac{1}{t}\right)}{P_G\left(-\frac{t+1}{t}\right)} - 1 \right\}. \tag{2.16}$$

The relation (i) follows by eliminating $P_G\left(-\frac{t+1}{t}\right) \Big/ P_{\overline{G}}\left(\frac{1}{t}\right)$ from (2.15) and (2.16). The relation (ii) is immediate from the definition (2.12). The third relation follows from (i) and (ii) when we express $G_1 \triangledown G_2$ as the complement of $\overline{G_1} \dot\cup \overline{G_2}$. $\qquad\square$

## 2.2 Coalescence and related graph compositions

Here we discuss further examples of characteristic polynomials of graphs constructed using various graph operations or modifications. The formulae obtained may be seen as reduction procedures for calculating the characteristic polynomials concerned. In these formulae, the characteristic polynomial of an empty graph should be interpreted as 1.

The proof of the first result is left as an exercise in evaluating determinants.

**Theorem 2.2.1.** *Let $G_j$ denote the graph obtained from $G$ by adding a pendant edge at the vertex $j$. Then*

$$P_{G_j}(x) = x P_G(x) - P_{G-j}(x). \tag{2.17}$$

By iterating formula (2.17), the characteristic polynomial of a tree can easily be computed. We may also apply Theorem 2.2.1 to the graph $G_j^n$ obtained from the connected graph $G$ by adding a path of length $n$ at the vertex $j$. We know from Chapter 1 that if $G$ is connected then $\lambda_1(G_u) > \lambda_1(G)$, because $G$ is a vertex-deleted subgraph of the connected graph $G_u$. Thus if $\rho_{jn}$ is the index of $G_j^n$ then we have

$$\lambda_1(G) < \rho_{j1} < \rho_{j2} < \rho_{j3} < \cdots,$$

while $\rho_{jn} \le \Delta(G_j^n) \le \Delta(G) + 1$ by Proposition 1.1.1. Hence the sequence $\rho_{j1}, \rho_{j2}, \rho_{j3}, \ldots$ converges to some limit $\rho_j > \lambda_1(G)$.

**Theorem 2.2.2** [Hof8]. *Let $G_j^n$ be the graph obtained from the connected graph $G$ by adding a path of length $n$ at the vertex $j$, and let $\rho_{jn}$ be the index of $G_j^n$. Suppose that $\rho_{jn} \to \rho_j > 2$ as $n \to \infty$. Then $\rho_j$ is the largest positive solution of the equation*

$$\frac{1}{2}(x + \sqrt{x^2 - 4}) P_G(x) - P_{G-j}(x) = 0.$$

**Proof.** For fixed $j$, let $f_n(x)$ be the characteristic polynomial of $G_j^n$. Thus $f_0(x) = P_G(x)$ and by Theorem 2.2.1 we have

$$f_n(x) = x f_{n-1}(x) - f_{n-2}(x) \ (n \ge 2), \quad f_1(x) = x P_G(x) - P_{G-j}(x).$$

The solution of this linear recurrence relation is given by

$$(\alpha(x) - \beta(x)) f_n(x) = (\alpha(x) P_G(x) - P_{G-j}(x)) \alpha(x)^n$$
$$- (\beta(x) P_G(x) - P_{G-j}(x)) \beta(x)^n,$$

where $\alpha(x), \beta(x) = \frac{1}{2}(x \pm \sqrt{x^2 - 4})$. If we divide this equation by $\alpha(x)^n$, set $x = \rho_{jn}$ and let $n \to \infty$ then we obtain the result. $\square$

We extend our deliberations to any graph with a cutvertex $w$. Such a graph may be regarded as a *coalescence* $G \cdot H$ of two graphs $G$ and $H$, obtained from $G \stackrel{.}{\cup} H$ by identifying a vertex $u$ of $G$ with a vertex $v$ of $H$. (Formally, $V(G \cdot H) = V(G - u) \stackrel{.}{\cup} V(H - v) \stackrel{.}{\cup} \{w\}$ with two vertices in $G \cdot H$ adjacent if they are adjacent in $G$ or $H$, or if one is $w$ and the other is a neighbour of $u$ in $G$ or a neighbour of $v$ in $H$.)

**Theorem 2.2.3.** *Let $G \cdot H$ be the coalescence in which the vertex $u$ of $G$ is identified with the vertex $v$ of $H$. Then*

$$P_{G \cdot H}(x) = P_G(x) P_{H-v}(x) + P_{G-u}(x) P_H(x) - x P_{G-u}(x) P_{H-v}(x).$$

$$(2.18)$$

**Proof.** The graph $G \cdot H$ has adjacency matrix $\begin{pmatrix} A' & \mathbf{r} & O \\ \mathbf{r}^\top & 0 & \mathbf{s}^\top \\ O & \mathbf{s} & B' \end{pmatrix}$, where $\begin{pmatrix} A' & \mathbf{r} \\ \mathbf{r}^\top & 0 \end{pmatrix}$ is the adjacency matrix of $G$ and $\begin{pmatrix} 0 & \mathbf{s}^\top \\ \mathbf{s} & B' \end{pmatrix}$ is the adjacency matrix of $H$. Now

$$P_{G \cdot H}(x) = \begin{vmatrix} xI - A' & -\mathbf{r} & O \\ -\mathbf{r}^\top & x & -\mathbf{s}^\top \\ O & -\mathbf{s} & xI - B' \end{vmatrix} =$$

$$\begin{vmatrix} xI - A' & -\mathbf{r} & O \\ -\mathbf{r}^\top & x & -\mathbf{s}^\top \\ O & 0 & xI - B' \end{vmatrix} + \begin{vmatrix} xI - A' & 0 & O \\ -\mathbf{r}^\top & x & -\mathbf{s}^\top \\ O & -\mathbf{s} & xI - B' \end{vmatrix} - \begin{vmatrix} xI - A' & 0 & O \\ -\mathbf{r}^\top & x & -\mathbf{s}^\top \\ O & 0 & xI - B' \end{vmatrix},$$

and the result follows. □

We may consider a graph with a bridge as a special case of Theorem 2.2.3. Let $GuvH$ be the graph obtained from $G \stackrel{.}{\cup} H$ by adding an edge joining the vertex $u$ of $G$ to the vertex $v$ of $H$.

**Theorem 2.2.4.** *The characteristic polynomial of $GuvH$ is given by*

$$P_{GuvH}(x) = P_G(x) P_H(x) - P_{G-u}(x) P_{H-v}(x). \qquad (2.19)$$

**Proof.** We regard $GuvH$ as a coalescence of $G_u$ and $H$. Using Theorems 2.2.3 and 2.2.1 in turn, we obtain:

$$P_{GuvH}(x) = P_{G_u}(x) P_{H-v}(x) + P_G(x) P_H(x) - x P_G(x) P_{H-v}(x)$$
$$= \{x P_G(x) - P_{G-u}(x)\} P_{H-v}(x) + P_G(x) P_H(x)$$
$$- x P_G(x) P_{H-v}(x),$$

and the result follows. □

The next result deals with a special type of graph with several cutvertices. Let $G$ be a graph with $n$ vertices, and let $H$ be a graph with $m$ vertices. The *corona* $G \circ H$ is the graph with $n + mn$ vertices obtained from $G$ and $n$ copies of $H$ by joining the $i$-th vertex of $G$ to each vertex in the $i$-th copy of $H$ ($i = 1, \ldots, n$).

**Theorem 2.2.5.** *Let $G$ be a graph with $n$ vertices, and let $H$ be an $r$-regular graph with $m$ vertices. The characteristic polynomial of the corona $G \circ H$ is given by*

$$P_{G \circ H}(x) = P_G \left( x - \frac{m}{x - r} \right) (P_H(x))^n .$$

**Proof.** We may express $P_{G \circ H}(x)$ in the form

$$\begin{vmatrix} xI - A & -J_1 & -J_2 & \cdots & -J_n \\ -J_1^\top & xI - B & & & \\ -J_2^\top & & xI - B & & \\ \vdots & & & \ddots & \\ -J_n^\top & & & & xI - B \end{vmatrix}$$

where (i) $A$, $B$ are the adjacency matrices of the graphs $G$, $H$ respectively, (ii) $J_k$ is the $n \times m$ matrix in which each entry of the $k$-th row is 1 and all other entries are 0.

For each $k = 1, \ldots, n$ let $s_k$ be the sum of rows $n + (k - 1)m + 1$, $n + (k - 1)m + 2, \ldots, n + (k - 1)m + m$, and subtract $(x - r)^{-1}s_k$ from the $k$-th row. We find $P_{G \circ H}(x) =$

$$\begin{vmatrix} \left(x - \frac{m}{x-r}\right)I - A & O & O & \cdots & O \\ -J_1^\top & xI - B & & & \\ -J_2^\top & & xI - B & & \\ \vdots & & & \ddots & \\ -J_n^\top & & & & xI - B \end{vmatrix} = P_G \left( x - \frac{m}{x - r} \right) (P_H(x))^n .$$

$\square$

As a special case of this result , we have $P_{G \circ K_1}(x) = x^n P_G \left( x - \frac{1}{x} \right)$. Thus if $\lambda_1, \ldots, \lambda_n$ are the eigenvalues of $G$, then $\frac{1}{2} \left( \lambda_i \pm \sqrt{\lambda_i^2 + 4} \right)$ ($i = 1, \ldots, n$) are the eigenvalues of $G \circ K_1$.

We now turn our attention to the vertex-deleted subgraphs which feature in Theorems 2.2.1 to 2.2.4. The formulae there can be refined by using graph angles, introduced in Section 1.3.

**Proposition 2.2.6.** *Let* $G - j$ *be the graph obtained from G by deleting the vertex j and all edges containing j. Then*

$$P_{G-j}(x) = P_G(x) \sum_{i=1}^{m} \frac{\alpha_{ij}^2}{x - \mu_i}.$$

**Proof.** Since the adjoint of $xI - A$ is $\det(xI - A)(xI - A)^{-1}$, we have

$$\text{adj}(xI - A) = P_G(x) \sum_{i=1}^{m} \frac{1}{x - \mu_i} P_i.$$

The result follows by equating diagonal entries in this matrix equation. $\quad\square$

Thus, given the spectrum of $G$, knowledge of the characteristic polynomials of its vertex-deleted subgraphs is equivalent to knowledge of the angles of $G$. Also, Theorem 2.2.1 yields:

$$P_{G_j}(x) = P_G(x) \left( x - \sum_{i=1}^{m} \frac{\alpha_{ij}^2}{x - \mu_i} \right), \tag{2.20}$$

while from Theorem 2.2.2 we obtain:

**Proposition 2.2.7** [CvRo1]. *Let* $G_j^n$ *be the graph obtained from G by adding a path of length n at vertex j, and let* $\rho_{jn}$ *be the index of* $G_j^n$. *Suppose that* $\rho_{jn} \to \rho_j > 2$ *as* $n \to \infty$. *Then* $\rho_j$ *is the largest positive solution of the equation*

$$\frac{1}{2}(x + \sqrt{x^2 - 4}) - \sum_{i=1}^{m} \frac{\alpha_{ij}^2}{x - \mu_i} = 0.$$

**Proof.** By Theorem 2.2.2, $\rho_j$ is the largest positive solution of the equation

$$\frac{1}{2} \left( x + \sqrt{x^2 - 4} \right) P_G(x) - P_{G-j}(x) = 0.$$

Moreover $P_G(\rho_j) \neq 0$ since $\rho_{j1} > \mu_1$. The result therefore follows immediately from Proposition 2.2.6. $\quad\square$

Restatements of Theorems 2.2.3 and 2.2.4 in terms of angles are left as exercises (see Exercises 2.8 and 2.9).

As in the previous section, we may use walk generating functions for $G$ in place of angles of $G$ when the spectrum of $G$ is known. Let $H_j^G(t)$ be the generating function for the number of closed walks of length $k$ in $G$ starting (and terminating) at the vertex $j$. Thus $H_j^G(t) = \sum_{k=0}^{\infty} a_{jj}^{(k)} t^k$ where $A^k = \left( a_{ij}^{(k)} \right)$. From Equation (1.10) we obtain

$$a_{jj}^{(k)} = \sum_{i=1}^{m} \alpha_{ij}^2 \mu_i^k, \tag{2.21}$$

and so

$$H_j^G(t) = \sum_{k=0}^{\infty} t^k \sum_{i=1}^{m} \alpha_{ij}^2 \mu_j^k = \sum_{i=1}^{m} \alpha_{ij}^2/(1 - \mu_i t).$$

Now, for example, we have

$$P_{G-j}(x) = \frac{1}{x} P_G(x) H_j^G\left(\frac{1}{x}\right),$$

and hence also

$$H_j^G(t) = P_{G-j}(1/t)/t P_G(1/t).$$

Before we establish a general formula for the characteristic polynomial of a graph modified by the addition of a vertex, we rewrite two of the results already described. The formula (2.20) can be written in the form

$$P_{G_j}(x) = P_G(x)\left(x - \sum_{i=1}^{m} \frac{\|P_i \mathbf{e}_j\|^2}{x - \mu_i}\right), \tag{2.22}$$

while (2.8) can be written in the form

$$P_{K_1 \triangledown G}(x) = P_G(x)\left(x - \sum_{i=1}^{m} \frac{\|P_i \mathbf{j}\|^2}{x - \mu_i}\right). \tag{2.23}$$

These are special cases of the following result:

**Theorem 2.2.8** [Row7]. *Let $G$ be a graph whose adjacency matrix $A$ has spectral decomposition $A = \sum_{i=1}^{m} \mu_i P_i$. Let $\emptyset \neq S \subseteq V(G) = \{1, 2, \ldots, n\}$ and let $G^*$ be the graph obtained from $G$ by adding one new vertex whose neighbours are the vertices in $S$. Then*

$$P_{G^*}(x) = P_G(x)\left(x - \sum_{i=1}^{m} \frac{\sigma_i^2}{x - \mu_i}\right), \quad \text{where } \sigma_i = \|\sum_{k \in S} P_i \mathbf{e}_k\|.$$

**Proof.** Let $\mathbf{r}$ be the characteristic vector of $S$; that is, $\mathbf{r} = \sum_{j \in S} \mathbf{e}_j$. Since $\mathrm{adj}(xI - A) = \det(xI - A)(xI - A)^{-1}$, we have

$$P_{G^*}(x) = \begin{vmatrix} x & -\mathbf{r}^\top \\ -\mathbf{r} & xI - A \end{vmatrix} = x \det(xI - A) - \mathbf{r}^\top \mathrm{adj}(xI - A)\mathbf{r}$$

$$= P_G(x)\left(x - \sum_{i=1}^{m} \frac{\|P_i \mathbf{r}\|^2}{x - \mu_i}\right).$$

$\square$

## 2.3 General reduction procedures

In the previous section we considered graphs constructed in prescribed fashion from smaller graphs. Here, for an arbitrary graph $G$, we discuss relations between $P_G(x)$ and the characteristic polynomials of proper subgraphs of $G$.

**Theorem 2.3.1** [Clar]. *For any graph $G$, with $V(G) = \{1, \ldots, n\}$, the derivative of $P_G(x)$ is given by*

$$P'_G(x) = \sum_{j=1}^{n} P_{G-j}(x). \qquad (2.24)$$

**Proof.** The derivative of an $n \times n$ determinant is the sum of $n$ determinants, obtained by differentiating each row in turn. Let $A$ be the adjacency matrix of $G$, and $A_j$ the matrix obtained from $A$ by deleting the $j$-th row and the $j$-th column. Row-by-row differentiation of $\det(xI - A)$ yields

$$P'_G(x) = \sum_{j=1}^{n} \det(xI - A_j) = \sum_{j=1}^{n} P_{G-j}(x).$$

$\square$

Some remarks are in order (see also Section 8.3):

(i) It follows from Theorem 2.3.1 that if we know the polynomials $P_{G-j}(x)$ $(j \in V(G))$, then we can determine $P_G(x)$ to within some constant $c$. We can determine $c$ if we also know one eigenvalue $\lambda$ of $G$. In particular, if some $P_{G-j}(x)$ has a repeated root $\lambda$, then by the Interlacing Theorem, $\lambda$ is an eigenvalue of $G$.

(ii) It is known (see [CvLe2]) that if $G$ is a tree then $P_G(x)$ is determined by the polynomials $P_{G-j}(x)$ $(j \in V(G))$.

(iii) It is known (see [Tut1] or [LauSc, Section 10.3]) that, for any graph $G$, $P_G(x)$ is determined by the vertex-deleted subgraphs $G - j$ $(j \in V(G))$.

We mention without proof an algorithm for the recursive computation of the characteristic polynomial of a multigraph $G$ (where loops and multiple edges are allowed). Let $G - [uv]$ denote the graph obtained from $G$ by deleting all edges between $u$ and $v$, and let $G^*$ be the graph obtained from $G - [uv]$ by amalgamating $u$ and $v$. If $m$ is the number of edges between $u$ and $v$ then (see [Row3]):

$$P_G(x) = P_{G-[uv]}(x) + m P_{G^*}(x) + m(x - m) P_{G-u-v}(x) - m P_{G-u}(x)$$
$$- m P_{G-v}(x).$$

*Graph operations and modifications*

This equation is called the *deletion-contraction algorithm*. Note that if $G$ is a graph then $G^*$ will have multiple edges precisely when $u$ and $v$ have a common neighbour in $G$; hence the multigraph setting. Once again, the equation is established by expanding the determinant which defines the characteristic polynomial. For subsequent results we need to relate the coefficients of $P_G(x)$ to the structure of $G$, and our starting point is the following result. Here an *elementary graph* is a graph in which each component is $K_2$ or a cycle.

**Theorem 2.3.2** [Har1]. *If $G$ is a graph with $n$ vertices and adjacency matrix $A$, then*

$$\det(A) = (-1)^n \sum_{H \in \mathcal{H}} (-1)^{p(H)} 2^{c(H)},$$

*where $\mathcal{H}$ is the set of elementary spanning subgraphs of $G$, $p(H)$ denotes the number of components of $H$ and $c(H)$ denotes the number of cycles in $H$.*

**Proof.** Consider a term $\mathrm{sgn}(\pi) a_{1,\pi(1)} a_{2,\pi(2)} \cdots a_{n,\pi(n)}$ in the expansion of $\det(A)$. If this term is non-zero then $j \sim \pi(j)$ for all $j = 1, 2, \ldots, n$. Thus $\pi$ is fixed-point-free and can be expressed as a composition $\gamma_1 \gamma_2 \cdots \gamma_t$ of disjoint cyclic permutations of length at least 2. This expression determines an elementary spanning subgraph $H$ in which the components isomorphic to $K_2$ are determined by the transpositions among the $\gamma_i$, and the cycles are determined by the remaining $\gamma_i$. The sign of $\pi$ is $(-1)^r$, where $r = \sum_{i=1}^{t} (\ell(\gamma_i) - 1)$ and $\ell(\gamma_i)$ is the length of $\gamma_i$. Since $t = p(H)$ and $\sum_{i=1}^{t} \ell(\gamma_i) = n$, we have $\mathrm{sgn}(\pi) = (-1)^{n-p(H)}$. Finally, $H$ arises from $2^{c(H)}$ permutations with the same sign as $\pi$, namely $\gamma_1^{\pm 1} \gamma_2^{\pm 1} \cdots \gamma_s^{\pm 1} \gamma_{s+1} \cdots \gamma_t$, where $s = c(H)$ and $\gamma_1, \gamma_2, \ldots, \gamma_s$ are the $\gamma_i$ of length $> 2$. $\square$

**Corollary 2.3.3** (Sachs' Coefficient Theorem [Sac2]). *Let $P_G(x) = x^n + c_1 x^{n-1} + \cdots + c_{n-1}x + c_n$, and let $\mathcal{H}_i$ be the set of elementary subgraphs of $G$ with $i$ vertices. Then*

$$c_i = \sum_{H \in \mathcal{H}_i} (-1)^{p(H)} 2^{c(H)} \quad (i = 1, \ldots, n).$$

**Proof.** The number $(-1)^i c_i$ is the sum of all $i \times i$ principal minors of $A$, and each such minor is the determinant of the adjacency matrix of an induced subgraph on $i$ vertices. An elementary subgraph with $i$ vertices is contained in exactly one such subgraph, and so the result follows by applying Theorem 2.3.2 to each minor. $\square$

**Theorem 2.3.4** [Sch2]. (i) *For any vertex u of the graph G,*

$$P_G(x) = x P_{G-u}(x) - \sum_{v \sim u} P_{G-u-v}(x) - 2 \sum_{Z \in \mathcal{C}(u)} P_{G-V(Z)}(x), \qquad (2.25)$$

*where $\mathcal{C}(u)$ denotes the set of all cycles containing u.*
(ii) *For any edge uv of the graph G,*

$$P_G(x) = P_{G-uv}(x) - P_{G-u-v}(x) - 2 \sum_{Z \in \mathcal{C}(uv)} P_{G-V(Z)}(x), \qquad (2.26)$$

*where $\mathcal{C}(uv)$ denotes the set of all cycles containing uv.*

**Proof.** (i) We follow the original proof of Schwenk by defining a one-to-one correspondence $H \leftrightarrow H'$ between elementary subgraphs that contribute to a coefficient on the left-hand side of (2.25), and those that contribute to a coefficient on the right-hand side. We distinguish three possibilities for an elementary subgraph $H$ of $G$ on $i$ vertices:

(a) if $u \notin V(H)$ then $H' = H$, regarded as a subgraph of $G - u$;
(b) if $u$ lies in a component $K = K_2$ of $H$, then $H' = H - V(K)$, regarded as a subgraph of $G - V(K)$;
(c) if $u$ lies in a cycle $Z$ of $H$, then $H' = H - V(Z)$, regarded as a subgraph of $G - V(Z)$.

Now, by applying Corollary 2.3.3 to each of the graphs that feature in (2.25), we can show that if $H$ contributes $c$ to the coefficient of $x^{n-i}$ on the left, then $H'$ contributes $c$ to the coefficient of $x^{n-i}$ on the right.

In case (a), $H'$ contributes $c$ to the coefficient of $x^{n-1-i}$ in $P_{G-u}(x)$, hence contributes $c$ to the coefficient of $x^{n-i}$ in $x P_{G-u}(x)$. (Note that $H'$ does not contribute to the coefficient of $x^{n-i}$ in the remaining terms.)

In case (b), $H'$ is an elementary spanning subgraph of exactly one graph $G - u - v$ with $v \sim u$, namely $G - V(K)$. Its contribution to the coefficient of $x^{(n-2)-(i-2)} (= x^{n-i})$ is $(-1)^{p(H')}2^{c(H')} = -(-1)^{p(H)}2^{c(H)} = -c$.

In case (c), $H'$ is an elementary spanning subgraph of exactly one graph $G - V(Z)$ with $Z \in \mathcal{C}(u)$. If $|V(Z)| = r$, then the contribution of $H'$ to the coefficient of $x^{(n-r)-(i-r)} (= x^{n-i})$ is $(-1)^{p(H')}2^{c(H')} = -\frac{1}{2}(-1)^{p(H)}2^{c(H)} = -\frac{1}{2}c$.

(ii) The proof, by exactly the same method, is left to the reader. $\qquad\square$

Finally, we mention without proof the following consequence of Jacobi's Theorem on the minors of an adjoint matrix (see [Pra, Section 2.5]):

**Theorem 2.3.5.** *Let u and v be vertices of the graph G, and let $\mathcal{P}_{uv}$ be the set of all u-v paths in G. Then*

$$P_{G-u}(x)P_{G-v}(x) - P_G(x)P_{G-u-v}(x) = \left( \sum_{P \in \mathcal{P}_{uv}} P_{G-V(P)}(x) \right)^2.$$

## 2.4 Line graphs and related operations

In this section we discuss the characteristic polynomials of line graphs and generalized line graphs, along with some related graph operations.

If $G$ is a regular graph, then the characteristic polynomial of $L(G)$ can be expressed in terms of the characteristic polynomial of $G$, as follows.

**Theorem 2.4.1.** *If G is a regular graph of degree r, with n vertices and m ($= \frac{1}{2}nr$) edges, then*

$$P_{L(G)}(x) = (x + 2)^{m-n} P_G(x - r + 2).$$

**Proof.** Recall that $BB^\top = A + rI$ and $B^\top B = A(L(G)) + 2I$, where $A = A(G)$ and $B$ is the incidence matrix of $G$. The theorem follows from the fact that $BB^\top$ and $B^\top B$ have the same non-zero eigenvalues. □

In the general case, we have:

$$BB^\top = A + D, \quad B^\top B = A(L(G)) + 2I, \tag{2.27}$$

where $D$ is the diagonal matrix of vertex degrees. From these relations we immediately obtain

$$P_{L(G)}(x) = (x + 2)^{m-n} Q_G(x + 2), \tag{2.28}$$

where $Q_G(x)$ is the characteristic polynomial of the signless Laplacian matrix $Q = A + D$. Properties of the matrix $Q$ and the corresponding spectrum will be discussed in Chapter 7.

The next theorem shows that a relation between $P_G(x)$ and $P_{L(G)}(x)$ can be established for certain non-regular graphs. Here we make use of the fact that if $M$ is a non-singular square matrix, then (writing $|M|$ for $\det(M)$) we have:

$$\begin{vmatrix} M & N \\ P & Q \end{vmatrix} = |M| \cdot |Q - PM^{-1}N|. \tag{2.29}$$

**Theorem 2.4.2** [Cve1]. *Let G be a semi-regular bipartite graph with $n_1$ independent vertices of degree $r_1$ and $n_2$ independent vertices of degree $r_2$, where $n_1 \geq n_2$. Then*

$$P_{L(G)}(x) = (x+2)^\beta P_G(\sqrt{\alpha_1\alpha_2}) \sqrt{\left(\frac{\alpha_1}{\alpha_2}\right)^{n_1-n_2}},$$

*where $\alpha_i = x - r_i + 2$ $(i = 1, 2)$ and $\beta = n_1 r_1 - n_1 - n_2$.*

**Proof.** We have

$$Q_G(x) = |xI - A - D| = \begin{vmatrix} (x-r_1)I_{n_1} & -K^\top \\ -K & (x-r_2)I_{n_2} \end{vmatrix},$$

where $K$ is an $n_2 \times n_1$ matrix. Making use of (2.29), we have:

$$|xI - A - D| = (x-r_1)^{n_1} \left| (x-r_2)I_{n_2} - K\frac{I_{n_1}}{x-r_1}K^\top \right|$$

$$= (x-r_1)^{n_1-n_2} \left| (x-r_1)(x-r_2)I_{n_2} - KK^\top \right|$$

$$= (x-r_1)^{n_1-n_2} P_{KK^\top}((x-r_1)(x-r_2)), \qquad (2.30)$$

where we write $P_M(x)$ for the characteristic polynomial of a matrix $M$. Now $P_{KK^\top}(x)$ can be expressed in terms of the characteristic polynomial of $A$. We have

$$A = \begin{pmatrix} O & K^\top \\ K & O \end{pmatrix}, \qquad A^2 = \begin{pmatrix} K^\top K & O \\ O & KK^\top \end{pmatrix}$$

and $P_{K^\top K}(x) = x^{n_1-n_2} P_{KK^\top}(x)$. Thus $P_{A^2}(x) = x^{n_1-n_2} P_{KK^\top}(x)^2$. On the other hand, since the eigenvalues of $A^2$ are the squares of the eigenvalues of $A$, and the latter are symmetric about 0 (see Exercise 1.6 and Theorem 3.2.3), we have $P_{A^2}(x^2) = P_A(x)^2$. Accordingly we obtain

$$P_{KK^\top}(x) = \sqrt{\frac{P_{A^2}(x)}{x^{n_1-n_2}}} = \sqrt{x^{n_2-n_1}}P_A(\sqrt{x}). \qquad (2.31)$$

Combining expressions (2.28), (2.30) and (2.31), we obtain the required formula. □

**Corollary 2.4.3.** *If G is a semi-regular bipartite graph with parameters $(n_1, n_2, r_1, r_2)$ $(n_1 \geq n_2)$ and if $\lambda_1, \lambda_2, \ldots, \lambda_{n_2}$ are the first $n_2$ largest eigenvalues of G, then*

$$P_{L(G)}(x) = (x - r_1 - r_2 + 2)(x - r_1 + 2)^{n_1-n_2}(x+2)^{n_1 r_1 - n_1 - n_2 + 1}$$

$$\times \prod_{i=2}^{n_2} ((x - r_1 + 2)(x - r_2 + 2) - \lambda_i^2).$$

**Proof.** The largest eigenvalue $\lambda_1$ is given by $\lambda_1 = \sqrt{r_1 r_2}$ because

$$\begin{pmatrix} O & K^\top \\ K & O \end{pmatrix} \begin{pmatrix} \sqrt{r_1}\mathbf{j}_{n_1} \\ \sqrt{r_2}\mathbf{j}_{n_2} \end{pmatrix} = \sqrt{r_1 r_2} \begin{pmatrix} \sqrt{r_1}\mathbf{j}_{n_1} \\ \sqrt{r_2}\mathbf{j}_{n_2} \end{pmatrix}.$$

Moreover, $G$ contains at least $n_1 - n_2$ eigenvalues equal to 0, because $K$ has rank at most $n_2$. Now the result follows from Theorem 2.4.2 when we expand $(x - r_1 + 2)(x - r_2 + 2) - \lambda_1^2$. □

Next we determine the characteristic polynomials of graphs obtained from regular graphs by other unary operations.

Recall that the subdivision graph $S(G)$ of a graph $G$ is the graph obtained by inserting a new vertex into every edge of $G$. The subdivision graph is a bipartite graph whose adjacency matrix is of the form

$$\begin{pmatrix} O & B^\top \\ B & O \end{pmatrix},$$

where $B$ is the incidence matrix of $G$. Using Equations (2.27) and (2.29) we have

$$P_{S(G)}(x) = \begin{vmatrix} xI_m & -B^\top \\ -B & xI_n \end{vmatrix} = x^m \left| xI_n - B\frac{1}{x}I_m B^\top \right| = x^{m-n} |x^2 I_n - BB^\top|.$$

If $G$ is $r$-regular then $BB^\top = A + rI$, and so we arrive at the following result:

**Theorem 2.4.4.** *If $G$ is a graph with $n$ vertices and $m$ edges then*

$$P_{S(G)}(x) = x^{m-n} Q_G(x^2).$$

*In particular, if $G$ is $r$-regular, then*

$$P_{S(G)}(x) = x^{m-n} P_G(x^2 - r). \tag{2.32}$$

Let $R(G)$ be the graph obtained from $G$ by adding, for each edge $uv$, a new vertex whose neighbours are $u$ and $v$. Thus the adjacency matrix of $R(G)$ is of the form

$$\begin{pmatrix} O & B^\top \\ B & A \end{pmatrix}.$$

**Theorem 2.4.5** [Cve4]. *If $G$ is a regular graph of degree $r$ with $n$ vertices and $m\left(= \frac{1}{2}nr\right)$ edges, then*

$$P_{R(G)}(x) = x^{m-n}(x+1)^n P_G\left(\frac{x^2 - r}{x+1}\right). \tag{2.33}$$

**Proof.** We have

$$P_{R(G)}(x) = \begin{vmatrix} xI_m & -B^\top \\ -B & xI_n - A \end{vmatrix} = x^m \cdot \left| xI_n - A - \frac{1}{x}BB^\top \right|$$

$$= x^{m-n} \cdot |x^2 I_n - xA - A - rI_n| = x^{m-n} \cdot |(x^2 - r)I_n - (x+1)A|$$

$$= x^{m-n}(x+1)^n P_G\left(\frac{x^2 - r}{x+1}\right).$$

$$\square$$

Next, let $Q(G)$ be the graph obtained from $G$ by inserting a new vertex into each edge of $G$, and joining by edges those pairs of new vertices which lie on adjacent edges of $G$. The adjacency matrix of $Q(G)$ is then of the form

$$\begin{pmatrix} O & B \\ B^\top & C \end{pmatrix},$$

where $C = A(L(G))$. Arguments similar to those above lead to the following result:

**Theorem 2.4.6** [Cve4]. *Let $G$ be a graph with $n$ vertices and $m$ edges. Then*

$$P_{Q(G)}(x) = x^{n-m}(x+1)^m P_{L(G)}\left(\frac{x^2 - 2}{x+1}\right). \tag{2.34}$$

In the case that $G$ is regular, we may apply Theorem 1.4.1 to obtain:

**Corollary 2.4.7.** *If $G$ is a regular graph of degree $r$ then*

$$P_{Q(G)}(x) = (x+2)^{m-n}(x+1)^n P_G\left(\frac{x^2 - (r-2)x - r}{x+1}\right). \tag{2.35}$$

Consideration of $Q(G)$, $R(G)$ and $S(G)$ leads us naturally to the investigation of total graphs: the *total graph* $T(G)$ of a graph $G$ is the graph whose vertices are the vertices and edges of $G$, with two vertices of $T(G)$ adjacent if and only if the corresponding elements of $G$ are adjacent or incident. Thus the adjacency matrix of $T(G)$ has the form

$$\begin{pmatrix} A & B \\ B^\top & C \end{pmatrix}.$$

If $G$ is $r$-regular with $n$ vertices and $m$ edges, we have

$$P_{T(G)}(x) = \begin{vmatrix} xI + rI - BB^\top & -B \\ -B^\top & xI + 2I - B^\top B \end{vmatrix}$$

$$= \begin{vmatrix} (x+r)I - BB^\top & -B \\ -(x+r+1)B^\top + B^\top BB^\top & (x+2)I \end{vmatrix}$$

$$= \begin{vmatrix} (x+r)I - BB^\top + \frac{1}{x+2}B(-(x+r+1)B^\top + B^\top BB^\top) & O \\ -(x+r+1)B^\top + B^\top BB^\top & (x+2)I \end{vmatrix}$$

$$= (x+2)^m \left| xI - A + \frac{1}{x+2}(A+rI)(A - (x+1)I) \right|$$

$$= (x+2)^{m-n} \left| A^2 - (2x - r + 3)A + (x^2 - (r-2)x - r)I \right|.$$

It follows that if $\lambda_1, \ldots, \lambda_n$ are the eigenvalues of $A$ then

$$P_{T(G)}(x) = (x+2)^{m-n} \prod_{i=1}^{n} \left( \lambda_i^2 - (2x - r + 3)\lambda_i + x^2 - (r-2)x - r \right)$$

$$= (x+2)^{m-n} \prod_{i=1}^{n} \left( x^2 - (2\lambda_i + r - 2)x + \lambda_i^2 + (r-3)\lambda_i - r \right).$$

Thus we have the following theorem.

**Theorem 2.4.8** [Cve3]. *Let $G$ be a regular graph of degree $r$ ($r > 1$) having $n$ vertices and $m$ edges. If the eigenvalues of $G$ are $\lambda_1, \ldots, \lambda_n$, then $T(G)$ has $m - n$ eigenvalues equal to $-2$ and the following $2n$ eigenvalues:*

$$\frac{1}{2} \left( 2\lambda_i + r - 2 \pm \sqrt{4\lambda_i + r^2 + 4} \right) \qquad (i = 1, \ldots, n).$$

In discussing the eigenvalues of $T(G)$ arising in Theorem 2.4.8, we shall consider only connected graphs. Note that $-r \le \lambda_i \le r$ ($i = 1, \ldots, n$), and consider the functions

$$f_1(x) = \frac{1}{2} \left( 2x + r - 2 + \sqrt{4x + r^2 + 4} \right),$$

$$f_2(x) = \frac{1}{2} \left( 2x + r - 2 - \sqrt{4x + r^2 + 4} \right).$$

Suppose first that $r > 2$. Both functions are increasing on the interval $[-r, r]$; the first one maps this interval onto $[-2, 2r]$, the second onto $[-r, r - 2]$. Thus the eigenvalues of $T(G)$ lie in the interval $[-r, 2r]$ (an observation that holds also for $r = 1$). The largest eigenvalue is, naturally, equal to $2r$, while $r - 2$ always lies in the spectrum. The smallest eigenvalue is equal to $-r$ if and only if $G$ is bipartite (see Theorem 3.2.4). The multiplicity of the eigenvalue $-2$ in $T(G)$ is equal to $m - n + m(-r) + m(-1)$, where $m(\lambda)$ is the multiplicity of the eigenvalue $\lambda$ in $G$.

Now suppose that $r = 2$. In this case the function $f_2(x)$ has a minimum at $x = -7/4$. Since $f_2(-7/4) = -9/4$, the smallest eigenvalue of $T(G)$ is greater than $-9/4$. Equality can never hold, since an eigenvalue of a graph cannot be rational non–integral number. But, since the eigenvalues of a connected regular graph $G$ of degree 2 with $n$ vertices are $2\cos\frac{2\pi}{n}i$ $(i = 1, 2, \ldots, n)$ (see Example 1.1.4), there exist graphs $G$ for which the smallest eigenvalue of $T(G)$ is arbitrarily close to the lower bound $-9/4$.

Lastly, the case $r = 1$ is quite simple: $G$ has eigenvalues $1, -1$, and $T(G)$ has eigenvalues $2, -1, -1$.

Turning now to generalized line graphs, we give a result which, in one special case, yields the whole spectrum. No general formula is known.

**Theorem 2.4.9.** *Let $G$ be a graph having vertex degrees $d_1, d_2, \ldots, d_n$. If $a_1, a_2, \ldots, a_n$ are non-negative integers such that $d_i + 2a_i = d$, $i = 1, 2, \ldots, n$, then*

$$P_{L(G;a_1,a_2,\ldots,a_n)}(x) = x^a(x + 2)^{m-n+a}P_G(x - d + 2), \quad \text{where } a = \sum_{i=1}^{n} a_i.$$

**Proof.** An incidence matrix of $L(G; a_1, a_2, \ldots, a_n)$ has the form

$$C = \begin{pmatrix} B & L_1 & L_2 & \ldots & L_n \\ O & M_1 & O & \ldots & O \\ O & O & M_2 & \ldots & O \\ \vdots & \vdots & \vdots & \ddots & \vdots \\ O & O & O & \ldots & M_n \end{pmatrix}$$

where $B$ is the incidence matrix of $G$; $L_i$ is an $n \times 2a_i$ matrix in which all entries of the $i$-th row are 1, and all other entries are 0; and $M_i$ is an $a_i \times 2a_i$ matrix of the form $(I \mid - I)$. We have $C^\top C = A + 2I$, where now $A$ is the adjacency matrix of $L(G; a_1, a_2, \ldots, a_n)$, and the theorem follows from the fact that $C^\top C$ and $CC^\top$ have the same non-zero eigenvalues. $\square$

# 2.5 Cartesian type operations

Next, we consider a very general graph operation called NEPS (*non-complete extended p-sum*) of graphs.

**Definition 2.5.1.** Let $\mathcal{B}$ be a set of non-zero binary $n$-tuples, i.e. $\mathcal{B} \subseteq \{0, 1\}^n \backslash \{(0, \ldots, 0)\}$. The NEPS of graphs $G_1, \ldots, G_n$ with basis $\mathcal{B}$ is the graph with vertex set $V(G_1) \times \cdots \times V(G_n)$, in which two vertices, say $(x_1, \ldots, x_n)$ and $(y_1, \ldots, y_n)$, are adjacent if and only if there exists an n-tuple $\beta = (\beta_1, \ldots, \beta_n) \in \mathcal{B}$ such that $x_i = y_i$ whenever $\beta_i = 0$, and $x_i$ is adjacent to $y_i$ (in $G_i$) whenever $\beta_i = 1$.

Clearly the NEPS construction generates many binary graph operations in which the vertex set of the resulting graph is the Cartesian product of the vertex sets of the graphs on which the operation is performed. We mention some special cases in which a graph is the NEPS of graphs $G_1, \ldots, G_n$ with basis $\mathcal{B}$. In particular, for $n = 2$ we have the following familiar operations:

(i) the *sum* $G_1 + G_2$, when $\mathcal{B} = \{(0, 1), (1, 0)\}$;
(ii) the *product* $G_1 \times G_2$, when $\mathcal{B} = \{(1, 1)\}$;
(iii) the *strong product* $G_1 * G_2$, when $\mathcal{B} = \{(0, 1), (1, 0), (1, 1)\}$.

(A variety of terms for these particular constructions can be found in the literature.)

The notion of NEPS arises in a natural way when studying spectral properties of graphs obtained by binary operations of the type mentioned above.

The adjacency matrix of a NEPS can be expressed in terms of the adjacency matrices of the constituent graphs by means of the *Kronecker product* of matrices. We define this product below, and note the properties which enable us to describe the spectrum of a NEPS.

**Definition 2.5.2.** The Kronecker product $A \otimes B$ of matrices $A = (a_{ij})_{m \times n}$ and $B = (b_{ij})_{p \times q}$ is the $mp \times nq$ matrix obtained from $A$ by replacing each element $a_{ij}$ with the block $a_{ij} B$.

Thus the entries $A \otimes B$ consist of all the $mnpq$ possible products of an entry of $A$ with an entry of $B$. The Kronecker product is an associative operation, and the following relations are well known (see, for example, [MaMi], p. 18 and p. 8). For square matrices $A$ and $B$, we have

$$\mathrm{tr}\,(A \otimes B) = \mathrm{tr}\,A \cdot \mathrm{tr}\,B, \qquad (2.36)$$

while

$$(A \otimes B) \cdot (C \otimes D) = (AC) \otimes (BD) \qquad (2.37)$$

whenever the products $AC$ and $BD$ exist.

Starting from (2.37) and using induction, we obtain

$$(A_1 \otimes \cdots \otimes A_n) \cdot (B_1 \otimes \cdots \otimes B_n) \cdots (M_1 \otimes \cdots \otimes M_n)$$
$$= (A_1 B_1 \cdots M_1) \otimes \cdots \otimes (A_n B_n \cdots M_n). \tag{2.38}$$

The proof of the next result is left as an exercise.

**Theorem 2.5.3.** *Let $A_1, \ldots, A_n$ be adjacency matrices of graphs $G_1, \ldots, G_n$, respectively. The NEPS $G$ with basis $\mathcal{B}$ of graphs $G_1, \ldots, G_n$ has as adjacency matrix the matrix $A$ given by*

$$A = \sum_{\beta \in \mathcal{B}} A_1^{\beta_1} \otimes \cdots \otimes A_n^{\beta_n}. \tag{2.39}$$

*Here $A_k^0$ is the identity matrix of the same size as $A_k$, and $A_k^1 = A_k$.*

One consequence of Theorem 2.5.3 is the following result.

**Theorem 2.5.4.** *If $\lambda_{i1}, \ldots, \lambda_{ik_i}$ are the eigenvalues of $G_i$ ($i = 1, \ldots, n$), then the spectrum of the NEPS of $G_1, \ldots, G_n$ with basis $\mathcal{B}$ consists of all possible values $\Lambda_{i_1, \ldots, i_n}$ where*

$$\Lambda_{i_1, \ldots, i_n} = \sum_{\beta \in \mathcal{B}} \lambda_{1i_1}^{\beta_1} \cdots \lambda_{ni_n}^{\beta_n} \quad (i_h = 1, \ldots, k_h; \ h = 1, \ldots, n). \tag{2.40}$$

**Proof.** Let $\mathbf{x}_{ij}$ ($j = 1, \ldots, k_i$) be linearly independent eigenvectors of $G_i$, with $A_i \mathbf{x}_{ij} = \lambda_{ij} \mathbf{x}_{ij}$ ($i = 1, 2, \ldots, n; j = 1, 2, \ldots, k_i$). Consider the vector

$$\mathbf{x} = \mathbf{x}_{1i_1} \otimes \cdots \otimes \mathbf{x}_{ni_n}.$$

Using Theorem 2.5.3, we see that $A\mathbf{x} = \Lambda_{i_1, \ldots, i_n} \mathbf{x}$. In this way, we find $k_1 k_2 \cdots k_n$ linearly independent eigenvectors, and hence all $k_1 k_2 \cdots k_n$ eigenvalues. $\qquad \square$

Thus if $\lambda_1, \ldots, \lambda_n$ and $\mu_1, \ldots, \mu_m$ are the eigenvalues of $G$ and $H$, respectively, then:

$\lambda_i + \mu_j$ ($i = 1, \ldots, n; \ j = 1, \ldots, m$) are the eigenvalues of $G + H$;

$\lambda_i \mu_j$ ($i = 1, \ldots, n; \ j = 1, \ldots, m$) are the eigenvalues of $G \times H$;

$\lambda_i + \mu_j + \lambda_i \mu_j$ ($i = 1, \ldots, n; \ j = 1, \ldots, m$) are the eigenvalues of $G * H$.

**Example 2.5.5.** We have $L(K_{m,n}) = K_m + K_n$. Since $K_n$ has spectrum $n - 1, (-1)^{n-1}$ we obtain $m + n - 2, (n - 2)^{m-1}, (m - 2)^{n-1}, (-2)^{(m-1)(n-1)}$ for the spectrum of $L(K_{m,n})$. $\qquad \square$

## 2.6 Spectra of graphs of particular types

In this section we shall determine the characteristic polynomials and spectra of certain graphs making use of the results described in this chapter. Some of the results of this section are well known in matrix theory, but we will deduce them using methods more consistent with the theory we have developed.

**1**. For the *null graph* $G$ with $n$ vertices, we see immediately that $P_G(x) = x^n$; in other words, the spectrum consists of $n$ eigenvalues equal to 0.

**2**. The *complete graph* $K_n$ with $n$ vertices is the complement of the graph of the previous example, and by Theorem 2.1.2 we have $P_{K_n}(x) = (x - n + 1)(x + 1)^{n-1}$, that is, the spectrum of $K_n$ consists of the eigenvalue $n - 1$ and $n - 1$ eigenvalues equal to $-1$.

**3**. Each component of a *regular graph $G$ of degree* 1 is isomorphic to the graph $K_2$, with characteristic polynomial $x^2 - 1$. If $G$ has $2k$ vertices, then by Theorem 2.1.1 we have $P_G(x) = (x^2 - 1)^k$.

**4**. The complement of the graph $kK_2$ above is the *regular graph $H$ of degree* $n - 2$ with $n = 2k$ vertices (i.e. the cocktail party graph $CP(k)$). By Theorem 2.1.2, its characteristic polynomial is $P_H(x) = (x - 2k + 2)x^k(x + 2)^{k-1}$.

**5**. For the *complete bipartite graph* $K_{n_1,n_2}$, we exploit the relation $K_{n_1,n_2} = G_1 \triangledown G_2$, where $G_1$, $G_2$ are graphs which consist of $n_1$, $n_2$ isolated vertices, respectively. Since $P_{G_1}(x) = x^{n_1}$ and $P_{G_2}(x) = x^{n_2}$, Theorem 2.1.4 yields $P_{K_{n_1,n_2}}(x) = (x^2 - n_1 n_2) \cdot x^{n_1+n_2-2}$. Thus the spectrum of the graph $K_{n_1,n_2}$ consists of $\sqrt{n_1 n_2}$, $-\sqrt{n_1 n_2}$ and $n_1 + n_2 - 2$ eigenvalues equal to 0. If $n_1 = n$ and $n_2 = 1$, we obtain a *star* with $n + 1$ vertices, and its characteristic polynomial is $P_{K_{1,n}}(x) = (x^2 - n)x^{n-1}$.

**6**. As already determined in Example 1.1.4, the spectrum of a *cycle* $C_n$ consists of the numbers $2\cos\dfrac{2\pi}{n}j$ $(j = 1, \ldots, n)$. Now $\cos\dfrac{2\pi}{n}j$ $(j = 1, ..., n)$ are the roots of $T_n(x) - 1$, where $T_n(x)$ is a Chebyshev polynomial of the first kind, defined by

$$\cos n\theta = T_n(\cos\theta).$$

Explicitly,

$$T_n(x) = \sum_{k=0}^{[n/2]} (-1)^k \frac{n}{n-k} \binom{n-k}{k} 2^{n-2k-1} x^{n-2k},$$

an expression which may be derived from the recurrence relation $T_{k+1}(x) = 2xT_k(x) - T_{k-1}(x)$ $(k \geq 1)$. Thus $P_{C_n}(x) = 2(T_n(x/2) - 1)$, that is,

$$P_{C_n}(x) = -2 + \sum_{k=0}^{[n/2]} (-1)^k \frac{n}{n-k} \binom{n-k}{k} x^{n-2k}.$$

**7**. By applying Theorem 2.2.3 we can deduce from the previous result the characteristic polynomial and spectrum of the *path $P_n$* with $n$ vertices. All vertex-deleted subgraphs of $C_n$ are isomorphic to the path $P_{n-1}$. Therefore, $P_{P_{n-1}}(x) = \dfrac{1}{n} P'_{C_n}(x)$, and so

$$P_{P_n}(x) = \sum_{k=0}^{[n/2]} (-1)^k \binom{n-k}{k} x^{n-2k}.$$

Chebyshev polynomials of the second kind are defined by

$$U_n(\cos\theta) = \frac{\sin(n+1)\theta}{\sin\theta}.$$

Thus $T'_n(\cos\theta) = nU_{n-1}(\cos\theta)$, and so $P_{P_n}(x) = U_n(x/2)$. It follows that the spectrum of the path $P_n$ consists of the numbers $2\cos\dfrac{\pi}{n+1}j$ $(j = 1, \ldots, n)$.

**8**. The *complete multipartite graph* $K_{n_1,\ldots,n_k}$ is the complement of the graph $G = K_{n_l} \mathbin{\dot\cup} \cdots \mathbin{\dot\cup} K_{n_k}$. We may extend the formula (2.5) to such a graph $G$, to obtain

$$P_{\overline{G}}(x) = (-1)^n P_G(-1-x) \left\{ 1 - \sum_{i=1}^{k} \frac{n_i}{(x+1) + (n_i - 1)} \right\}.$$

where $n = n_1 + \cdots + n_k$. Since

$$P_G(x) = (x+1)^{n-k}(x - n_1 + 1) \cdots (x - n_k + 1),$$

we readily obtain:

$$P_{K_{n_1,\ldots,n_k}}(x) = x^{n-k} \left( 1 - \sum_{i=1}^{k} \frac{n_i}{x + n_i} \right) \prod_{j=1}^{k} (x + n_j),$$

or

$$P_{K_{n_1,\ldots,n_k}}(x) = \sum_{i=0}^{k} (1 - i) S_i x^{n-i},$$

where $S_0 = 1$ and for $i \in \{1, \ldots, k\}$, $S_i$ is the $i$-th elementary symmetric function of the numbers $n_1, \ldots, n_k$.

**9**. Interesting graphs can be obtained if we consider the *sum of two paths*, or of *a path and a cycle*, or of *two cycles*.

The *sum of two paths* having $m$ and $n$ vertices respectively is the graph of an $m \times n$ lattice, represented in Fig. 2.1. According to Theorem 2.5.4, the spectrum of this graph consists of all numbers of the form

$$2\cos\frac{\pi}{m+1}j + 2\cos\frac{\pi}{n+1}k \quad (j = 1, \ldots, m; k = 1, \ldots, n).$$

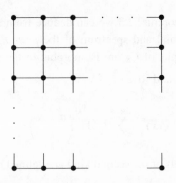

Figure 2.1 The sum of two paths.

The *sum of the cycle $C_m$ and the path $P_n$* gives the graph of an analogous lattice on a cylinder which can be obtained from the graph of Fig 2.1 (with $m + 1$ instead of $m$) by identifying the vertices of the first row with the corresponding vertices of the last row. The spectrum of this graph consists of the numbers

$$2\cos\frac{2\pi}{m}j + 2\cos\frac{\pi}{n+1}k \quad (j = 1, \ldots, m; k = 1, \ldots, n).$$

In similar fashion, the *sum of two cycles* is the graph of a square lattice on a torus, and its spectrum consists of the numbers

$$2\cos\frac{2\pi}{m}j + 2\cos\frac{2\pi}{n}k \quad (j = 1, \ldots, m; k = 1, \ldots, n).$$

If we consider the strong product instead of the sum, we obtain the graphs corresponding to modified square lattices, in which 'diagonals' are added to each 'square'. Again, the spectra can be easily determined.

**10**. The *graph of a k-dimensional (finite) lattice* is a graph $G$ whose vertices are all the $k$-tuples of numbers $1, \ldots, n$, with two $k$-tuples adjacent if and only if they differ in exactly one coordinate. For $n = 2$, $G$ is just the hypercube $Q_k$. For $k = 2$ the graph $G$ is just $L(K_{n,n})$, and for $k = 3$ we obtain the *cubic lattice graph*. In the general case, $G$ is the sum of $k$ graphs, each isomorphic to $K_n$. Now the sum $G_1 + \cdots + G_k$ is a NEPS whose basis consists of all $k$-tuples of the numbers $0, 1$ in which exactly one number $1$ appears. By Theorem 2.5.4, the eigenvalues of $G_1 + \cdots + G_k$ are the numbers $\lambda_{1i_1} + \cdots + \lambda_{ki_k}$, where $\lambda_{ji_j}$ is an eigenvalue of $G_j$. In the case that each $G_j$ is isomorphic to $K_n$, we find that the eigenvalues of $G$ are the numbers $\lambda_j = n(k - j) - k$ with multiplicity

$$p_j = \binom{k}{j}(n - 1)^j \quad (j = 0, 1, \ldots, k).$$

**11**. The *Möbius ladder* $M_n$ is the graph with $2n$ vertices $1, \ldots, 2n$ in which the following pairs of vertices are adjacent:

$$(j, j + 1), \quad j = 1, \ldots, 2n - 1,$$
$$(1, 2n),$$
$$(j, j + n), \quad j = 1, \ldots, n.$$

In other words, the adjacency matrix $A$ of $M_n$ is a circulant $2n \times 2n$ matrix whose entries in the first row are equal to 0 except for 1 in the second, $(n + 1)$-th, and $(2n)$-th columns. Thus $A = P + P^n + P^{2n-1}$, where $P$ is the permutation matrix determined by a cyclic permutation of length $2n$. Therefore (cf. Example 1.1.4) the spectrum of $M_n$ consists of numbers

$$\lambda_j = e^{\frac{2\pi j}{2n}i} + \left(e^{\frac{2\pi j}{2n}i}\right)^n + \left(e^{\frac{2\pi j}{2n}i}\right)^{2n-1} \quad (i = \sqrt{-1}; \; j = 1, \ldots, 2n),$$

that is,

$$\lambda_j = 2\cos\frac{\pi}{n}j + (-1)^j \quad (j = 1, \ldots, 2n),$$

a formula similar to those above, but obtained without invoking the results of this chapter.

## Exercises

**2.1** For the graph $G$ of Example 1.1.3, find $P_G(x)$ by using the fact that $G \cong K_{1,2,2}$.

**2.2** Show that the Petersen graph (Example 1.1.5) is isomorphic to $\overline{L(K_5)}$, and use this fact to determine its spectrum.

**2.3** Prove Proposition 2.1.4.

**2.4** Let $G^j$ be the multigraph obtained from $G$ by adding a loop at vertex $j$. Show that

$$P_{G^j}(x) = P_G(x)\left(1 - \sum_{i=1}^{m} \frac{\alpha_{ij}^2}{x - \mu_i}\right).$$

**2.5** Prove Theorem 2.2.1.

**2.6** Let $G$ be a graph with a pendant edge $uv$. Show that 0 has the same multiplicity as an eigenvalue of $G$ and $G - u - v$.

**2.7** Let $G_j$, $G_j'$ be the $B$-graphs obtained from the $B$-graph $G$ by adding, at the vertex $j$, a pendant edge and a petal, respectively. Show that

$$P_{L(G_j')}(x) = -2x\, P_{L(G_j)}(x) - 2x^2 P_{L(G)}(x).$$

**2.8** Let $F$ be the graph obtained from $G \cup H$ by introducing an edge between the vertex $u$ of $G$ and the vertex $v$ of $H$. Show that

$$P_F(x) = P_G(x)P_H(x)\left\{1 - \sum_{i=1}^{m}\sum_{k=1}^{p}\frac{\alpha_{iu}^2 \delta_{kv}^2}{(x-\mu_i)(x-\nu_k)}\right\},$$

where $\alpha_{1u}, \ldots, \alpha_{mu}$ are the angles of $G$ at $u$, and $\delta_{1v}, \ldots, \delta_{pv}$ are the angles of $H$ at $v$.

**2.9** Let $F$ be the coalescence of graphs $G$ and $H$ obtained by identifying the vertex $u$ of $G$ with the vertex $v$ of $H$. Show that (in the notation of Exercise 2.8)

$$P_F(x) = \frac{1}{x}P_G(x)P_H(x)\left\{1 - \left(1 - x\sum_{i=1}^{m}\frac{\alpha_{iu}^2}{x-\mu_i}\right)\left(1 - x\sum_{k=1}^{p}\frac{\gamma_{kv}^2}{x-\nu_k}\right)\right\}.$$

**2.10** Let $F \cdot G$ be the coalescence whose vertex $w$ is obtained by identifying the vertex $u$ of $F$ with the vertex $v$ of $G$. Show that

$$\frac{1}{H_w^{F \cdot G}(t)} = \frac{1}{H_u^F(t)} + \frac{1}{H_v^G(t)} - 1.$$

**2.11** Prove by induction on $k$ that the $k$-th derivative of the characteristic polynomial of a graph $G$ is given by the formula

$$P_G^{(k)}(x) = k!\sum_{|S|=k} P_{G-S}(x),$$

where the summation runs over all $k$-subsets $S$ of $V(G)$.

**2.12** Verify the deletion-contraction algorithm [Row3].

**2.13** Use the deletion-contraction algorithm to prove that if the graph $H$ is obtained from the graph $G$ by subdividing the edge $uv$ then

$$P_H(x) = P_G(x) + (x-1)P_{G-uv}(x) - P_{G-u}(x) - P_{G-v}(x) + P_{G-u-v}(x).$$

**2.14** Let $T$ be a tree with $2k$ vertices. Use Corollary 2.3.3 to show that the constant term in $P_T(x)$ is $(-1)^k$ or $0$ according as $T$ does or does not have a perfect matching.

**2.15** Prove Theorem 2.3.4(ii).

**2.16** Let $G$ be an $r$-regular graph with $n$ vertices such that both $G$ and $\overline{G}$ are connected. Show that

$$\frac{P_{G-j}(x)}{P_G(x)} + \frac{P_{\overline{G}-j}(-x-1)}{P_{\overline{G}}(-x-1)} = \frac{1}{(x-r)(x+n-r)}.$$

**2.17** Prove Theorem 2.5.3.

**2.18** Show that (i) if $G$ is a bipartite graph then $K_2 \times G = 2G$, (ii) if $G$ is a connected non-bipartite graph then $K_2 \times G$ is a connected bipartite graph.

**2.19** Let $\lambda$ be a simple eigenvalue of the graph $G$. Show that $\mathcal{E}(\lambda)$ is spanned by a vector $(x_1, \ldots, x_n)^\top$ such that $x_j^2 = |P_{G-j}(\lambda)|$ $(j = 1, \ldots, n)$ [CvRS9].

# Notes

A majority of the results in this chapter, some with different proofs, can be found in Chapter 2 of [CvDSa] or Chapter 4 of [CvRS2], along with references to the original papers. The characteristic polynomials of a join of graphs (Theorem 2.1.5) and a complete multipartite graph (Section 2.6) were originally derived by means of walk generating functions. Sachs' Coefficient Theorem (Corollary 2.3.3) was proved independently by Spialter [Spia] and Milić [Mil]. Theorems 2.2.1, 2.2.3 and 2.2.4 can be obtained as consequences of the deletion-contraction algorithm (see [Row3]). Formula (2.23) appears in [Cve4], while [Mnu] contains a generalization of Theorem 2.4.4 to the $k$-th subdivision graph $S_k(G)$ (obtained from $G$ by inserting $k$ vertices of degree 2 in each edge).

A survey of characteristic polynomials of modified graphs is given in [Row11]. Local modifications of a graph may be regarded as graph perturbations (see Section 8.1), and the resulting perturbations of eigenvalues are discussed in [Row5, Row6] and [CvRS2, Chapter 6]. Corollary 2.4.7 corrects Equation (2.37) of [CvDSa]. A review of results on NEPS can be found in [CvSi1].

# 3

# Spectrum and structure

In this chapter we consider various relations between the structure of a graph and its spectrum. We saw in Chapter 1 that the spectrum of a graph does not determine the graph up to isomorphism; nevertheless, often significant information on graph invariants or properties can be extracted from the spectrum. We consider constraints on certain eigenvalues as well as the role of further spectral invariants such as graph angles.

## 3.1 Counting certain subgraphs

The following result, noted in Chapter 1 as a consequence of Proposition 1.3.4, plays a basic role in counting certain subgraphs of a graph with spectrum $\lambda_1 \geq \lambda_2 \geq \cdots \geq \lambda_n$.

**Theorem 3.1.1.** *The number of closed walks of length k in a graph G is equal to $s_k$, where*

$$s_k = \sum_{i=1}^{n} \lambda_i^k, \tag{3.1}$$

the $k$-th spectral moment of $G$.

Clearly $s_1 = 0$ (equivalently, $G$ has no loops). If $G$ has $e$ edges and $t$ triangles, then $s_2 = 2e$ and $s_3 = 6t$. To see this, note first that a closed walk of length 2 traverses an edge, while the edge $ij$ accounts for two closed walks of length 2, namely $iji$ and $jij$. Secondly, a closed walk of length 3 traverses a triangle, and each triangle accounts for six closed walks of length 3 (there are three choices of starting point and two choices of orientation).

52

Accordingly, we have:

(i) the number of vertices is equal to the number of eigenvalues (with repetitions);
(ii) the number of edges is equal to $\frac{1}{2}s_2$;
(iii) the mean degree is $\frac{1}{n}s_2$;
(iv) the number of triangles is equal to $\frac{1}{6}s_3$;
(v) the average number of triangles containing a given vertex is $\frac{1}{2n}s_3$.

These observations explain why graphs are often ordered lexicographically by the sequence $(s_0, s_1, \ldots, s_{n-1})$, as in Tables A3 and A4 of the Appendix. (Recall that $s_0, s_1, \ldots, s_{n-1}$ determine the spectrum and hence all other spectral moments.)

When $k \geq 4$, a closed walk of length $k$ can trace more than one type of subgraph; for example, when $k = 4$ we have three possible types, namely $K_2$, $P_3$ and $C_4$. Moreover, when traversing $P_3$, the number of $v$-$v$ walks of length 4 depends on $v$. To take this fact into consideration, we use graph angles (which are graph invariants related to the vertices): Equation (1.10) yields the following 'local' counterpart of Theorem 3.1.1.

**Theorem 3.1.2.** *The number $n_k(j)$ of closed walks of length $k$ starting (and terminating) at vertex $j$ of a graph $G$ is given by*

$$n_k(j) = \sum_{i=1}^{m} \alpha_{ij}^2 \mu_i^k. \qquad (3.2)$$

An immediate consequence is that the degree of any vertex, and the number of triangles incident with any vertex, can be extracted from the eigenvalues and angles of a graph. In such situations we say that the corresponding invariant (or property) is $EA$-reconstructible.

**Theorem 3.1.3.** *The degree $d_j$ of vertex $j$, and the number $t_j$ of triangles containing vertex $j$ of a graph $G$, are given by*

$$d_j = \sum_{i=1}^{m} \alpha_{ij}^2 \mu_i^2, \quad t_j = \frac{1}{2} \sum_{i=1}^{m} \alpha_{ij}^2 \mu_i^3.$$

**Proof.** This follows from (3.2) since $n_2(j) = d_j$ and $n_3(j) = 2t_j$. $\qquad \square$

**Remark 3.1.4.** Let $f$ be the number of subgraphs of $G$ isomorphic to $P_3$. Counting pairs of edges containing a given vertex, we find that $f = \sum_{i=1}^{n} \binom{d_i}{2}$. Now it follows from Theorem 3.1.3 that $f$ is $EA$-reconstructible. $\qquad \square$

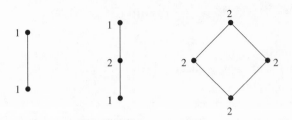

Figure 3.1 Graphs from Theorem 3.1.5.

The next two results show that the number of quadrangles (4-cycles), and the number of pentagons (5-cycles) are also $EA$-reconstructible.

**Theorem 3.1.5** [CvRo1]. *The number $q$ of quadrangles in a graph $G$ is given by*

$$q = \frac{1}{8} \sum_{i=1}^{m} \sum_{j=1}^{n} \alpha_{ij}^2 \mu_i^2 \left( \mu_i^2 + 1 - 2 \sum_{h=1}^{m} \alpha_{hj}^2 \mu_h^2 \right).$$

**Proof.** We first claim that $s_4 = 2e + 4f + 8q$, where $f$ (as above) is the number of paths of length 2 in $G$. To see this, note that the subgraph traversed by a closed walk of length 4 is $K_2$ or $P_3$ or $C_4$. For each of these graphs, Fig. 3.1 shows the number of closed walks of length 4 starting at each vertex (and traversing the graph). The total number of closed walks of length 4 traversing the graph is 2, 4 or 8 respectively.

Now $e$ and $s_4$ are determined by the spectrum of $G$, while $f$ is $EA$-reconstructible (see Remark 3.1.4). Accordingly, $q$ is $EA$-reconstructible, and the explicit formula is a matter of algebraic manipulation (Exercise 3.1). $\quad \square$

**Theorem 3.1.6** [CvRo1]. *The number $p$ of pentagons in a graph $G$ is given by*

$$p = \frac{1}{10} \sum_{i=1}^{m} \sum_{j=1}^{n} \alpha_{ij}^2 \mu_i^3 \left( \mu_i^2 + 5 - 5 \sum_{h=1}^{m} \alpha_{hj}^2 \mu_h^2 \right).$$

**Proof.** Arguing as in the proof of the previous theorem, we have $s_5 = 30t + 10s + 10p$, where $t$ is the number of triangles and $s$ the number of subgraphs consisting of a triangle and one pendant edge. Note that $s = \sum_{j=1}^{n} t_j(d_j - 2)$, where $d_j$ and $t_j$ are given by Theorem 3.1.3. The result now follows by algebraic manipulation (Exercise 3.2). $\quad \square$

## 3.2 Regularity and bipartiteness

We know from Chapter 1 that vertex degrees are not determined by the spectrum of a graph (see Fig. 1.3). On the other hand, we can tell from the spectrum whether or not all vertex degrees of $G$ are the same, and if they are, we can find the degree of regularity.

**Theorem 3.2.1** [ColSi]. *Let $\lambda_1$ be the index of the graph $G$, and let $\overline{d}$ and $\Delta$ be its average degree and maximum degree, respectively. Then*

$$\overline{d} \leq \lambda_1 \leq \Delta.$$

*Moreover, $\overline{d} = \lambda_1$ if and only if $G$ is regular. For a connected graph $G$, $\lambda_1 = \Delta$ if and only if $G$ is regular.*

**Proof.** For the first inequality, recall that the index of $G$ is given by Equation (1.6):

$$\lambda_1 = \sup\{\mathbf{x}^\top A\mathbf{x} \ : \ \mathbf{x} \in I\!\!R^n, \ ||\mathbf{x}|| = 1\},$$

where $A$ is the adjacency matrix of $G$. Taking $\mathbf{x} = \frac{1}{\sqrt{n}}(1, 1, \ldots, 1)^\top$, we see that $\lambda_1 \geq \overline{d}$. Moreover, by Rayleigh's Principle, equality holds if and only if $\mathbf{x} = \frac{1}{\sqrt{n}}(1, 1, \ldots, 1)^\top$ is an eigenvector of $G$. But the latter holds if and only if $G$ is regular (Proposition 1.1.2).

The second inequality follows from Proposition 1.1.1, while if $G$ is $r$-regular then $\lambda_1 = r \ (= \Delta)$ by Proposition 1.1.2.

Now suppose that $G$ is connected and $\lambda_1 = \Delta$. Let $\mathbf{x} = (x_1, x_2, \ldots, x_n)^\top$ be an eigenvector corresponding to $\lambda_1$. By Theorem 1.3.6 we may assume that all entries of $\mathbf{x}$ are positive. Let $x_u = \max_i\{x_i\}$. Now the equation

$$\Delta x_u = \sum_{v \sim u} x_v \tag{3.3}$$

shows that $\deg(u) = \Delta$ and $x_v = x_u$ for all $v \sim u$. Repetition of the argument shows that that all vertices have degree $\Delta$ (and that $G$ has the all-1 vector as an eigenvector). Thus $G$ is regular. $\qquad\square$

Since $n\overline{d} = \text{tr}(A^2)$ we immediately obtain the following:

**Corollary 3.2.2.** *A graph $G$ is regular (of degree $\lambda_1$) if and only if*

$$n\lambda_1 = \lambda_1^2 + \lambda_2^2 + \cdots + \lambda_n^2.$$

Thus regularity can be recognized from the spectrum. Next we show that the same is true of bipartiteness. If $G$ is bipartite on $U \ \dot\cup \ V$, then $G$ has an adjacency matrix of the form

$$A = \begin{pmatrix} O & P \\ Q & O \end{pmatrix},$$

where $Q = P^\top$; here, the non-zero row entries of $P$ correspond to edges incident with vertices from $U$, while the non-zero row entries of $Q$ correspond to edges incident with vertices from $V$. Now suppose that $\mu$ is an eigenvalue of $G$, and that

$$\mathbf{x} = \begin{pmatrix} \mathbf{y} \\ \mathbf{z} \end{pmatrix}$$

is an arbitrary eigenvector from $\mathcal{E}(\mu)$. Consequently, we have $P\mathbf{z} = \mu\mathbf{y}$ and $Q\mathbf{y} = \mu\mathbf{z}$. Consider next the vector

$$\mathbf{x}' = \begin{pmatrix} \mathbf{y} \\ -\mathbf{z} \end{pmatrix}.$$

We have

$$A\mathbf{x}' = \begin{pmatrix} O & P \\ Q & O \end{pmatrix} \begin{pmatrix} \mathbf{y} \\ -\mathbf{z} \end{pmatrix} = \begin{pmatrix} -P\mathbf{z} \\ Q\mathbf{y} \end{pmatrix} = \begin{pmatrix} -\mu\mathbf{y} \\ \mu\mathbf{z} \end{pmatrix} = -\mu\mathbf{x}'.$$

This shows not only that $-\mu$ is an eigenvalue of $G$, but also that $\mathcal{E}(-\mu)$ and $\mathcal{E}(\mu)$ have the same dimension. Thus we have proved (in answer to Exercise 1.6) that the spectrum of a bipartite graph is symmetric about 0.

We will now prove that the converse is true. Accordingly, let $G$ be a graph whose spectrum is symmetric about 0. Then all the odd spectral moments of $G$ are zero; in particular, $G$ has no cycles of odd length (by Theorem 3.1.1). Hence $G$ is bipartite, and we have the following result (rediscovered many times in the literature).

**Theorem 3.2.3.** *A graph $G$ is bipartite if and only if its spectrum is symmetric with respect to the origin.*

For connected graphs we have a substantially stronger result:

**Theorem 3.2.4.** *A connected graph $G$ is bipartite if and only if $\lambda_1 = -\lambda_n$.*

**Proof.** In the light of Theorem 3.2.3, it remains to prove that if $\lambda_1 = -\lambda_n$ then $G$ is bipartite. This is a consequence of a theorem of Frobenius [Gan, Vol. 2, p. 53], but we can also argue as follows.

The largest eigenvalue of $A^2$ is $\lambda_1^2$, and it is not a simple eigenvalue. By Theorem 1.3.6, $A^2$ is reducible, say with bipartition $U \cup V$; then $G$ has no $U$-$V$ walks of length 2. Suppose by way of contradiction that $U$ has adjacent vertices $u_1$ and $u_2$, and let $v \in V$. Let $w_0 w_1 \cdots w_m$ be a shortest path from $u_1$ to $v$, and let $k$ be least such that $w_{k+1} \in V$. If $k > 0$ then $w_{k-1} w_k w_{k+1}$

is a $U$-$V$ walk of length 2. If $k = 0$ then $u_2 w_0 w_1$ is a $U$-$V$ walk of length 2, a contradiction. Therefore, $U$ is independent; similarly $V$ is independent. This completes the proof. □

We conclude this section by discussing cycles of shortest length. We define the *odd-girth* of $G$, denoted by $og(G)$, as the length of the shortest odd cycle.

The following theorem is stated in terms of the characteristic polynomial $P_G(x)$. (Although knowledge of $P_G(x)$ is equivalent to knowledge of the spectrum of $G$, computational considerations can make for differences in practice.)

**Theorem 3.2.5** [Sac2]. *Let $x^n + c_1 x^{n-1} + c_2 x^{n-2} + \cdots + c_{n-1} x + c_n$ be the characteristic polynomial of a graph $G$. Then the odd girth of $G$ is equal to the index of the first non-zero coefficient from the sequence $c_1, c_3, c_5, \ldots$; the number of cycles of this length is equal to $-\frac{1}{2} c_h$, where $h = og(G)$.*

**Proof.** Recall from Corollary 2.3.3 that

$$c_i = \sum_{H \in \mathcal{H}_i} (-1)^{p(H)} 2^{c(H)} \quad (i = 1, 2, \ldots, n),$$

where $\mathcal{H}_i$ is the set of all elementary subgraphs on $i$ vertices (subgraphs of $G$ whose components are either cycles or isomorphic to $K_2$), $p(H)$ is the number of components of $H$, and $c(H)$ is the number of cycles in $H$.

Thus if $og(G) = 2k + 1$ then $c_{2l+1} = 0$ whenever $l < k$ because then no elementary subgraph has an odd number of vertices. In the case that $k = l$, an elementary subgraph must be an odd cycle, and so $c_{2k+1} = -2s(G)$, where $s(G)$ is the number of cycles of length $og(G)$. The result follows. □

A natural question now arises. Is it possible to identify (from the characteristic polynomial) the length of the shortest even cycle, and to find the number of such cycles? The answer is no. To see this, consider again the smallest pair of cospectral graphs shown in Fig. 1.3(a): $K_{1,4}$ has no cycle, while $C_4 \overset{.}{\cup} K_1$ has just one, which is even.

However, the following theorem of Sachs can sometimes be of use. Observe first that if $G$ has girth $g$ then for $i < g$ we have

$$c_i = \begin{cases} 0 & \text{if } i \text{ is odd} \\ (-1)^q b_q & \text{if } i = 2q, \end{cases}$$

where $b_q$ is the number of elementary subgraphs consisting of $q$ disjoint copies of $K_2$.

For $i = g$, elementary subgraphs can be of two types, either disjoint copies of $K_2$ (arising only when $g$ is even) or one copy of $C_g$. Accordingly, we define

$$\hat{c}_i = \begin{cases} c_i & \text{if } i \text{ is odd} \\ c_i - (-1)^q b_q & \text{if } i = 2q \end{cases}, \tag{3.4}$$

for $i = 1, 2, \ldots, n$. Then $\hat{c}_i = 0$ for $i < g$, and $-\hat{c}_g$ is equal to twice the number of cycles of length $g$. Thus we have proved:

**Theorem 3.2.6.** *If $\hat{c}_i$ is given by (3.4) then the girth $g$ of $G$ is equal to the index of the first non-zero coefficient from the sequence $\hat{c}_1, \hat{c}_2, \hat{c}_3, \ldots$; the number of cycles of this length is equal to $-\frac{1}{2}\hat{c}_g$.*

For regular graphs we can say more. As observed in [Sac1], if $G$ is $r$-regular with $n$ vertices and girth $g$ then for $q < g$, $b_q$ can be expressed in terms of $q$, $n$ and $r$. Therefore we have:

**Theorem 3.2.7** [Sac2]. *If $G$ is a regular graph, then the girth of $G$ is determined by its characteristic polynomial (and hence by the spectrum).*

With a more detailed analysis, we can obtain the following result, stated without proof.

**Theorem 3.2.8** [Sac2]. *Let $G$ be an $r$-regular graph with $n$ vertices and girth $g$. If $h \leq \min\{n, 2g - 1\}$ then the number of cycles in $G$ of length $h$ is determined by $r$ and the coefficients $c_1, c_2, \ldots, c_h$ in the characteristic polynomial of $G$.*

## 3.3 Connectedness and metric invariants

In general, connectedness is a property not determined by the spectrum of a graph. (For instance, $K_{1,4}$ is connected, while $C_4 \cup K_1$ is not.) Nevertheless, for some classes of graphs, we can deduce whether or not $G$ is connected. Indeed, this is true for regular graphs, as we now demonstrate.

We have already seen in Corollary 3.2.2 that regularity can be recognized from the spectrum. Moreover the degree of a regular graph $G$ is just the index of $G$ (Proposition 1.1.2). By Theorem 1.3.6, each component of $G$ contributes one to the multiplicity of $r$, and so we have the following result:

**Theorem 3.3.1.** *If $G$ is $r$-regular then its index is equal to $r$, and the number of components of $G$ is equal to the multiplicity of $r$.*

In the general case, it follows from Theorem 1.3.6 that the number of components with index $\lambda_1$ is equal to the multiplicity of $\lambda_1$. In Corollary 1.3.8 we noted a general condition for an arbitrary graph to be connected: a graph is connected if and only if its index is of multiplicity one with a positive eigenvector. This result can be seen in the context of angles as follows

**Lemma 3.3.2.** *Vertices belonging to the components whose index coincides with the index of a graph are EA-reconstructible.*

**Proof.** We show that the vertices in question are precisely those vertices $j$ for which $\alpha_{1j} \neq 0$. First, if $j$ does not lie in a component with index $\lambda_1$ then every vector in $\mathcal{E}(\lambda_1)$ has $j$-th entry 0, and so $\mathbf{e}_j$ is orthogonal to $\mathcal{E}(\lambda_1)$, equivalently, $\alpha_{1j} = 0$. On the other hand, if $j$ does lie in a component with index $\lambda_1$ then by Theorem 1.3.6 there exists an eigenvector corresponding to $\lambda_1$ whose $j$-th entry is non-zero, and so $\alpha_{1j} \neq 0$. □

**Theorem 3.3.3.** *The property of a graph being connected, or disconnected, is EA-reconstructible.*

**Proof.** By the previous lemma we can reconstruct all vertices belonging to the components with index $\lambda_1$. If the number of these vertices is less than the number of vertices of the whole graph, then the graph is is not connected; otherwise, the same conclusion holds when the index is not a simple eigenvalue. Only in the remaining case is the graph connected. □

**Remark 3.3.4.** In view of Equation (1.9), we may now reformulate Corollary 1.3.8 as follows: a graph is connected if and only if $\sum_{j=1}^{n} \alpha_{1j}^2 = 1$ and $\alpha_{1j} \neq 0$ $(j = 1, \ldots, n)$. □

If we restrict ourselves to connected graphs we can ask more: for example, we can ask how large is the diameter, and we can pose the same question for the eccentricities of the vertices. (Recall that the *diameter* diam($G$) of a connected graph $G$ is the maximum distance between two vertices of $G$, while the *eccentricity* ecc($u$) of a vertex $u$ is the maximum distance of a vertex from $u$.)

**Theorem 3.3.5.** *If $G$ is a connected graph with precisely $m$ distinct eigenvalues then*

$$\mathrm{diam}(G) \leq m - 1.$$

**Proof.** Assume the contrary, so that $G$ has vertices $s$ and $t$ at distance $m$. The adjacency matrix $A$ of $G$ has minimal polynomial of degree $m$, and so we may write $A^m = \sum_{k=0}^{m-1} a_k A^k$. This yields the required contradiction because the $(s, t)$-entry on the right is zero, while the $(s, t)$-entry on the left is non-zero. □

For upper bounds on the eccentricities in a connected graph, we make use of the angle matrix $(\alpha_{ij})$:

**Theorem 3.3.6.** *Let $u$ be a vertex in the connected graph $G$. If $m(u)$ is the number of non-zero entries in the $u$-th column of the angle matrix of $G$, then*

$$\text{ecc}(u) \le m(u) - 1.$$

**Proof.** Suppose by way of contradiction that $e \ge m(u)$, where $e = \text{ecc}(u)$. From the spectral decomposition of the adjacency matrix $A$ of $G$ we have

$$A^k = \mu_1^k P_1 + \mu_2^k P_2 + \cdots + \mu_m^k P_m \quad (k = 0, 1, 2, \ldots). \tag{3.5}$$

Suppose that $v$ is a vertex of $G$ at distance $e$ from $u$. Then the $(u, v)$-entry of $A^k$ is zero for all $k \in \{0, 1, \ldots, e - 1\}$. Let $x_j$ be the $(u, v)$-entry of $P_j$ ($j = 1, 2, \ldots, m$). Comparing $(u, v)$-entries in (3.5) (for $k = 0, 1, \ldots, e - 1$) we obtain a system of $e$ equations in the $m$ unknowns $x_1, x_2, \ldots, x_m$, which reads

$$\sum_{j=1}^{m} \mu_j^k x_j = 0 \quad (k = 0, 1, \ldots, e - 1).$$

Note next that $x_j = (P_j \mathbf{e}_u)^\top (P_j \mathbf{e}_v)$, which is zero if $\alpha_{ju} = 0$. Accordingly, the above system reduces to a system of $e$ equations in $m(u)$ unknowns. The system consisting of the first $m(u)$ equations has a Vandermonde determinant, and so all the remaining $x_j$ are also zero. From (3.5), we see that the $(u, v)$-entry of $A^k$ is zero for all $k$. Hence $G$ is not connected, a contradiction. □

## 3.4 Line graphs and related graphs

We saw in Chapter 1 that the spectrum of any generalized line graph is bounded from below by $-2$. However not every graph with this spectral property is a generalized line graph (see Exercise 3.8), and an early problem in spectral graph theory was to describe all the graphs whose spectrum lies in $[-2, \infty)$. This problem has received much attention from researchers over the years, and the graphs in question are now very well understood.

**Definition 3.4.1.** An exceptional graph is a connected graph, other than a generalized line graph, with least eigenvalue $\ge -2$.

In this section we explain (without proofs of all the details) why there are only finitely many exceptional graphs. We go on to discuss the multiplicity of $-2$ as an eigenvalue of a generalized line graph, and to describe the graphs with least eigenvalue greater than $-2$.

Let $A$ be the adjacency matrix of a graph with $n$ vertices and least eigenvalue $\geq -2$. Suppose that the multiplicity (possibly zero) of $-2$ is $n - r$, so that $A + 2I$ is a positive semi-definite matrix of rank $r$. As we saw in Section 1.3, $A + 2I = Q^\top Q$ for some $r \times n$ matrix $Q$. In other words, if $Q = (\mathbf{q}_1 | \cdots | \mathbf{q}_n)$, then $A + 2I$ is the Gram matrix of the vectors $\mathbf{q}_1, \ldots, \mathbf{q}_n$. Note that $\|\mathbf{q}_i\| = \sqrt{2}$ and for $i \neq j$, we have

$$\mathbf{q}_i^\top \mathbf{q}_j = 1 \text{ if } i \sim j, \quad \mathbf{q}_i^\top \mathbf{q}_j = 0 \text{ if } i \not\sim j.$$

Thus if $\ell_i$ is the line (1-dimensional subspace) in $I\!R^r$ spanned by $\mathbf{q}_i$ then the angle between $\ell_i$ and $\ell_j$ ($i \neq j$) is $60°$ if $i \sim j$, and $90°$ if $i \not\sim j$.

Accordingly, we should investigate sets of lines at angles of $60°$ and $90°$ (through the origin) in Euclidean space, and we call such sets *line systems*. For fixed $r$, any line system in $I\!R^r$ is finite. This can be seen as follows, by considering the points at which the lines intersect the unit sphere centred at the origin: the distance between any two such points cannot be less than 1, and so these points have disjoint neighbourhoods of fixed positive area on the surface of the sphere.

A line system $\mathcal{L}$ is *decomposable* if it can be partitioned into two subsets $\mathcal{L}_1$ and $\mathcal{L}_2$ such that every line in $\mathcal{L}_1$ is orthogonal to every line in $\mathcal{L}_2$; otherwise, $\mathcal{L}$ is *indecomposable*. Note that, with the notation above, the system $\{\ell_1, \ldots, \ell_n\}$ is indecomposable if and only if $G$ is connected.

A *star* is a set of three coplanar coincident lines such that the angle between any pair of them is $60°$. A system $\mathcal{L}$ of lines is *star-closed* if for any two lines $\ell, \ell'$ in $\mathcal{L}$, the third line from the star determined by $\ell$ and $\ell'$ also lies in $\mathcal{L}$.

**Theorem 3.4.2.** *Any line system in $I\!R^r$ is contained in a star-closed line system in $I\!R^r$.*

**Proof.** Let $\mathcal{L}$ be a line system in $I\!R^r$, and consider any pair of lines in $\mathcal{L}$ at $60°$, say $\langle \mathbf{x} \rangle, \langle \mathbf{y} \rangle$ where $\mathbf{x}^\top \mathbf{x} = \mathbf{y}^\top \mathbf{y} = 2$ and $\mathbf{x}^\top \mathbf{y} = -1$. We show that if $\langle \mathbf{x} + \mathbf{y} \rangle \notin \mathcal{L}$ then we may add $\langle \mathbf{x} + \mathbf{y} \rangle$ to obtain a larger line system (necessarily also in $I\!R^r$); note that $(\mathbf{x} + \mathbf{y})^\top (\mathbf{x} + \mathbf{y}) = 2$. If $\langle \mathbf{u} \rangle$ is any line of $\mathcal{L}$ other than $\langle \mathbf{x} \rangle, \langle \mathbf{y} \rangle$ then we may choose $\mathbf{u}$ so that $\mathbf{u}^\top \mathbf{u} = 2$ and $\mathbf{u}^\top (\mathbf{x} + \mathbf{y}) \in \{0, 1, 2\}$. If however $\mathbf{u}^\top (\mathbf{x} + \mathbf{y}) = 2$ then $(\mathbf{u} - \mathbf{x} - \mathbf{y})^\top (\mathbf{u} - \mathbf{x} - \mathbf{y}) = 0$ and we have the contradiction $\mathbf{u} = \mathbf{x} + \mathbf{y}$. Thus $\langle \mathbf{u} \rangle$ makes an angle of $60°$ or $90°$ with every line of $\mathcal{L}$ and so it may be added to $\mathcal{L}$ to form a line system $\mathcal{L}'$. If $\mathcal{L}'$ is not star-closed then the procedure may be repeated. In view of the finiteness property noted above, we obtain a star-closed line system in $I\!R^r$ after finitely many steps. □

For any line $\ell$ in a line system $\mathcal{L}$ of size $n$, there are two vectors of length $\sqrt{2}$ lying along $\ell$. The set of $2n$ such vectors arising from $\mathcal{L}$ is called a *root system*, a term borrowed from the theory of Lie algebras. For any root system $\mathcal{R}$, we write $\overline{\mathcal{R}}$ for the line system determined by $\mathcal{R}$. In 1976, Cameron, Goethals, Seidel and Shult [CamGSS] classified the indecomposable star-closed systems of lines as follows.

**Theorem 3.4.3.** *To within an orthogonal transformation, the only indecomposable star-closed line systems are* $\overline{A_n}$, $\overline{D_n}$, $\overline{E_6}$, $\overline{E_7}$ *and* $\overline{E_8}$, *where*

(i) $A_n = \{\mathbf{e}_i - \mathbf{e}_j : \mathbf{e}_i, \mathbf{e}_j \in \mathbb{R}^{n+1}, i \neq j, 1 \le i, j \le n+1\}$ $(n = 2, 3, \ldots)$,

(ii) $D_n = \{\pm \mathbf{e}_i \pm \mathbf{e}_j : \mathbf{e}_i, \mathbf{e}_j \in \mathbb{R}^n, i \neq j, 1 \le i, j \le n\}$ $(n = 2, 3, \ldots)$,

(iii) $E_8 = D_8 \cup \{\frac{1}{2} \sum_{i=1}^{8} \epsilon_i \mathbf{e}_i : \epsilon_i = \pm 1, \prod_{i=1}^{8} \epsilon_i = 1\}$,

(iv) $E_7 = \{\mathbf{u} \in E_8 : \mathbf{u} \text{ is orthogonal to a fixed vector in } E_8 \}$,

(v) $E_6 = \{\mathbf{u} \in E_8 : \mathbf{u} \text{ is orthogonal to a fixed star in } E_8 \}$.

Several remarks are in order. First, the group of symmetries of $E_8$ acts transitively on vectors and on stars, and so $E_7$ and $E_6$ are well defined. Secondly, $A_{n-1} \subseteq D_n$ $(n = 2, 3, \ldots)$. Also, $|A_n| = n(n + 1)$, $|D_n| = 2n(n - 1)$, $|E_8| = 240$, $|E_7| = 126$ and $|E_6| = 72$.

We say that a graph $G$ is *represented* in the root system $R$ if its adjacency matrix $A$ satisfies $A + 2I = Q^\top Q$ where the columns of $Q$ lie in $\mathcal{R}$. Proofs of the next two theorems are left as exercises.

**Theorem 3.4.4.** *A graph has a representation in* $A_n$ *if and only if it is the line graph of a bipartite graph with* $n + 1$ *vertices.*

**Theorem 3.4.5.** *A graph has a representation in* $D_n$ *if and only if it is a generalized line graph.*

As a consequence, we have:

**Theorem 3.4.6.** *All exceptional graphs are representable in the root system* $E_8$.

It follows that there are only finitely many exceptional graphs. In principle, they can be found by identifying all the subsets $S$ of $E_8$ with the property that any two vectors in $S$ have scalar product 0 or 1. (It suffices to find the maximal subsets with this property, since every exceptional graph is an induced subgraph of a graph determined by such a maximal subset.) This forbidding computation was circumvented by Cvetković, Lepović, Rowlinson and Simić who used the star complement technique described in Chapter 5 to determine the maximal exceptional graphs. There are only 473 such graphs and they are described in the paper [CvLRS2], published in 2002. The regular exceptional graphs are discussed in Chapter 4.

An exceptional graph without $-2$ as an eigenvalue has at most 8 vertices because it has a representation in $E_8$ and $A + 2I$ is non-singular. It turns out that there are 573 such graphs: 20 with 6 vertices, 110 with 7 vertices, and 443 with 8 vertices. We denote these families of graphs by $\mathcal{G}_6$, $\mathcal{G}_7$, $\mathcal{G}_8$ respectively; they are listed in [CvRS7, Appendix A2]. In order to complete the description of all graphs with least eigenvalue greater than $-2$, we go on to investigate the eigenspace of $-2$ for generalized line graphs.

Recall from Section 1.2 that if $A$ is the adjacency matrix of the generalized line graph $L(\hat{H})$ then $A + 2I = C^\top C$ where $C$ is a vertex–edge incidence matrix of the $B$-graph $\hat{H}$. It follows (Exercise 3.9) that the eigenspace of $L(\hat{H})$ corresponding to $-2$ is just the nullspace of $C$:

**Lemma 3.4.7.** *Let $C$ be a vertex–edge incidence matrix of $\hat{H}$. The non-zero vector $\mathbf{x}$ is an eigenvector for $L(\hat{H})$ corresponding to $-2$ if and only if $C\mathbf{x} = \mathbf{0}$.*

Consequently $-2$ is the least eigenvalue of $L(\hat{H})$ if and only if $C\mathbf{x} = \mathbf{0}$ for some non-zero vector $\mathbf{x}$. More generally, the multiplicity of $-2$ as an eigenvalue of $L(\hat{H})$ is just the nullity of $C$. In the case of line graphs we can therefore use the following result:

**Lemma 3.4.8 [Sac3, Nuff].** *Let $B$ be the vertex–edge incidence matrix of a connected graph $H$ with $n$ vertices, $n > 1$. Then*

$$\text{rank}(B) = \begin{cases} n - 1 & \text{if } H \text{ is bipartite,} \\ n & \text{if } H \text{ is non-bipartite.} \end{cases} \tag{3.6}$$

**Proof.** Let $B = (b_{ij})$, with rows $B_1, \ldots, B_n$, and assume that the rows are linearly dependent, say

$$c_1 B_1 + \cdots + c_n B_n = \mathbf{0} \quad \text{and} \quad (c_1, \ldots, c_n) \neq (0, \ldots, 0). \tag{3.7}$$

If two vertices $v_s$ and $v_t$ are joined by the edge $e_j$ then $b_{sj} = b_{tj} = 1$, while $b_{kj} = 0$ for all $k \neq s, t$. Consequently, from (3.7) we obtain $c_s = -c_t$.

It follows that for any path $i_1 i_2 \ldots i_k$ starting at a vertex for which $c_{i_1} = c \neq 0$, the coefficients $c_{i_1}, c_{i_2}, \ldots, c_{i_k}$ are alternately $c$ and $-c$. Since $H$ is connected we deduce that $H$ is bipartite and that $\dim\{\mathbf{x} \in I\!\!R^n : \mathbf{x}^\top B = \mathbf{0}\} = 1$. The result follows. $\square$

Let $m_G(\lambda)$ denote the multiplicity of $\lambda$ as an eigenvalue of the graph $G$. From Lemmas 3.4.7 and 3.4.8 we have the following result.

**Theorem 3.4.9.** *Let $H$ be a connected graph with $n$ vertices and $m$ edges. Then*

$$m_{L(H)}(-2) = \begin{cases} m - n + 1 & \text{if } H \text{ is bipartite,} \\ m - n & \text{if } H \text{ is non-bipartite.} \end{cases} \tag{3.8}$$

Recall that we write $\lambda(G)$ for the smallest eigenvalue of a graph $G$. It follows from Theorem 3.4.9 that if $H$ is a unicyclic graph with cycle $Z$ then $\lambda(H) = -2$ if $Z$ has even length, and $\lambda(H) > -2$ if $Z$ has odd length. Thus we have:

**Corollary 3.4.10** [Doo1]. *Let $H$ be a connected graph. Then $\lambda(L(H)) > -2$ if and only if $H$ is a tree or an odd-unicyclic graph.*

**Corollary 3.4.11.** *Let $H$ be a (connected) graph with diameter $d$. Then*

$$-2 \leq \lambda(L(H)) \leq -2\cos\frac{\pi}{d+1},$$

*and these bounds are best possible.*

**Proof.** It remains to consider the second inequality. Since the diameter of $L(H)$ is not less than $d - 1$, $L(H)$ has a path $P_d$ as an induced subgraph. By the Interlacing Theorem we have $\lambda(L(H)) \leq \lambda(P_d) = -2\cos\frac{\pi}{d+1}$, and equality holds when $H = P_{d+1}$. □

To obtain an analogue of (3.8) for generalized line graphs, we can proceed in the same way as above:

**Lemma 3.4.12.** *Suppose that $C$ is an incidence matrix of a connected $B$-graph $H(a_1, a_2, \ldots, a_n)$ for which $(a_1, a_2, \ldots, a_n) \neq (0, 0, \ldots, 0)$. Then*

$$\text{rank}(C) = n + \sum_{i=1}^{n} a_i. \tag{3.9}$$

**Proof.** Let $C = (c_{ij})$, with rows $C_1, \ldots, C_r$, where $r = n + \sum_{i=1}^{n} a_i$. To show that these rows are linearly independent, suppose that $c_1 C_1 + \cdots + c_r C_r = \mathbf{0}$. Our multigraph contains vertices $h$ and $i$ joined by two edges, say $j$-th and $k$-th. From Section 1.2 we know that, without loss of generality, $c_{hj} = c_{hk} = 1$, $c_{ij} = -c_{ik} = 1$ and $c_{lj} = c_{lk} = 0$ for all $l \neq i, h$. It follows that $c_h = c_i = 0$. Tracing paths from $h$ as in the proof of Lemma 3.4.8, we find that $c_1 = \cdots = c_r = 0$. The lemma follows. □

Now from Lemmas 3.4.7 and 3.4.12 we obtain the following analogue of Theorem 3.4.9.

**Theorem 3.4.13** [CvDS2]. *Suppose that $H$ is a connected graph with $n$ vertices and $m$ edges. If $\hat{H} = H(a_1, a_2, \ldots, a_n)$, where $(a_1, a_2, \ldots, a_n) \neq (0, 0, \ldots, 0)$, then*

$$m_{L(\hat{H})}(-2) = m - n + \sum_{i=1}^{n} a_i. \tag{3.10}$$

It follows that if $G$ is a generalized line graph without $-2$ as an eigenvalue, then either $G$ is a line graph (and Corollary 3.4.10 applies) or $G = L(\hat{H})$ where $\hat{H}$ is a tree with just one petal attached. Now we can complete the description of graphs whose eigenvalues lie in the interval $(-2, \infty)$:

**Theorem 3.4.14** [DooCv]. *If $H$ is a connected graph with least eigenvalue greater than $-2$ then one of the following holds:*

(a) $H = L(K)$ *where $K$ is a tree, or a tree with a single petal attached, or an odd-unicyclic graph;*

(b) $H$ *is one of the 573 graphs in $\mathcal{G}_6 \cup \mathcal{G}_7 \cup \mathcal{G}_8$.*

## 3.5 More on regular graphs

In this section we discuss the significance of the second largest eigenvalue, and the eigenvalue with second largest modulus, as invariants of regular graphs. We also consider the Hoffman polynomial of a regular graph and the mean degree of an arbitrary induced subgraph of a regular graph.

### 3.5.1 The second largest eigenvalue

The second largest eigenvalue of a connected regular graph plays an important role in determining the graph structure. This phenomenon was observed in 1976, in respect of connected cubic graphs, by Bussemaker, Čobeljić, Cvetković and Seidel [BuČCS]. For each $n \le 14$, the connected cubic graphs with $n$ vertices were ordered lexicographically by their spectrum $(\lambda_1, \lambda_2, \ldots, \lambda_n)$; since $\lambda_1 = 3$ throughout, $\lambda_2$ plays the primary role. It can be observed from Table A5 in the Appendix that for small values of $\lambda_2$ the graphs have a more 'round' shape (smaller diameter, higher connectivity and girth), while for large values of $\lambda_2$ the graphs have a more 'path-like' shape (larger diameter, lower connectivity and girth). A partial explanation of these empirical observations was offered in 1978 by Cvetković:

**Theorem 3.5.1** [Cve7]. *Let $G$ be an $r$-regular graph on $n$ vertices. Let $v$ be any vertex of $G$ and let $\bar{d}$ be the average vertex degree of the subgraph induced by the vertices not adjacent to $v$. Then*

$$\bar{d} \le r \frac{\lambda_2^2 + \lambda_2(n - r)}{\lambda_2(n - 1) + r}.$$

**Proof.** We partition $V(G)$ into three parts, consisting of $v$, the vertices adjacent to $v$, and the vertices not adjacent to $v$. If we partition the adjacency matrix $A$ of $G$ into corresponding blocks then the average row sums in the blocks form the matrix

$$B = \begin{pmatrix} 0 & r & 0 \\ 1 & r - \nu - 1 & \nu \\ 0 & r - \overline{d} & \overline{d} \end{pmatrix},$$

where $\nu$ is the mean number of edges from a vertex adjacent to $v$ to vertices not adjacent to $v$. Counting in two ways the total number of such edges, we have $r\nu = (n - 1 - r)(r - \overline{d})$. By Corollary 1.3.13, the eigenvalues of $B$ interlace those of $A$. Since $B$ has characteristic polynomial $(x - r)(x^2 - (\overline{d} - \nu - 1)x - \overline{d})$, and this must be non-positive at $x = \lambda_2$, we have

$$\lambda_2^2 - (\overline{d} - \nu - 1)\lambda_2 - \overline{d} \geq 0.$$

Now the result follows by substituting $(n - 1 - r)(r - \overline{d})/r$ for $\nu$ in this inequality.                                                                          □

In Theorem 3.5.1, the upper bound for $\overline{d}$ decreases as $\lambda_2$ decreases. Now a decrease in $\overline{d}$ reduces the number of edges in the subgraph $H_1$ induced by the vertices not adjacent to $v$ (and hence brings edges closer to $v$). Moreover, when $\overline{d}$ decreases, so does $r - \nu - 1$, the average vertex degree in the subgraph $H_2$ induced by the neighbours of $v$. Thus we have fewer edges in $H_1$ and $H_2$, and more edges between these subgraphs: this phenomenon corresponds intuitively to the graph assuming a more 'round' shape.

The cubic graphs for which the second largest eigenvalue is maximal were identified in [BGI]; for each even $n \geq 4$, there is a unique such graph $G_n$ with $n$ vertices. The graph $G_4$ is necessarily $K_4$, $G_6$ is the prism $K_3 + K_2$, and $G_8$ is the first graph with 8 vertices in Table A5 of the Appendix. The graphs $G_n$ ($n \geq 10$) are illustrated in Fig. 3.2.

Finally we note without proof a result of Nilli [Nil]: if $G$ is a connected $r$-regular graph which contains two edges whose distance apart is at least $2k + 2$ then

$$\lambda_2(G) \geq 2\sqrt{r - 1}\left(1 - \frac{1}{k + 1}\right) + \frac{1}{k + 1}. \tag{3.11}$$

(The distance between two edges is the length of a shortest path whose terminal vertices are vertices of the edges in question.)

Figure 3.2 (a) The graph $G_n$ for $n \equiv 2 \pmod 4$.

Figure 3.2 (b) The graph $G_n$ for $n \equiv 0 \pmod 4$.

### 3.5.2 The eigenvalue with second largest modulus

Here we discuss a relation between $\Lambda(G)$, the second largest modulus of an eigenvalue of a connected regular graph $G$, and an expansion property of $G$. For $X \subseteq V(G)$ let $N(X)$ be the set of vertices of $G$ adjacent to some vertex of $X$. The expansion of a graph is defined in many ways in the literature, but the essential requirement of $G$ as a 'good' expander is that for any $X \subseteq V(G)$, $N(X)$ should be suitably 'large' in comparison with $|X|$. This concept is made precise (for arbitrary graphs) in Chapter 7; here we establish a lower bound for $|N(X)|/|X|$ in regular graphs.

For any $X, Y \subseteq V$, let $e(X, Y)$ be the number of ordered edges with the first endvertex in $X$ and the other in $Y$. Thus

$$e(X, Y) = |\{(u, v) \in V(G)^2 : u \sim v, \ u \in X \text{ and } v \in Y\}|,$$

and edges whose endvertices are both in $X \cap Y$ are counted twice.

**Lemma 3.5.2.** *Let $G$ be a connected $r$-regular graph on $n$ vertices, with eigenvalues $\lambda_1(=r), \lambda_2, \ldots, \lambda_n$. Let $X, Y$ be subsets of $V(G)$ with $|X| = \alpha n$ and $|Y| = \beta n$. If $\Lambda = \max_i \{|\lambda_i| : \lambda_i \neq \pm r\}$ then*

$$|e(X, Y) - \alpha\beta rn| \leq \Lambda n\sqrt{(\alpha - \alpha^2)(\beta - \beta^2)}.$$

**Proof.** Let $\mathbf{x}$ and $\mathbf{y}$ be the characteristic vectors of the sets $X$ and $Y$, and let $A$ be the adjacency matrix of $G$. Note that $\mathbf{x}^\top A\mathbf{y} = e(X, Y)$. Define $\mathbf{v} = \mathbf{x} - \alpha\mathbf{j}$ and $\mathbf{w} = \mathbf{y} - \beta\mathbf{j}$, where $\mathbf{j}$ is the all-1 vector in $\mathbb{R}^n$. Since each of $(\alpha\mathbf{j})^\top A\mathbf{y}$, $\mathbf{x}^\top A(\beta\mathbf{j})$, $(\alpha\mathbf{j})^\top A(\beta\mathbf{j})$ is equal to $\alpha\beta rn$, we have

$$\mathbf{v}^\top A\mathbf{w} = e(X, Y) - \alpha\beta rn.$$

On the other hand, since the vectors $\mathbf{v}$, $\mathbf{w}$ lie in $\mathbf{j}^\perp$ we have

$$|\mathbf{v}^\top A\mathbf{w}| \leq \Lambda |\mathbf{v}^\top \mathbf{w}| \leq \Lambda \|\mathbf{v}\|\|\mathbf{w}\| = \Lambda n\sqrt{(\alpha - \alpha^2)(\beta - \beta^2)}.$$

This completes the proof.                                                      $\square$

**Theorem 3.5.3** [Tan]. *Let $G$ be a connected $r$-regular graph on $n$ vertices, with eigenvalues $\lambda_1(=r), \lambda_2, \ldots, \lambda_n$. If $\Lambda = \max_i \{|\lambda_i| : \lambda_i \neq \pm r\}$ then for any $X \subseteq V(G)$,*

$$\frac{|N(X)|}{|X|} \geq \frac{r^2}{\Lambda^2 + (r^2 - \Lambda^2)\frac{|X|}{n}}.$$

**Proof.** We apply Lemma 3.5.2 to $X$ and $Y$, where $Y = V(G) \setminus N(X)$. Note that $e(X, Y) = 0$, and so if $|X| = \alpha n$, $|Y| = \beta n$, then $\alpha\beta rn \leq \Lambda n\sqrt{(\alpha - \alpha^2)(\beta - \beta^2)}$. Hence

$$\alpha\beta r^2 \leq \Lambda^2(1 - \alpha - \beta + \alpha\beta),$$

equivalently,

$$\beta \leq \frac{\Lambda^2(1 - \alpha)}{\Lambda^2 + (r^2 - \Lambda^2)\alpha},$$

Now we have

$$|N(X)| = (1 - \beta)n \geq \frac{|X|r^2}{\Lambda^2 + (r^2 - \Lambda^2)\alpha},$$

and we are done.                                                              $\square$

It follows that if $\Lambda$ is small compared with $r$ then $G$ is a good expander. How small can $\Lambda$ be? A good indication is provided by the following result of Alon and Boppana (cf. [LuPS, Proposition 4.2]). For fixed $r > 1$, let $(G_m)_{m \in \mathbb{N}}$ be a family of connected $r$-regular graphs such that $|(V(G_m)| \to \infty$ as $m \to \infty$. Then

$$\liminf_{m \to \infty} \Lambda(G_m) \geq 2\sqrt{r - 1}. \tag{3.12}$$

This explains the importance of the following class of graphs:

**Definition 3.5.4.** A Ramanujan graph is a connected $r$-regular graph $G$ for which $\Lambda(G) \leq 2\sqrt{r - 1}$.

An infinite family of Ramanujan graphs $\{X^{p,q}\}$ was first constructed by Lubotzky, Phillips and Sarnak [LuPS] in 1988. Here, $p$ and $q$ are distinct primes, both congruent to 1 mod 4, such that $p$ is non-square mod $q$. The graph $X^{p,q}$ is realized as a certain vertex-transitive bipartite graph of degree

$p + 1$ with $q(q^2 - 1)$ vertices. In fact, infinite families of Ramanujan graphs of degree $r$ exist whenever $r - 1$ is a prime power (see [Mor2]). Finally we note that the inequality (3.11) restricts the diameter of a Ramanujan graph $G$ for which $\Lambda(G) < 2\sqrt{r - 1}$.

### 3.5.3 Miscellaneous results

Here we note two properties of regular graphs required in subsequent sections. For any graph $G$ with adjacency matrix $A$, the *adjacency algebra* of $G$ consists of all matrices of the form $f(A)$, where $f(x)$ is a polynomial with real coefficients. Hoffman identified the following characteristic property of the adjacency algebra of a regular connected graph.

**Theorem 3.5.5** [Hof3]. *The all-1 matrix $J$ belongs to the adjacency algebra of the graph $G$ if and only if $G$ is regular and connected.*

**Proof.** Suppose first that $J$ lies in the adjacency algebra $\mathcal{A}$ of $G$. Then $AJ = JA$, and so $G$ is regular. If $G$ is not connected then consider vertices $u$ and $v$ lying in different components of $G$. By Proposition 1.3.3, the $(u, v)$-entry of $f(A)$ is zero for all $f(x) \in \mathbb{R}[x]$; hence $J \notin \mathcal{A}$, a contradiction.

Conversely, suppose that $G$ is $r$-regular and connected. Then $G$ has index $r$ and the minimal polynomial of $A$ has the form $(x - r)g(x)$. Since $Ag(A) = rg(A)$, each column of $g(A)$ lies in the eigenspace $\mathcal{E}(r)$. Since $G$ is connected, $\mathcal{E}(r)$ is spanned by the all-1 vector $\mathbf{j}$ (cf. Theorem 1.3.5), and so $g(A)$ has the form $(c_1\mathbf{j}| \cdots |c_n\mathbf{j})$. Since $g(A)$ is a symmetric matrix, $c_1 = \cdots = c_n$. Thus $g(A) = cJ$ for some $c$, and the result follows. $\qquad\square$

The above proof shows that $h(A) = J$ where $h(x) = c^{-1}g(x)$; the polynomial $h(x)$ is called the *Hoffman polynomial* of $G$. If $\mu_1 = r, \mu_2, \ldots, \mu_m$ are the distinct eigenvalues of $G$, then the only non-zero eigenvalue of $g(A)$ is $\prod_{i=2}^{m}(r - \mu_i)$. Hence $\prod_{i=2}^{m}(r - \mu_i) = cn$, and so

$$h(x) = n \prod_{i=2}^{m} \frac{x - \mu_i}{r - \mu_i},$$

and we have the following formula for $J$:

**Corollary 3.5.6.** *If $G$ is an $r$-regular connected graph on $n$ vertices, with distinct eigenvalues $\mu_1 = r, \mu_2, \ldots, \mu_m$ then*

$$J = n \prod_{i=2}^{m} \frac{A - \mu_i I}{r - \mu_i}.$$

The final result of this section can be regarded as a generalization of Theorem 3.5.1.

**Theorem 3.5.7.** *Let $G$ be an $r$-regular graph with eigenvalues $r(= \lambda_1) \geq \lambda_2 \geq \cdots \geq \lambda_n$. Let $G_1$ be an induced subgraph of $G$ with $n_1$ vertices and mean degree $d_1$. Then*

$$\frac{n_1(r - \lambda_n)}{n} + \lambda_n \leq d_1 \leq \frac{n_1(r - \lambda_2)}{n} + \lambda_2. \tag{3.13}$$

**Proof.** We partition $V(G)$ into $V(G_1)$ and its complement, and consider the corresponding blocking of the adjacency matrix of $G$. The average row sums in the blocks form the matrix

$$B = \begin{pmatrix} d_1 & r - d_1 \\ \dfrac{(r - d_1)n_1}{n - n_1} & r - \dfrac{(r - d_1)n_1}{n - n_1} \end{pmatrix}.$$

The eigenvalues of $B$ are $r$ and $d_1 - (r - d_1)n_1/(n - n_1)$. By Corollary 1.3.13 we have $\lambda_n \leq d_1 - (r - d_1)n_1/(n - n_1)$, and the first inequality in (3.13) follows.

In order to prove the second inequality, we consider the complements $\overline{G}$, $\overline{G}_1$. The graph $\overline{G}$ is a regular graph on $n$ vertices of degree $n - 1 - d$, and by Theorem 2.1.2 its least eigenvalue is $-\lambda_2 - 1$. The graph $\overline{G}_1$ is an induced subgraph of $\overline{G}$ with $n_1$ vertices and mean degree $n - 1 - d_1$. If we now apply the first inequality of (3.13) to $\overline{G}$ and $\overline{G}_1$ we obtain the second inequality in (3.13). $\qquad\square$

## 3.6 Strongly regular graphs

Recall from Chapter 1 that a strongly regular graph with parameters $(n, r, e, f)$ is an $r$-regular graph on $n$ vertices in which any two adjacent vertices have exactly $e$ common neighbours and any two non-adjacent vertices have exactly $f$ common neighbours. Strongly regular graphs are important in relation to algorithms designed to determine whether or not two graphs are isomorphic (the 'graph isomorphism problem'), since they often represent the hardest case to deal with. At the same time, they are very well suited to investigation by spectral techniques, not least because (as we show below) knowledge of their spectrum is equivalent to knowledge of their parameters.

To exclude the complete graphs and their complements, we assume throughout that $0 < r < n - 1$. We have seen that the Petersen graph is a strongly regular graph with parameters $(10, 3, 0, 1)$. Some examples of infinite families of strongly regular graphs are given below.

**Examples 3.6.1.** (i) For $n > 3$, the *triangular graph* $T(n) = L(K_n)$ is strongly regular with parameters $\left(\frac{1}{2}n(n-1), 2n-4, n-2, 4\right)$.
(ii) For $n > 1$, the *lattice graph* $L(K_{n,n})$ is a strongly regular graph with parameters $(n^2, 2n-2, n-2, 2)$. □

**Example 3.6.2.** Let $GF(q)$ be a field with $q$ elements, where $q \equiv 1 \bmod 4$. The *Paley graph* $P(q)$ is the graph whose vertices are the elements of $GF(q)$, with $u \sim v$ if and only if $u - v$ is a square in $GF(q)$. (Note that the condition $q \equiv 1 \bmod 4$ ensures that $u - v$ is a square if and only if $v - u$ is a square.) The graph $P(q)$ is strongly regular with parameters $\left(q, \frac{1}{2}(q-1), \frac{1}{4}(q-5), \frac{1}{4}(q-1)\right)$. □

**Example 3.6.3.** Let $\Gamma$ be a finite group of permutations of the set $V$. Then $\Gamma$ has a natural action on $V^2$, given by

$$\gamma : (u, v) \mapsto (\gamma(u), \gamma(v)) \quad (\gamma \in \Gamma).$$

We say that $(\Gamma, V)$ is a permutation group of *rank s* if $\Gamma$ has $s$ orbits on $V^2$. (The orbits include the 'diagonal' orbit $D = \{(v, v) : v \in V\}$, and the permutation groups of rank 2 are precisely the doubly transitive groups.) Suppose that $\Gamma$ is of even order with rank 3, and let $D, O_1, O_2$ be the orbits of $(\Gamma, V^2)$. Since $|\Gamma|$ is even, $\Gamma$ contains an involution $\tau$. Let $a, b$ be points of $V$ interchanged by $\tau$. Without loss of generality, $(a, b) \in O_1$. Then $(b, a) \in O_1$ and it follows that $(u, v) \in O_1$ if and only if $(v, u) \in O_1$. Now we may define a graph $G$ with $V(G) = V$ and $u \sim v$ if and only if $(u, v) \in O_1$. It is easy to see that $G$ is strongly regular, with $\Gamma$ as a subgroup of its automorphism group. Such a graph is called a *rank 3 graph*. Note that the graph obtained in the same way from $O_2$ is just the complement of $G$. For an explicit example, we may take $\Gamma$ to be the alternating group on $\{1, 2, 3, 4, 5\}$ and $V$ to be the set of 10 unordered pairs in $\{1, 2, 3, 4, 5\}$. Then without loss of generality, $O_1$ consists of disjoint pairs, and $O_2$ consists of intersecting pairs; in this case, $G$ is the Petersen graph and $\overline{G} = L(K_5)$. □

It is a simple matter (Exercise 3.11) to check that if $G$ is a strongly regular graph with parameters $(n, r, e, f)$ then its complement $\overline{G}$ is strongly regular with parameters $(\bar{n}, \bar{r}, \bar{e}, \bar{f})$, where

$$\bar{n} = n, \quad \bar{r} = n - r - 1, \quad \bar{e} = n - 2 - 2r + f, \quad \bar{f} = n - 2r + e.$$

A strongly regular graph $G$ is *primitive* if both $G$ and $\overline{G}$ are connected; otherwise $G$ is *imprimitive*. It is straightforward to show (Exercise 3.12) that a strongly regular graph $G$ is imprimitive if and only if $G$ or $\overline{G}$ is a complete multipartite graph of the form $K_{m,m,\ldots,m}$.

The parameters of strongly regular graphs are not independent. Indeed, if we consider a fixed vertex $u$ and count in two ways the edges $vw$ such that $u$ is adjacent to $v$ but not to $w$, then we find

$$r(r - e - 1) = (n - r - 1)f. \tag{3.14}$$

Some other conditions on parameters will be discussed later.

From the definition of strongly regular graphs, we see that the adjacency matrix $A$ satisfies

$$A^2 = eA + f(J - A - I) + rI, \tag{3.15}$$

or equivalently

$$A^2 + (f - e)A + (f - r)I = fJ. \tag{3.16}$$

Since $AJ = rJ$, it follows from (3.16) that

$$(A - rI)(A^2 + (f - e)A + (f - r)I) = O. \tag{3.17}$$

The following theorem of Shrikhande and Bhagwandas gives a spectral characterization of the strongly regular graphs.

**Theorem 3.6.4 [ShrBh].** *Let $G$ be a connected regular graph of degree $r > 0$. Then $G$ is strongly regular if and only if it has exactly three distinct eigenvalues, say $\mu_1 = r$, $\mu_2 = s$ and $\mu_3 = t$. In this situation,*

$$e = r + s + t + st, \quad f = r + st, \quad n = \frac{(r - s)(r - t)}{r + st}.$$

**Proof.** Suppose that $G$ is strongly regular. If $\mu_2$ is the only eigenvalue different from $r$ then $r + (n - 1)\mu_2 = 0$. Thus $\mu_2$ is rational and hence an integer. But $0 < r < n - 1$, and so we have a contradiction. From (3.17), we know that the minimal polynomial of $G$ has degree at most 3, and so $G$ has exactly three distinct eigenvalues.

Conversely, suppose that $G$ is a connected $r$-regular graph, with exactly three distinct eigenvalues $r, s, t$. By Theorem 3.5.6 we have a relation of the form

$$aA^2 + bA + cI = J \quad (a \neq 0) \tag{3.18}$$

where $s, t$ are the roots of the quadratic $ax^2 + bx + c$. It follows that the number of walks of length 2 between vertices $i$ and $j$ is $\frac{1-b}{a}$ if $i \sim j$, and $\frac{1}{a}$ if $i \not\sim j$.

Therefore $G$ is strongly regular. From Theorem 3.5.6 we have

$$a = \frac{n}{(r-s)(r-t)}, \quad b = \frac{-n(s+t)}{(r-s)(r-t)}, \quad c = \frac{nst}{(r-s)(r-t)}. \quad (3.19)$$

Equating diagonal entries in (3.18), we find that $ar + c = 1$, and so

$$n(r+st) = (r-s)(r-t). \quad (3.20)$$

The formulae for $e$ and $f$ now follow from (3.17) and (3.18) since $e = \frac{1-b}{a}$ and $f = \frac{1}{a}$. $\quad\square$

Theorem 3.6.4 gives the parameters of a strongly regular graph in terms of eigenvalues. In the reverse direction, we have:

**Theorem 3.6.5.** *The distinct eigenvalues of a connected strongly regular graph with parameters $(n, r, e, f)$ are $r, s, t$, where*

$$s, t = \frac{1}{2}(e - f) \pm \sqrt{\Delta} \quad \text{and} \quad \Delta = (e - f)^2 + 4(r - f).$$

*Their respective multiplicities are $1, k, l$ where*

$$k, l = \frac{1}{2}\left\{n - 1 \mp \frac{2r + (n-1)(e-f)}{\sqrt{\Delta}}\right\}.$$

**Proof.** Since $G$ is connected and $r$-regular, $r$ is an eigenvalue of multiplicity 1. Eigenvectors corresponding to other eigenvalues are orthogonal to the all-1 vector, and so from (3.17) we see that $s, t$ are the roots of the quadratic $x^2 + (f - e)x + (f - r)$. Their multiplicities $k, l$ are determined from the equations

$$1 + k + l = n, \quad r + ks + lt = 0.$$

Here the first equation is obtained by counting eigenvalues, and the second by summing eigenvalues. $\quad\square$

Theorem 3.6.5 provides a nice feasibility condition for the parameters of a strongly regular graph: the parameters must be such that $k$ and $l$ are positive integers. (In practice, this condition turns out to be very powerful.) Further, if $\Delta$ is not a perfect square, then $k = l$ since $2r + (n-1)(e-f)$ is necessarily 0; in this situation, a strongly regular graph is called a *conference graph*. For example, the Paley graph of Example 3.6.3 is a conference graph. Since $\sqrt{\Delta} = s - t$, we have:

**Theorem 3.6.6.** *If $G$ is a strongly regular graph with parameters $(n, r, e, f)$ and eigenvalues $r, s, t$, then one of the following holds:*

(a) *G is a conference graph;*
(b) *each eigenvalue of G is an integer and* $(e - f)^2 + 4(r - f) = (s - t)^2.$

So far a complete characterization of parameters of strongly regular graphs is not known. We conclude this section by giving a further condition on multiplicities and a further condition on eigenvalues. In each case we use the fact that, in the light of (3.15), the matrices $I, A, J - I - A$ form a basis for the adjacency algebra.

**Theorem 3.6.7.** *Let G be a primitive strongly regular graph on n vertices, with eigenvalue multiplicities* $1, k, l$. *Then*

$$n \leq \min \left\{ \tfrac{1}{2}k(k + 3), \tfrac{1}{2}l(l + 3) \right\}.$$

**Proof.** Let $P$ represent the orthogonal projection of $I\!R^n$ onto the eigenspace of dimension $k$. From (1.5) we know that $P$ is a quadratic polynomial in $A$; hence, using (3.16), we can express $P$ in the form

$$P = \alpha I + \beta A + \gamma (J - I - A)$$

for some $\alpha, \beta, \gamma \in I\!R$. Using Theorem 3.6.5, we find (Exercise 3.14):

$$\alpha = \frac{k}{n}, \quad \beta = \frac{ks}{nr}, \quad \gamma = \frac{-k(s + 1)}{n(n - r - 1)}. \tag{3.21}$$

In particular, $\alpha \neq \beta$ and $\alpha \neq \gamma$. Since $P$ has spectrum $1^k, 0^{n-k}$, we may write $P = H^\top H$, where $H$ has size $k \times n$ and rank $k$. Thus if $H$ has columns $\mathbf{h}_1, \ldots, \mathbf{h}_n$ then

$$\mathbf{h}_i^\top \mathbf{h}_j = \begin{cases} \alpha & \text{if } i = j \\ \beta & \text{if } i \sim j \\ \gamma & \text{if } i \not\sim j, \ i \neq j. \end{cases}$$

Now let $\Omega$ be the sphere in $I\!R^k$ with equation $\|\mathbf{x}\| = \alpha$, and define $f_i : \Omega \to I\!R$ by

$$f_i(\mathbf{x}) = \frac{(\mathbf{h}_i^\top \mathbf{x} - \beta)(\mathbf{h}_i^\top \mathbf{x} - \gamma)}{(\alpha - \beta)(\alpha - \gamma)} \quad (i = 1, \ldots, n).$$

Each $f_i$ lies in $V_1 \oplus V_2$, where $V_1$ is the space of all homogeneous linear functions $\Omega \to I\!R$ and $V_2$ is the space of all homogenoeus quadratic functions $\Omega \to I\!R$. Note that the constant functions lie in $V_2$ because $\alpha^2 = x_1^2 + \cdots + x_n^2$ for all $(x_1, \ldots, x_n)^\top \in \Omega$. Also, $\dim(V_1 \oplus V_2) = k + \left(k + \binom{k}{2}\right) = \frac{1}{2}k(k + 3)$.

The functions $f_1, \ldots, f_n$ are linearly independent because $f_i(\mathbf{h}_j) = \delta_{ij}$. It follows that $n \leq \frac{1}{2}k(k + 3)$. Similarly, $n \leq \frac{1}{2}l(l + 3)$. $\qquad\square$

The bound for $n$ in Theorem 3.6.7 is known as the *absolute bound* for strongly regular graphs, since it is independent of $\alpha, \beta, \gamma$ (cf. [Sei4]). Graphs that attain the bound are called *extremal* strongly regular graphs. Only five such graphs $G$ are known: in these cases, $G$ or $\overline{G}$ is one of $C_5$, $Sch_{16}$, $McL_{112}$. Here, $Sch_{16}$ is the Schläfli graph of Example 1.2.5, and $McL_{112}$ is the McLaughlin graph, the unique strongly regular graph with parameters $(275, 112, 30, 56)$. This last graph is a rank 3 graph (and the corresponding rank 3 group has order $1, 796, 256, 000$); it was first constructed in [McL], and an alternative construction is described in [CamLi, Chapter 4].

For the next result, we require the following observation: since the adjacency algebra $\mathcal{A}$ of a strongly regular graph $G$ has $\{I, A, J - I - A\}$ as a basis, $\mathcal{A}$ is closed under Hadamard multiplication. (If the matrices $(x_{ij})$, $(y_{ij})$ have the same size then their Hadamard product is $(x_{ij} y_{ij})$, denoted by $(x_{ij}) \circ (y_{ij})$.) In the notation of Chapter 1, the projection matrices $P_1, P_2, P_3$ form a basis for $\mathcal{A}$ (cf. Equation (1.10)), and so

$$P_i \circ P_j = \sum_{i=1}^{3} q_{ijk} P_k$$

for some $q_{ijk} \in \mathbb{R}$. The real numbers $q_{ijk}$ are called the *Krein parameters* of $G$; note that $q_{ij1}, q_{ij2}, q_{ij3}$ are eigenvalues of $P_i \circ P_j$. Since $P_i \circ P_j$ is a principal submatrix of the positive semi-definite matrix $P_i \otimes P_j$, it too is a positive semi-definite matrix. Thus $q_{ijk} \geq 0$ for all $i, j, k$.

With notation as in the proof of Theorem 3.6.7, for $P = P_2$ we have

$$P \circ P = \alpha^2 I + \beta^2 A + \gamma^2 (J - I - A),$$

with eigenvalues

$$q_{221} = \alpha^2 + \beta^2 r + \gamma^2 (n - r - 1),$$
$$q_{222} = \alpha^2 + \beta^2 s + \gamma^2 (-s - 1),$$
$$q_{223} = \alpha^2 + \beta^2 t + \gamma^2 (-t - 1).$$

These may be expressed in terms of $n, r, s$ and $k$ using (3.21); we find that

$$q_{222} = \frac{k^2}{n^2} \left\{ 1 + \frac{s^3}{r^2} - \frac{(s + 1)^3}{(n - r - 1)^2} \right\}.$$

From (3.20) we have $n - r - 1 = -\dfrac{r(s + 1)(t + 1)}{r + st}$ and so

$$n^2 r^2 (t + 1)^2 q_{222} = k^2 (r - s)\{r(t^2 + 2t - s) + s(t^2 - 2st - s)\}. \quad (3.22)$$

It follows from (3.21) that $r(t^2 + 2t - s) + s(t^2 - 2st - s) \geq 0$: this is the first inequality of Theorem 3.6.8 below. The second inequality is derived by interchanging $s$ and $t$ (i.e. by taking $P = P_3$).

**Theorem 3.6.8[Sco].** *The eigenvalues $r, s, t$ of any primitive strongly regular graph of degree $r$ satisfy the inequalities*

(i) $(r + s)(t + 1)^2 \geq (s + 1)(r + s + 2st)$,
(ii) $(r + t)(s + 1)^2 \geq (t + 1)(r + t + 2st)$.

The inequalities of Theorem 3.6.8 are known as the *Krein inequalities*. To describe the implications of equality here, we define the subconstituents associated with a vertex $u$ of a strongly regular graph $G$: the *first subconstituent* is the regular subgraph of $G$ induced by the neighbours of $u$, and the *second subconstituent* of $u$ is the regular subgraph of $G$ induced by the non-neighbours of $u$. It can be shown that if the first bound of Theorem 3.6.8 is attained then $r = k$, while if the second is attained then $r = l$. In either case, one of the following holds: (a) $G$ is a 5-cycle; (b) in $G$ or $\overline{G}$, all the first subconstituents are null graphs, and all the second subconstituents are strongly regular; (c) all subconstituents of $G$ are strongly regular.

## 3.7 Distance-regular graphs

Let $G$ be a connected graph of diameter $d$, and for $i \in \{0, 1, \ldots, d\}$, let $\Gamma_i(u)$ denote the set of vertices at distance $i$ from the vertex $u$. We say that $G$ is *distance-regular* if there exist non-negative integers $b_0, b_1, \ldots, b_{d-1}$ and $c_1, c_2, \ldots, c_d$ such that for any two vertices $u, v$ at distance $i$,

$$b_i = |\Gamma_{i+1}(u) \cap \Gamma_1(v)| \ (i = 0, \ldots, d - 1),$$
$$c_i = |\Gamma_{i-1}(u) \cap \Gamma_1(v)| \ (i = 1, \ldots, d).$$

Thus $v$ has exactly $b_i$ neighbours at distance $i + 1$ from $u$, and $c_i$ neighbours at distance $i - 1$ from $u$. The array

$$\{b_0, b_1, \ldots, b_{d-1}; c_1, c_2, \ldots, c_d\}$$

is called the *intersection array* for $G$. Note that $c_1 = 1$ and $G$ is regular of degree $b_0$; we write $r = b_0$. Hence the number of neighbours of $v$ at distance $i$ from $u$ is $a_i$, where

$$a_i = r - b_i - c_i \ (i = 1, \ldots, d - 1) \quad \text{and} \quad a_d = r - c_d.$$

Note also that the distance-regular graphs of diameter 2 are precisely the connected strongly regular graphs.

The class of distance-regular graphs clearly includes the *distance-transitive* graphs: these are the connected graphs with the property that for any vertices $u, v, u', v'$ with $d(u, v) = d(u', v')$ there exists an automorphism which maps $u$ to $u'$ and $v$ to $v'$.

**Examples 3.7.1.** (i) The Petersen graph is distance-transitive with intersection array $\{3, 2; 1, 1\}$. The skeleta of the Platonic solids are also distance-transitive: the arrays are $\{3; 1\}$ for the tetrahedron, $\{4, 1; 1, 4\}$ for the octahedron, $\{3, 2, 1; 1, 2, 3\}$ for the cube, $\{5, 2, 1; 1, 2, 5\}$ for the icosahedron, and $\{3, 2, 1, 1, 1; 1, 1, 1, 2, 3\}$ for the dodecahedron.

(ii) The *Johnson graph* $J(n, m)$ has as its vertices the $m$-subsets of an $n$-set $X$; two such subsets are adjacent in $J(n, m)$ if they have exactly $m - 1$ elements in common. Thus $J(n, m) = J(n, n - m)$, $J(n, 1) = K_n$ and $J(n, 2) = L(K_n)$. The graph $J(n, m)$ is distance-transitive with diameter $d = \min\{m, n-m\}$ and parameters

$$b_i = (m - i)(n - m - i) \ (i = 0, \ldots, d - 1), \quad c_i = i^2 \ (i = 1, \ldots, d).$$

(iii) An example of a distance-regular graph that is not distance-transitive is the strongly regular graph defined as follows. The vertices are $u_1, \ldots, u_{13}, v_1, \ldots, v_{13}$ and the edges are given by:

$$u_i \sim u_j \text{ if and only if } |i - j| \equiv 1, 3 \text{ or } 4 \bmod 13,$$
$$v_i \sim v_j \text{ if and only if } |i - j| \equiv 2, 5 \text{ or } 6 \bmod 13,$$
$$u_i \sim v_j \text{ if and only if } |i - j| \equiv 0, 1, 3 \text{ or } 9 \bmod 13.$$

The intersection array is $\{10, 6; 1, 4\}$. $\qquad\qquad\qquad\qquad\qquad\qquad\square$

If we let $|\Gamma_i(u)| = k_i$ and count in two ways the edges between $\Gamma_i(u)$ and $\Gamma_{i+1}(u)$, we find that

$$k_0 = 1, \quad k_1 = r, \quad k_{i+1} = \frac{k_i b_i}{c_{i+1}} \ (i = 1, 2, \ldots, d - 1). \qquad (3.23)$$

We may illustrate these parameters in a diagram as shown in Fig. 3.3.

The parameters in an intersection array are subject to a number of constraints, the simplest of which are the following:

**Proposition 3.7.2.** *For any distance-regular graph with intersection array* $\{r, b_1, \ldots, b_{d-1}; 1, c_2, \ldots, c_d\}$, *we have*

(i) $1 \leq c_2 \leq c_3 \leq \cdots \leq c_d$,

(ii) $r \geq b_1 \geq b_2 \geq \cdots \geq b_{d-1}$,

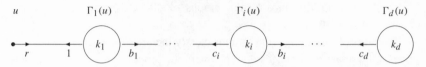

Figure 3.3 A representation of a distance-regular graph.

(iii) *for each* $j \in \{2, \ldots, d\}$, $rb_1 \cdots b_{j-1}/c_2 c_3 \cdots c_j$ *is an integer* ($=$ $|\Gamma_j(v)|$).

**Proof.** (i) Let $d(u, v) = i + 1 \le d$, and consider a path $uw \ldots v$ of length $i + 1$. Since $\Gamma_{i-1}(w) \cap \Gamma_1(v) \subseteq \Gamma_i(u) \cap \Gamma_1(v)$, we have $c_i \le c_{i+1}$.
(ii) Let $d(u, v) = i < d$, and consider a path $uw \ldots v$ of length $i$. Since $\Gamma_{i+1}(u) \cap \Gamma_1(v) \subseteq \Gamma_i(w) \cap \Gamma_1(v)$, we have $b_i \le b_{i-1}$.
(iii) The third assertion follows from (3.23) by induction on $i$.  □

For any graph we can define *distance matrices* $A_0, A_1, \ldots$ as follows: the $(i, j)$-entry of $A_h$ is 1 if $d(i, j) = h$, and 0 otherwise. (Thus $A_0 = I$ and $A_1 = A$.) For a distance-regular graph $G$, it is straightforward to show by induction on $k$ that, for each non-negative integer $k$, the $(i, j)$-entry of $A^k$ depends only on $d(i, j)$ (Exercise 3.16). In other words, each $A^k$ is a linear combination of $A_0, A_1, \ldots, A_d$; equivalently, the adjacency algebra $\mathcal{A}$ of $G$ has $\{A_0, A_1, \ldots, A_d\}$ as a basis. Since $I, A, A^2, \ldots, A^d$ are linearly independent, $\{I, A, \ldots, A^d\}$ is another basis for $\mathcal{A}$. Thus the minimal polynomial $m_A$ of $A$ has degree $d + 1$ and $G$ has precisely $d + 1$ distinct eigenvalues.

Now consider the linear transformation $\tau$ of $\mathcal{A}$ given by: $X \mapsto XA$ ($X \in \mathcal{A}$). The matrix of $\tau$ with respect to $\{I, A, \ldots, A^d\}$ is just the companion matrix of $m_A$; but the matrix $B$ of $\tau$ with respect to $\{A_0, A_1, \ldots, A_d\}$ has a tridiagonal form, because

$$A_i A = b_{i-1} A_{i-1} + a_i A_i + c_{i+1} A_{i+1} \quad (0 < i < d) \quad \text{and}$$
$$A_0 A = 0 A_0 + 1 A_1, \quad A_d A = b_{d-1} A_{d-1} + a_d A_d. \tag{3.24}$$

It follows from (3.24) that

$$B = \begin{pmatrix} 0 & 1 & & & & & \\ r & a_1 & c_2 & & & & \\ & b_1 & a_2 & . & & & \\ & & b_2 & . & . & & \\ & & & . & . & . & \\ & & & & . & . & c_{d-1} \\ & & & & . & a_{d-1} & c_d \\ & & & & & b_{d-1} & a_d \end{pmatrix}.$$

Note first that $A$ and $B$ share the same distinct eigenvalues $\mu_1, \ldots, \mu_{d+1}$, because $m_B = m_\tau = m_A$. Secondly, if the polynomials $v_0(x), \ldots, v_d(x)$ are defined recursively by:

$$v_0(x) = 1, \quad v_1(x) = x,$$

$$c_{i+1}v_{i+1}(x) + (a_i - x)v_i(x) + b_{i-1}v_{i-1}(x) = 0 \quad (0 < i < d) \qquad (3.25)$$

then it follows from (3.24) by induction on $i$ that $A_i = v_i(A)$ ($i = 0, 1, \ldots, d$). In other words, $B$ determines the transition matrix from $\{A_0, A_1, \ldots, A_d\}$ to $\{I, A, A^2, \ldots, A^d\}$. Hence $B$ determines the inverse transition matrix $(w_{hk})$, where $A^k = \sum_{h=0}^{d} w_{hk}A_h$ ($k = 0, 1, \ldots, d$). Since $\mathrm{tr}(A_0) = n$, while $\mathrm{tr}(A_h) = 0$ for $h \in \{1, \ldots, d\}$, we have $\sum_{i=1}^{d+1} m(\mu_i)\mu_i^k = nw_{0k}$ ($k = 0, 1, \ldots, d$), where $m(\mu_i)$ denotes the multiplicity of $\mu_i$ as an eigenvalue of $A$. It is clear from these $d + 1$ equations that the multiplicities $m(\mu_i)$ are determined by $B$. Consequently we have:

**Theorem 3.7.3.** *The spectrum of a distance-regular graph $G$ is determined by the intersection array for $G$.*

We shall determine the multiplicity $m(\mu_i)$ explicitly in terms of $k_0, k_1, \ldots, k_d$ and an eigenvector of $B$ corresponding to $\mu_i$. (Note that all the eigenspaces of $B$ are one-dimensional.)

**Lemma 3.7.4.** *For $j = 1, 2, \ldots, d + 1$, let $\mathbf{v}_j = (v_0(\mu_j), v_1(\mu_j), \ldots, v_d(\mu_j))^\top$ and $\mathbf{u}_j = K^{-1}\mathbf{v}_j$, where $K = \mathrm{diag}(k_0, k_1, \ldots, k_d)$. Then $\mathbf{v}_j$ is an eigenvector of $B$ and $\mathbf{u}_j$ is an eigenvector of $B^\top$, both corresponding to the eigenvalue $\mu_j$.*

**Proof.** We have directly from (3.24) that $B\mathbf{v}_j = \mu_j\mathbf{v}_j$. From (3.22) we have $BK = KB^\top$, and so $B^\top\mathbf{u}_j = B^\top K^{-1}\mathbf{v}_j = K^{-1}B\mathbf{v}_j = K^{-1}(\mu_j\mathbf{v}_j) = \mu_j\mathbf{u}_j$. $\square$

**Theorem 3.7.5.** *With the notation above, the eigenvalue $\mu_i$ of a distance-regular graph on $n$ vertices has multiplicity*

$$m(\mu_i) = \frac{n}{\mathbf{v}_i^\top K^{-1}\mathbf{v}_i}.$$

**Proof.** It follows from Lemma 3.7.4 that $\mathbf{u}_i^\top \mathbf{v}_j = 0$ when $i \neq j$, because $\mu_i\mathbf{u}_i^\top\mathbf{v}_j = \mathbf{u}_i^\top B\mathbf{v}_j = \mu_j\mathbf{u}_i^\top\mathbf{v}_j$. Now let $\mathbf{u}_i = (u_{i0}, u_{i1}, \ldots, u_{id})^\top$ and $\mathbf{v}_i = (v_{i0}, v_{i1}, \ldots, v_{id})^\top$. Note that $u_{i0} = 1$ because $k_0 = 1$ and $v_{i0} = v_0(\mu_i) = 1$. We calculate in two ways the trace of the matrix

$$M_i = \sum_{j=0}^{d} u_{ij}A_j.$$

First, since $A_1, \ldots, A_d$ have zero diagonal, we have $\mathrm{tr}(M_i) = u_{i0}\mathrm{tr}(A_0) = n$. Secondly, since $A_j = v_j(A)$, the eigenvalues of $A_j$ are $v_j(\mu_1), \ldots, v_j(\mu_{d+1})$, with multiplcities $m(\mu_1), \ldots, m(\mu_{d+1})$, and so $\mathrm{tr}(A_j) = \sum_{k=1}^{d+1} m(\mu_k)v_j(\mu_k)$. Hence

$$\mathrm{tr}(M_i) = \sum_{j=0}^{d} u_{ij} \sum_{k=1}^{d+1} m(\mu_k)v_{kj} = \sum_{k=1}^{d+1} m(\mu_k)\mathbf{u}_i^\top \mathbf{v}_k = m(\mu_i)\mathbf{u}_i^\top \mathbf{v}_i,$$

and the result follows.                                                        □

Since $m(\mu_j)$ is a positive integer, Theorem 3.7.5 imposes a further restriction on the parameters of an intersection array. For example [Big2, p. 168] there is no distance-regular graph with intersection array {3, 2, 1; 1, 1, 3}, an array not excluded by Theorem 3.7.2. Further necessary conditions on the parameters of an intersection array, analogous to the Krein inequalities of Theorem 3.6.8, arise from the fact that $\mathcal{A}$ is closed under Hadamard multiplication. For these and other constraints the reader is referred to the monograph by Brouwer, Cohen and Neumaier [BroCN].

We note that, in contrast to the situation for strongly regular graphs, the property of distance-regularity (of diameter > 2) cannot in general be identified from the spectrum. Haemers and Spence [HaeSp] show that while there is a unique distance-regular graph with intersection array {13, 6, 1; 1, 6, 13}, it is one of no fewer than 515 graphs of diameter 3 which share the same spectrum, namely $\{13^1, \sqrt{13}^7, (-1)^{13}, (-\sqrt{13})^7\}$. On the other hand there are four nonisomorphic distance-regular graphs with intersection array {7, 6, 4; 1, 3, 7}; they have spectrum $\{7^1, 2^{14}, (-2)^{14}, (-7)^1\}$ and they are the only graphs with this spectrum.

Finally we mention the Bannai–Ito conjecture [BanIt, p. 237], which asserts that for fixed $r > 2$ there are only finitely many distance-regular graphs of degree $r$. This has been confirmed for $r \in \{3, 4\}$ by Bannai and Ito themselves, and for $r \in \{5, 6, 7\}$ by Koolen and Moulton [KooMo2].[1]

## 3.8 Automorphisms and eigenspaces

Recall that an automorphism of a graph $G$ is a permutation $\pi$ of $V(G)$ such that $u \sim v$ if and only if $\pi(u) \sim \pi(v)$. The group of all automorphisms of $G$ is denoted by $\mathrm{Aut}(G)$, and the order of $\mathrm{Aut}(G)$ is a measure of the symmetry of $G$. Vertices in the same orbit of $\mathrm{Aut}(G)$ are said to be *similar*.

---

[1] The Babbai–Ito conjecture has now been confirmed for all $r > 2$ by Bang, Koolen and Moulton.

Symmetries are commonly used as a tool in the study of mathematical structures, and the symmetric features of graphs are often revealed by an appropriate geometric representation. For example, the graph $C_5$ can be viewed as a regular pentagon, whose symmetries consist of five rotations and five reflections. Although the full group of automorphisms of the Petersen graph $P$ (the group $S_5$) is not readily identified from a single diagrammatic representation, the standard drawing of $P$ (Fig. 1.2) shows that it too has a symmetry of order 5. For both $C_5$ and $P$, the presence of an automorphism of order 5 guarantees the existence of a multiple eigenvalue. This follows from Theorem 3.8.4 below and illustrates the flavour of the results in this section.

As usual, let $G$ be a graph with vertex-set $V(G) = \{1, 2, \ldots, n\}$ and adjacency matrix $A = (a_{ij})$. For any permutation $\pi$ of $\{1, 2, \ldots, n\}$, let $P(\pi)$ be the permutation matrix $(\delta_{\pi(i)j})$. Note that $P(\pi)^{-1} = P(\pi)^{\top}$ and that the map $\pi \mapsto P(\pi)$ is a monomorphism from the symmetric group $S_n$ into the multiplicative group of orthogonal $n \times n$ matrices.

Let $A'$ be the adjacency matrix of $G$ obtained when the vertices $1, 2, \ldots, n$ are relabelled $\pi(1), \pi(2), \ldots, \pi(n)$. Then $A' = P(\pi)^{\top} A P(\pi)$ because the $(i, j)$-entry of $P(\pi)^{\top} A P(\pi)$ is

$$\sum_h \sum_k \delta_{\pi(h)i} a_{hk} \delta_{\pi(k)j} = a_{\pi^{-1}(i)\pi^{-1}(j)}.$$

Since $\pi$ is an automorphism of $G$ if and only if $A' = A$, we have:

**Proposition 3.8.1.** *The permutation $\pi$ is an automorphism of $G$ if and only if $A = P(\pi)^{\top} A P(\pi)$, equivalently $P(\pi)A = A P(\pi)$.*

It follows that if $\lambda$ is an eigenvalue of $G$ and $\mathbf{x} \in \mathcal{E}_A(\lambda)$ then for each automorphism $\pi$ of $G$ we have

$$A P(\pi)\mathbf{x} = P(\pi)A\mathbf{x} = \lambda P(\pi)\mathbf{x}.$$

Thus each eigenspace is $P(\pi)$-invariant for every $\pi \in \mathrm{Aut}(G)$; we say simply that the eigenspaces are invariant under the automorphism group. Clearly, if $\mathbf{x}$ and $P(\pi)\mathbf{x}$ are linearly independent eigenvectors, then $\lambda$ is a multiple eigenvalue. This simple observation is crucial to what follows.

**Lemma 3.8.2.** *If $\lambda$ is a simple eigenvalue of $G$, and if $\mathbf{x}$ is an eigenvector corresponding to $\lambda$ then $P(\pi)\mathbf{x} = \pm\mathbf{x}$ for each $\pi \in \mathrm{Aut}(G)$.*

**Proof.** Since $\mathbf{x}$ and $P(\pi)\mathbf{x}$ are linearly dependent eigenvectors in $\mathbb{R}^n$, we have $P(\pi)\mathbf{x} = c\mathbf{x}$ for some $c \in \mathbb{R}$. Since $\| P(\pi)\mathbf{x} \| = \|\mathbf{x}\|$, we have $c = \pm 1$. $\qquad\square$

**Theorem 3.8.3** [PeSa2]. *Let $G$ be a vertex-transitive graph with $n$ vertices of degree $r$, and let $\lambda$ be a simple eigenvalue of $G$. If $n$ is odd then $\lambda = r$; if $n$ is even then $\lambda = 2k - r$ for some $k \in \{0, 1, \ldots, r\}$.*

**Proof.** Let $\mathbf{x} = (x_1, x_2, \ldots, x_n)^\top$ be an eigenvector of $G$ corresponding to $\lambda$. If $\pi$ is an automorphism of $G$ such that $\pi(j) = i$ then $x_i$ is the $j$-th entry of $P(\pi)\mathbf{x}$. By Lemma 3.8.2, $x_i = \pm x_j$. Since $G$ is vertex-transitive, it follows that all entries of $\mathbf{x}$ have the same absolute value.

Suppose first that $n$ is odd. If $\lambda \neq r$ then $\mathcal{E}(\lambda) \perp \mathcal{E}(r)$ and so $\sum_{i=1}^n x_i = 0$. But this sum cannot vanish under the established conditions, and consequently $\lambda = r$ is the only possibility.

Assume now that $n$ is even. For a fixed vertex $i$, suppose that $i$ has $k$ neighbours $j$ such that $x_j = x_i$, and $r - k$ neighbours $j$ such that $x_j = -x_i$. From the $i$-th eigenvalue equation, we have

$$\lambda x_i = \sum_{j \sim i} x_j = k x_i - (r - k) x_i,$$

whence $\lambda = 2k - r$, as required. This completes the proof. □

**Theorem 3.8.4** [Mow, PeSa2]. *If $G$ is a graph with an automorphism of order greater than 2, then $G$ has a multiple eigenvalue.*

**Proof.** Suppose by the way of contradiction that all eigenvalues are simple. If $\mathbf{x}$ is an eigenvector of $G$, then, by Lemma 3.8.2, $P(\pi)^2\mathbf{x} = \mathbf{x}$ for every automorphism $\pi$ of $G$. Since $\mathbb{R}^n$ has a basis of eigenvectors, we have $P(\pi)^2 = I$. Hence $\pi^2$ is the identity permutation for every automorphism $\pi$, contrary to assumption. □

The proof of Theorem 3.8.4 shows that if all eigenvalues of $G$ are simple, then every non-identity automorphism has order 2, equivalently $\mathrm{Aut}(G)$ is an elementary abelian 2-group. To describe the general situation, let $U$ be an orthogonal matrix such that $U^\top A U = D = \mathrm{diag}(\lambda_1, \ldots, \lambda_n)$, and let $O(k)$ denote the multiplicative group of $k \times k$ orthogonal matrices. If $\pi \in \mathrm{Aut}(G)$ then $U^\top P(\pi) U$ commutes with $D$. Hence if the distinct eigenvalues $\mu_1, \ldots, \mu_m$ have multiplicities $k_1, \ldots, k_m$ then $U^\top P(\pi) U$ has the block-diagonal form $X_1(\pi) \dotplus \cdots \dotplus X_m(\pi)$, where $X_i(\pi) \in O(k_i)$. Accordingly, we have:

**Proposition 3.8.5.** *If $G$ has eigenvalue multiplicities $k_1, \ldots, k_m$ then $\mathrm{Aut}(G)$ is isomorphic to a subgroup of $O(k_1) \times \cdots \times O(k_m)$.*

When some eigenvalues are simple, we can obtain some additional information on $\mathrm{Aut}(G)$ by counting the non-real eigenvalues of $P(\pi)$. We denote the number of such eigenvalues by $w(\pi)$. Thus if $\pi$ is a $t$-cycle then $w(\pi)$ is $t-1$ if $t$ is odd, and $t-2$ if $t$ is even. If $\pi = \pi_1\pi_2\cdots\pi_j$ as a product of disjoint cycles then $w(\pi) = \sum_{i=1}^{k} w(\pi_i)$. Now suppose that $G$ has $n-r$ simple eigenvalues $(0 \le r < n)$. With appropriate ordering of the columns of $U$, $U^\top P(\pi)U$ has the block-diagonal form $X(\pi) \dotplus e_{r+1}(\pi) \dotplus \cdots \dotplus e_n(\pi)$, where each $e_i(\pi)$ is $\pm 1$. The non-real eigenvalues of $P(\pi)$ are necessarily eigenvalues of $X(\pi)$ and so number at most $r$. Hence $w(\pi) \le r$ and each constituent cycle $\pi_k$ of $\pi$ has length at most $r+2$. We deduce the following result:

**Proposition 3.8.6** [Row1]. *If $G$ has $n-r$ simple eigenvalues then the order of any automorphism of $G$ divides the least common multiple of* $2, 3, 4, \ldots, r+2$.

Finally, we mention without proof two upper bounds for the number $s$ of simple eigenvalues of a graph on vertices. First, $s$ is at most the largest power of 2 that divides $n$ [SaSt]; secondly, if $\mathrm{Aut}(G)$ has no orbit on which it acts as an elementary abelian 2-group then $s \le \frac{5}{9}n$ [Row1].

## 3.9 Equitable partitions, divisors and main eigenvalues

Equitable partitions and divisors represent a powerful tool in spectral graph theory. In particular we shall see how to exploit regularity properties of a graph to obtain part of the spectrum, including the main eigenvalues.

**Definition 3.9.1.** Given a graph $G$, the partition $V(G) = V_1 \,\dot\cup\, V_2 \,\dot\cup\cdots\dot\cup\, V_k$ is an equitable partition if every vertex in $V_i$ has the same number of neighbours in $V_j$, for all $i, j \in \{1, 2, \ldots, k\}$.

Clearly, every graph has a trivial equitable partition, in which each cell is a singleton. For the existence of a non-trivial equitable partition, some local regularity is required. For example, in a complete multipartite graph the usual colouring gives rise to an equitable partition in which the cells are the colour classes. In general, it is often convenient to assign different colours to the cells of an equitable partition. Then the subgraphs induced by the vertices of the same colour are regular, while edges joining the vertices from two different cells give rise to a semi–regular bipartite graph. In view of this colouring, an equitable partition is sometimes called a *colouration*.

Suppose now that $\Pi$ is an equitable partition $V(G) = V_1 \,\dot\cup\, V_2 \,\dot\cup\cdots\dot\cup\, V_k$, and that each vertex in $V_i$ has $b_{ij}$ neighbours in $V_j$ $(i, j \in \{1, 2, \ldots, k\})$. Let $D_\Pi$ be the directed multigraph with vertices $V_1, V_2, \ldots, V_k$ and $b_{ij}$ arcs from $V_i$ to $V_j$. We call $D_\Pi$ the *divisor* of $G$ with respect to $\Pi$. The matrix $(b_{ij})$ is called the *divisor matrix* of $\Pi$, denoted by $B_\Pi$.

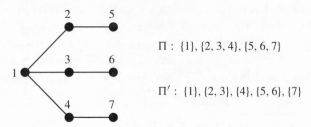

$$\Pi : \{1\}, \{2, 3, 4\}, \{5, 6, 7\}$$

$$\Pi' : \{1\}, \{2, 3\}, \{4\}, \{5, 6\}, \{7\}$$

Figure 3.4 Two equitable partitions of a graph.

**Example 3.9.2.** Fig. 3.4 shows a graph and two of its equitable partitions. The corresponding divisor matrices are:

$$B_\Pi = \begin{pmatrix} 0 & 3 & 0 \\ 1 & 0 & 1 \\ 0 & 1 & 0 \end{pmatrix}, \quad B_{\Pi'} = \begin{pmatrix} 0 & 2 & 1 & 0 & 0 \\ 1 & 0 & 0 & 1 & 0 \\ 1 & 0 & 0 & 0 & 1 \\ 0 & 1 & 0 & 0 & 0 \\ 0 & 0 & 1 & 0 & 0 \end{pmatrix}.$$

It is clear from Fig. 3.3 that, for any vertex $u$ in a distance-regular graph of diameter $d$, the sets $\Gamma_i(u)$ ($i = 0, 1, \ldots, d$) form an equitable partition. (The corresponding divisor matrix is the transpose of the matrix $B$ considered in Section 3.7.) Also, for any graph $G$, the orbits of $\text{Aut}(G)$, or of any subgroup of $\text{Aut}(G)$, form an equitable partition (Exercise 3.18); the first partition in Example 3.9.2 is such a partition.

For any partition $\Pi$ of $V(G)$ with cells $V_1, V_2, \ldots, V_k$, let $C_\Pi$ be the $n \times k$ matrix whose columns are the characteristic vectors of $V_1, V_2, \ldots, V_k$. We call $C_\Pi$ the *characteristic matrix* of $\Pi$. Note that $C_\Pi^\top C_\Pi = \text{diag}(|V_1|, |V_2|, \ldots, |V_k|)$.

**Proposition 3.9.3.** *Let $G$ be a graph with adjacency matrix $A$. If $\Pi$ is an equitable partition of $G$, with divisor matrix $B$ and characteristic matrix $C$, then*

$$AC = CB, \quad B = (C^\top C)^{-1} C^\top AC.$$

**Proof.** It suffices to note that if $i \in V_h$ then the $(i, j)$-entry of both $AC$ and $CB$ is $b_{hj}$. □

The following theorem characterizes the equitable partitions:

**Theorem 3.9.4.** *Let $G$ be a graph with adjacency matrix $A$, and let $\Pi$ be a partition of $G$ with characteristic matrix $C$. Then $\Pi$ is an equitable partition of $V(G)$ if and only if the column space of $C$ is $A$-invariant.*

**Proof.** If $\Pi$ is an equitable partition then, in the notation of Proposition 3.9.3, $AC = CB$, and so the column space of $C$ is $A$-invariant. Conversely, if the column space of $C$ is $A$-invariant then there exists a matrix $B = (b_{ij})$ such that $AC = CB$. Equating entries in this matrix equation, we find that each vertex in the $h$-th cell of $\Pi$ is adjacent to $b_{hj}$ vertices in the $j$-th cell. $\square$

We can now prove the first of the two main results on divisors.

**Theorem 3.9.5.** *The characteristic polynomial of any divisor of a graph divides the characteristic polynomial of the graph.*

**Proof.** We use the notation of Proposition 3.9.3. Let $C^*$ be an $n \times (n - k)$ matrix whose columns are vectors which extend the columns of $C$ to a basis of $\mathbb{R}^n$. Then there exists matrices $X$ and $Y$ such that

$$AC^* = CX + C^*Y.$$

From this equation and Proposition 3.9.3 we obtain:

$$A\,(C \mid C^*) = (C \mid C^*)\begin{pmatrix} B & X \\ O & Y \end{pmatrix}.$$

Since $(C \mid C^*)$ is invertible, it follows that $\det(xI - A) = \det(xI - B)\det(xI - Y)$. $\square$

**Remark 3.9.6.** In the situation of Theorem 3.9.5, we have $AC = CB$ and hence

$$f(A)C = Cf(B) \tag{3.26}$$

for any polynomial $f(x) \in \mathbb{R}[x]$. In particular, we have

$$(xI - A)C\mathbf{v} = C(xI - B)\mathbf{v}$$

for all $\mathbf{v} \in \mathbb{R}^n$. Since $C\mathbf{x} = \mathbf{0}$ if and only if $\mathbf{x} = \mathbf{0}$, it follows that $\mathbf{v}$ is an eigenvector of $B$ if and only if $C\mathbf{v}$ is an eigenvector of $A$.

Since the column space of $C$ is $A$-invariant, it has a basis consisting of $k$ eigenvectors of $A$. Each such eigenvector has the form $C\mathbf{v}$, and so its entries are constant on each cell of the underlying equitable partition $\Pi$. We may choose $n - k$ further eigenvectors of $A$ orthogonal to the column space of $C$, and the entries of such vectors sum to zero on each cell of $\Pi$. $\square$

The second main result on divisors of a graph $G$ concerns the main eigenvalues of $G$. Recall that the eigenvalue $\mu_i$ is a *main* eigenvalue of $G$ if $\mathcal{E}(\mu_i)$ is not orthogonal to the all-1 vector $\mathbf{j}$. In the notation of Chapter 1, this is equivalent to the condition $P_i\mathbf{j} \neq \mathbf{0}$.

**Definition 3.9.7.** Let $G$ be a graph whose distinct eigenvalues are $\mu_1, \ldots, \mu_m$. The main part of the spectrum of $G$ is the subset $\mathcal{M}$ of $\{\mu_1, \ldots, \mu_m\}$ consisting of the main eigenvalues of $G$, and we define

$$M_G(x) = \prod_{\mu_i \in \mathcal{M}} (x - \mu_i).$$

Note that cospectral graphs need not have the same main part of the spectrum; for example the graphs $K_{1,4}$, $C_4 \mathbin{\dot\cup} K_1$ are cospectral and $-2$ is a main eigenvalue of the first graph but not the second. On the other hand, it follows from Proposition 2.1.3 that if $G_1$ and $G_2$ are cospectral graphs with cospectral complements then $G_1$ and $G_2$ share the same main part of the spectrum.

**Lemma 3.9.8.** *Let* $f(x) \in I\!\!R[x]$. *Then* $f(A)\mathbf{j} = \mathbf{0}$ *if and only if* $M_G(x)$ *divides* $f(x)$.

**Proof.** We may use the spectral decomposition of $A$ to express $f(A)\mathbf{j}$ as an element of $\mathcal{E}(\mu_1) \oplus \mathcal{E}(\mu_2) \oplus \cdots \oplus \mathcal{E}(\mu_m)$:

$$f(A)\mathbf{j} = f(\mu_1)P_1\mathbf{j} + f(\mu_2)P_2\mathbf{j} + \cdots + f(\mu_m)P_m\mathbf{j},$$

where the $i$-th summand is $\mathbf{0}$ if $\mu_i \notin \mathcal{M}$. Hence $f(A)\mathbf{j} = \mathbf{0}$ if and only if $f(\mu_i) = 0$ for each $\mu_i \in \mathcal{M}$. The result follows. $\square$

The second main result on divisors is the following.

**Theorem 3.9.9.** *The characteristic polynomial of any divisor of a graph $G$ is divisible by* $M_G(x)$.

**Proof.** Let $B$ be a divisor matrix, with characteristic polynomial $f(x)$. By Equation (3.26), we have $f(A)C\mathbf{j}_k = Cf(B)\mathbf{j}_k$, where $\mathbf{j}_k$ is the all-1 vector in $I\!\!R^k$. Now $C\mathbf{j}_k = \mathbf{j}$, while $f(B) = O$ (by the Cayley–Hamilton Theorem). Hence $f(A)\mathbf{j} = \mathbf{0}$, and the result follows from Lemma 3.9.8. $\square$

**Corollary 3.9.10.** *If* $\mathrm{Aut}(G)$ *has $s$ orbits on $V(G)$ then $G$ has at most $s$ main eigenvalues.*

**Proof.** The orbits of $\mathrm{Aut}(G)$ constitute an equitable partition $\Pi$ for which $\det(xI - B_\Pi)$ has degree $s$. By Theorem 3.9.9, $\det(xI - B_\Pi)$ is divisible by $M_G(x)$. The result follows because $M_G(x)$ has degree $|\mathcal{M}|$. $\square$

The largest eigenvalue $\mu_1$ always belongs to $\mathcal{M}$ because $\mathcal{E}(\mu_1)$ contains an eigenvector whose entries are non-negative. Hence we have:

**Corollary 3.9.11.** *Any divisor of a graph $G$ has the index of $G$ as an eigenvalue.*

In an $r$-regular graph, every eigenspace other than $\mathcal{E}(r)$ is orthogonal to $\mathbf{j}$, and so we have:

**Corollary 3.9.12.** *The graphs with exactly one main eigenvalue are precisely the regular graphs.*

It is an open problem to determine the graphs with exactly $s$ main eigen-values, where $s > 1$. To describe one of the known results for the case $s = 2$, recall that a *harmonic* graph is a graph for which the vector $\mathbf{d}$ of vertex-degrees is an eigenvector. If $A\mathbf{d} = \mu\mathbf{d}$ then we say that $G$ is $\mu$-harmonic. In this situation, $\mu = \mu_1 \in \mathbb{Z}$ because $\mu$ is rational, while the entries of $\mathbf{d}$ are non-negative. Further, if $G$ has no isolated vertices, then for any vertex $v$ of $G$, $\mu$ is the mean degree of the neighbours of $v$. The graph of Fig. 3.4 is 2-harmonic.

**Proposition 3.9.13.** *Let $G$ be a non-trivial connected graph with index $\mu$. Then $G$ is harmonic and non-regular if and only if the main eigenvalues of $G$ are $\mu$ and $0$.*

**Proof.** Note that $\mathbf{d} = A\mathbf{j} \neq \mathbf{0}$, while the relation $A\mathbf{d} = \mu\mathbf{d}$ may be written as $(A^2 - \mu A)\mathbf{j} = \mathbf{0}$. Hence $G$ is harmonic and non-regular if and only if $(A^2 - \mu A)\mathbf{j} = \mathbf{0}$ and $A\mathbf{j} \neq \mu\mathbf{j}$. By Lemma 3.9.8, $G$ is harmonic and non-regular if and only if $M_G(x) = x(x - \mu)$. $\qquad\square$

## 3.10 Spectral bounds for graph invariants

In this section we give some further bounds on non-spectral invariants in terms of graph eigenvalues. The existence of such bounds provides some justification for ordering graphs lexicographically by spectrum: small changes in eigen-values will restrict changes to the relevant structural invariants. (See also the remarks in Section 3.5.)

Here we discuss the stability number, the clique number and the chromatic number (all defined below). Spectral bounds for these invariants are of interest in the context of complexity: the problem of determining each of the invariants is NP-complete, whereas the spectral bounds can be determined in polynomial time.

The *stability number* (or *independence number*) of a graph $G$ is denoted by $\alpha(G)$: this is the largest number of pairwise non-adjacent vertices in $G$.

**Theorem 3.10.1.** *Let $G$ be a graph on $n$ vertices. Let $n^+$ and $n^-$ denote the number of positive and negative eigenvalues of $G$ respectively. Then*

$$\alpha(G) \leq \min\{n - n^+, n - n^-\}.$$

**Proof.** A set of $s$ independent vertices in $G$ induces a null subgraph $H$. By the Interlacing Theorem (applied to the adjacency matrix $A$ of $G$) we have

$$\lambda_{n-s+i}(G) \leq \lambda_i(H) \leq \lambda_i(G) \quad (i = 1, 2, \ldots, s).$$

It follows that $0 \leq \lambda_s(G)$, and so $n^- \leq n - s$. If we apply the same argument to the matrix $-A$ in place of $A$, we find that $n^+ \leq n - s$. Thus $s \leq \min\{n - n^+, n - n^-\}$, as required. $\qquad\square$

Note that the bound in Theorem 3.10.1 is attained by a complete graph. For regular graphs the following bound was obtained (but not published) by Hoffman.

**Theorem 3.10.2.** *If $G$ is a regular graph with spectrum $\lambda_1 \geq \cdots \geq \lambda_n$, then*

$$\alpha(G) \leq n \frac{-\lambda_n}{\lambda_1 - \lambda_n}.$$

**Proof.** From Theorem 3.5.7 (the left-hand inequality), we have

$$|V(H)| \leq n \frac{\bar{d} - \lambda_n}{\lambda_1 - \lambda_n}$$

for any induced subgraph $H$ with mean degree $\bar{d}$. If $H$ is a null graph then $\bar{d} = 0$ and the result follows. $\qquad\square$

The *clique number* of $G$, denoted by $\omega(G)$, is the number of vertices in the largest clique of $G$. Thus $\omega(G) = \alpha(\overline{G})$.

**Theorem 3.10.3.** *Let $m^-$, $m^0$, $m^+$ denote the number of eigenvalues of a graph $G$ which are less than, equal to, or greater than $-1$, respectively. Let $s = \min\{m^- + m^0 + 1, m^0 + m^+, 1 + \rho\}$, where $\rho$ is the index of $G$. Then $\omega(G) \leq s$. If $s = m^- + m^0 + 1$ and the eigenvalues greater than $-1$ exceed $m^- + m^0$ then $\omega(G) \leq s - 1$.*

**Proof.** Suppose that $G$ contains a clique on $k$ vertices. Then by the Interlacing Theorem we have

$$\lambda_{n-k+1} \leq k - 1 \leq \lambda_1 = \rho, \tag{3.27}$$

$$\lambda_{n-k+i} \leq -1 \leq \lambda_i \ (i = 2, \ldots, k). \tag{3.28}$$

From (3.28) we have $k \leq m^- + m^0 + 1$ and $k \leq m^0 + m^+$. From the right-hand side of (3.27) we have $k \leq 1 + \rho$. Hence $k \leq s$. If $k = s = m^- + m^0 + 1$ then (again by interlacing) $k - 1 \geq \lambda_*$, where $\lambda_*$ denotes the least eigenvalue greater than $-1$. In this situation $\lambda_* \leq m^- + m^0$, and the last assertion of the theorem follows. $\qquad\qquad\qquad\qquad\qquad\qquad\qquad\qquad\qquad\qquad\qquad\square$

Note that the upper bound $s$ in Theorem 3.10.3 is attained in the complete multipartite graph $\overline{m K_n}$ (for which $m^- = m-1$, $m^0 = 0$ and $m^+ = mn - m + 1$). For a spectral lower bound on $\omega(G)$, we make use of the Motzkin–Straus inequality:

**Lemma 3.10.4** [MotSt]. *If $G$ is a graph with adjacency matrix $A$ then*

$$\max\{\mathbf{x}^\top A\mathbf{x} : \mathbf{x} \geq \mathbf{0}, \ \mathbf{j}^\top \mathbf{x} = 1\} = 1 - \frac{1}{\omega(G)}.$$

**Proof.** Let $\mathbf{x} = (x_1, \ldots, x_n)^\top$ and let $S$ be the simplex $\{\mathbf{x} \in \mathbb{R}^n : \mathbf{x} \geq \mathbf{0}, \ \mathbf{j}^\top \mathbf{x} = 1\}$. We write $F(\mathbf{x}) = \mathbf{x}^\top A\mathbf{x}$ and $f(G) = \max\{F(\mathbf{x}) : \mathbf{x} \in S\}$. If the vertices $1, \ldots k$ induce a largest clique, and if we set $x_1 = \cdots = x_k = 1/k$, $x_{k+1} = \cdots = x_n = 0$, then

$$f(G) \geq 2\binom{k}{2}\frac{1}{k^2} = 1 - \frac{1}{k} = 1 - \frac{1}{\omega(G)}.$$

The reverse inequality is proved by induction on $n$. If $n = 1$ then $f(G) = 0$ and $\omega(G) = 1$. Now suppose that $n > 1$ and the result holds for graphs with $n - 1$ vertices. If the maximum $f(G)$ is attained on a hyperplane $x_i = 0$ then, applying the induction hypothesis to $G' = G - i$, we have

$$f(G) = f(G') = 1 - \frac{1}{\omega(G')} \leq 1 - \frac{1}{\omega(G)}.$$

Otherwise, the maximum $f(G)$ is attained at a point $\mathbf{c} = (c_1, c_2, \ldots, c_n)^\top$ with all $c_i > 0$. If we apply the method of Lagrange multipliers (with multiplier $\theta$) to the function

$$F(x_1, x_2, \ldots, x_n) - \theta(x_1 + x_2 + \cdots + x_n - 1),$$

we find that $F_1(\mathbf{c}) = F_2(\mathbf{c}) = \cdots = F_n(\mathbf{c}) = \theta$, where $F_i(\mathbf{x}) = \partial F/\partial x_i$. If $G$ is not complete, say vertices 1 and 2 are non-adjacent, then for any $c \in \mathbb{R}$, we have

$$F(x_1 - c, x_2 + c, x_3, \ldots, x_n) = F(\mathbf{x}) - c(F_1(\mathbf{x}) - F_2(\mathbf{x})).$$

Taking $c = c_1$ we find that

$$F(0, c_1 + c_2, c_3, \ldots, c_n) = F(\mathbf{c}).$$

Thus the maximum $f(G)$ is attained on $x_1 = 0$ and the result follows as before. Finally, if $G$ is complete then $n = \omega(G)$ and the Cauchy–Schwarz inequality yields:

$$F(\mathbf{x}) = (x_1 + \cdots + x_n)^2 - x_1^2 - \cdots - x_n^2 = 1 - \|\mathbf{x}\|^2 \leq 1 - \tfrac{1}{n}. \qquad \square$$

**Theorem 3.10.5** [Nik1]. *If G is a graph with n vertices and m edges then*

$$\lambda_1(G) \leq \sqrt{2m \frac{\omega(G) - 1}{\omega(G)}}, \quad equivalently \quad \omega(G) \geq \frac{2m}{2m - \lambda_1^2}.$$

**Proof.** Let $\mathbf{y} = (y_1, \ldots, y_n)$ be a unit eigenvector of $G$ corresponding to $\lambda_1(G)$. By the Cauchy–Schwarz inequality, we have

$$\lambda_1(G)^2 = \left( 2 \sum_{i \sim j} y_i y_j \right)^2 \leq 4m \sum_{i \sim j} y_i^2 y_j^2.$$

Applying Lemma 3.10.4 to the vector $\mathbf{x} = (y_1^2, \ldots, y_n^2)^\top$, we have

$$2 \sum_{i \sim j} y_i^2 y_j^2 \leq \frac{\omega(G) - 1}{\omega(G)},$$

and the result follows. $\qquad \square$

We mention without proof a related result of Bollobás and Nikiforov [BolNi]: if $G$ has $n$ vertices and $k_s(G)$ denotes the number of cliques in $G$ with $s$ vertices then

$$k_{r+1}(G) \geq \left( \frac{\lambda_1(G)}{n} - 1 + \frac{1}{r} \right) \frac{r(r-1)}{r+1} \left( \frac{n}{r} \right)^{r+1}.$$

A *k-colouring* of the graph $G$ is an assignment of $k$ colours to the vertices of $G$ such that adjacent vertices have different colours. The *chromatic number* of $G$, denoted by $\chi(G)$, is the smallest $k$ for which $G$ has a $k$-colouring. The spectral upper bound for $\chi(G)$ which follows is an improvement on the well-known inequality $\chi(G) \leq 1 + \Delta(G)$.

**Theorem 3.10.6** [Wilf]. *For any graph G we have $\chi(G) \leq 1 + \lambda_1(G)$.*

**Proof.** Suppose that $k = \chi(G)$. We may delete vertices from $G$ as necessary to obtain an induced subgraph $H$ such that $\chi(H) = k$ and $\chi(H - v) = k - 1$ for any vertex $v$ of $H$. In a $(k - 1)$-colouring of $H - v$, all $k - 1$ colours are represented among the neighbours of $v$ (for otherwise the $(k - 1)$-colouring of $H - v$ may be extended to a $(k - 1)$-colouring of $H$). Thus $\delta(H) \geq k - 1$. Using Theorem 3.2.1 and interlacing, we have

$$k \leq \delta(H) + 1 \leq \lambda_1(H) + 1 \leq \lambda_1(G) + 1,$$

and the result follows. $\qquad \square$

**Theorem 3.10.7** [Hof6]. *Let $G$ be a graph with $n$ vertices and at least one edge. Then*

$$\chi(G) \geq 1 + \frac{\lambda_1(G)}{|\lambda_n(G)|}.$$

**Proof.** Let $k = \chi(G)$ and consider a partition of $V(G)$ into $k$ colour classes. Each colour class is an independent set, and so (with an appropriate labelling of vertices) the adjacency matrix $A(G)$ has a block form in which all the diagonal blocks $A_{ii}$ are zero matrices. By Corollary 1.3.17 we have

$$\lambda_1(G) + (k-1)\lambda_n(G) \leq \sum_{i=1}^{k} \lambda_{\max}(A_{ii}) = 0.$$

Since $G$ has at least one edge, we have $\lambda_n(G) < 0$ (for example by interlacing). The result follows on division by $|\lambda_n(G)|$. $\qquad\square$

The bound in Theorem 3.10.7 is attained in any non-trivial complete graph. Note that always $|V(G)| \leq \chi(G)\alpha(G)$, and so for regular graphs, Theorem 3.10.7 follows from Theorem 3.10.2.

Since $\chi(G) \geq \omega(G)$, Theorem 3.10.5 provides another lower bound for $\chi(G)$. Finally, we mention without proof a further bound from [Nik4]: for any graph $G$ with $n$ vertices,

$$\chi(G) \geq 1 + \frac{\lambda_1(G)}{\nu_1(G) - \lambda_n(G)},$$

where $\nu_1(G)$ is the largest eigenvalue of the Laplacian matrix of $G$.

## 3.11 Constraints on individual eigenvalues

We have already seen that, in general, the spectrum of a graph does not determine the graph completely. Nevertheless it can often happen that just a single eigenvalue can provide considerable structural information. In what follows we examine some such situations, with a focus on the largest and second largest eigenvalues. Graphs with least eigenvalue $\geq -2$ were investigated in Section 3.4; the general relationship between graph structure and a single eigenvalue is discussed in Chapter 5.

### 3.11.1 The largest eigenvalue

The largest eigenvalue of a graph is always non-negative. For a connected graph $G$, the largest eigenvalue is equal to 0 if and only if $G = K_1$; it is

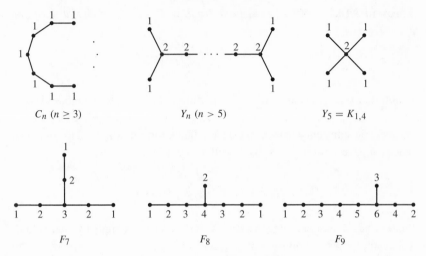

$C_n \ (n \geq 3)$          $Y_n \ (n > 5)$          $Y_5 = K_{1,4}$

$F_7$                    $F_8$                    $F_9$

Figure 3.5  The Smith graphs.

equal to 1 if and only if $G = K_2$; and it is equal to $\sqrt{2}$ if and only if $G = K_{1,2}$. All these conclusions follow from the Interlacing Theorem. In addition, there is no graph whose largest eigenvalue lies in the intervals $(0, 1)$ and $(1, \sqrt{2})$. On the other hand, there are infinitely many graphs whose largest eigenvalue lies in the interval $(\sqrt{2}, 2)$. We show that these graphs are proper subgraphs of the *Smith graphs*, i.e. the graphs whose largest eigenvalue is equal to 2.

**Theorem 3.11.1** [Smi]. *The connected graphs whose largest eigenvalue does not exceed* 2 *are precisely the induced subgraphs of the graphs shown in Fig. 3.5, where the graphs are labelled with a subscript that denotes the number of vertices.*

**Proof.** In Fig. 3.5, the vertices of each graph are labelled with the entries of an eigenvector corresponding to the eigenvalue 2. Since all these entries are positive each graph in Fig. 3.5 has 2 as the largest eigenvalue.

Any connected graph may be constructed from $K_1$ by adding vertices successively and maintaining connectedness at each stage. As we saw in Proposition 1.3.9, $\lambda_1$ increases strictly with the addition of each vertex. Hence if $G$ is a connected graph with $\lambda_1(G) \leq 2$ then $G$ is either a cycle $C_n$ or a tree; moreover $K_{1,4}$ is the only possible tree with a vertex of degree greater than 3. If the maximum degree is 3, then either $G$ is $Y_n$ or $G$ has a unique vertex of degree 3 with three paths attached. In the second case, either $G$ is $F_7$ or one of the three paths has length 1. If one path has length 1 then either $G$ is

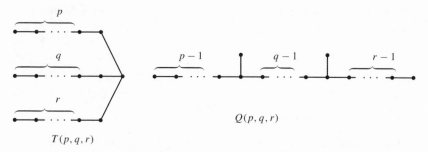

Figure 3.6 The graphs $T(p,q,r)$ and $Q(p,q,r)$.

$F_8$ or a second path has length less than 3. In the latter case, $G$ is an induced subgraph of $F_9$ or $F_8$. Finally, if the maximum degree of a vertex in $G$ is 2 then $G$ is a path and hence an induced subgraph of some $C_n$. This completes the proof. □

It is interesting to see what happens if the upper bound for $\lambda_1$ is extended a little beyond 2. The next bound considered in the literature is $\sqrt{2+\sqrt{5}} \approx 2.05817$, and then the structure of the graphs in question is still relatively simple, as we now describe.

Let $T(p,q,r)$ and $Q(p,q,r)$ be the graphs depicted in Fig. 3.6. Then we have:

**Theorem 3.11.2** [BroNe, CvDG]. *If $G$ is a connected graph whose largest eigenvalue lies in the interval $(2, \sqrt{2+\sqrt{5}}\,)$ then $G$ is one of the following graphs:*

(a) $T(p,q,r)$ *for* $p = 1, q = 2, r > 5$, *or* $p = 1, q > 2, r > 3$, *or* $p = 2$, $q = 2, r > 2$, *or* $p = 2, q = 3, r = 3$;

(b) $Q(p,q,r)$ *for* $(p,q,r) \in \{(2,1,3), (3,4,3), (3,5,4), (4,7,4), (4,8,5)\}$, *or* $p > 1, r > 1, q \geq q^*(p,r)$, *where* $(p,r) \neq (2,2)$ *and*

$$q^*(p,r) = \begin{cases} p+r & \text{if } p > 3, \\ 2+r & \text{if } p = 3, \\ -1+r & \text{if } p = 2. \end{cases}$$

It is also worth mentioning that, while $\sqrt{2+\sqrt{5}}$ cannot be an eigenvalue of any graph, any real number $\alpha$ greater than $\sqrt{2+\sqrt{5}}$ is a limit point for graph indices. In other words, there is a sequence of graphs $G_1, G_2, \ldots$ such that the sequence $\lambda_1(G_1), \lambda_1(G_2), \ldots$ converges to $\alpha$ (see [She]).

More recently, the number $\frac{3}{2}\sqrt{2} \approx 2.12312$ was considered by Woo and Neumaier as the 'next' bound for the index:

**Theorem 3.11.3** [WoNe2]. *If $G$ is a graph with index at most $\frac{3}{2}\sqrt{2}$ then $G$ is one of the following:*

(a) *a tree of maximum degree 3 such that all vertices of degree 3 lie on a path,*
(b) *a unicyclic graph of maximum degree 3 such that all vertices of degree 3 lie on a cycle,*
(c) *a tree of maximum degree 4 such that all vertices of degree 2 lie on a path.*

Note that the converse of Theorem 3.11.3 is false in general. An interesting bound beyond $\frac{3}{2}\sqrt{2}$ has not yet been identified.

## 3.11.2 The second largest eigenvalue

In this subsection we give a survey (mostly without proofs) of results that describe, for various values of $\alpha$, the graphs $G$ such that $\lambda_2(G) \leq \alpha$. Always $\lambda_2(G) \geq -1$, with equality if and only if $G$ is complete. Indeed, if $G$ is not complete then $G$ has $K_{1,2}$ as an induced subgraph, and we have $\lambda_2(G) \geq 0$ by interlacing.

**Proposition 3.11.4** [Smi]. *The non-trivial connected graphs $G$ with $\lambda_2(G) = 0$ are precisely the complete multipartite graphs other than the graphs $K_n$ ($n > 1$).*

**Proof.** Let $G$ be a connected graph which is not complete. If $G$ is not a complete multipartite graph then $G$ has $K_2 \,\dot\cup\, K_1$ as an induced subgraph $H$. Considering a shortest path in $G$ between the two components of $H$, we see that $G$ has $K_1 \triangledown (K_2 \,\dot\cup\, K_1)$ or $P_4$ as an induced subgraph. Since both of these graphs have second largest eigenvalue greater than 0 (see Table A1), we have $\lambda_2(G) > 0$ by interlacing.

If $G$ is a complete multipartite graph then we can use the Courant–Weyl inequalities to show that $\lambda_2(G) = 0$. By Theorem 1.3.15 we have $\lambda_2(G) + \lambda_n(\overline{G}) \leq -1$; the claim follows since the components of $\overline{G}$ are complete graphs, and one of them is non-trivial. □

We state the following result without proof:

**Theorem 3.11.5** [CaoHo]. *The connected graphs $G$ with $0 < \lambda_2(G) < \frac{1}{3}$ are the graphs $H_n = (n-3)K_1 \triangledown (K_2 \,\dot\cup\, K_1)$ ($n \geq 4$).*

This remarkable result shows that the graphs $H_n$ are determined by the second largest eigenvalue. Note also that $\lim_{n\to\infty} \lambda_2(H_n) = 1/3$, while $1/3$ itself is not a graph eigenvalue.

The question arises as to whether there are any wider classes of graphs whose structure is, to some extent, determined by larger upper bounds on the second largest eigenvalue. It turns out that $\sqrt{2} - 1$ is a good choice of upper bound in this respect, because the graphs which arise can be described explicitly. They were found independently by Li [Li] and Petrović [Pet2]; details appear in [PetRa, Chapter 3]. The next bound, a more natural one, is the golden section $\sigma = \frac{\sqrt{5}-1}{2}$. Since $\lambda_2(P_4) = \sigma$ and $\lambda_2(2K_2) = 1$, neither $P_4$ nor $2K_2$ is an induced subgraph of a graph $G$ for which $\lambda_2(G) < \sigma$. We denote by $\mathcal{C}$ the class of graphs without $P_4$ or $2K_2$ as an induced subgraph. If $G \in \mathcal{C}$ then either $G$ has an isolated vertex or $\overline{G}$ is not connected (Exercise 3.25). It follows that $\mathcal{C}$ can be defined recursively as follows:

(i) $K_1 \in \mathcal{C}$;
(ii) if $G \in \mathcal{C}$ then $G \dot{\cup} K_1 \in \mathcal{C}$;
(iii) if $G_1, G_2 \in \mathcal{H}$ then $G_1 \triangledown G_2 \in \mathcal{C}$.

Now we introduce some more terminology. The graphs $G$ with $\lambda_2(G) \leq \sigma$ (the $\sigma$-*property*) will be called $\sigma$-*graphs*. The graphs $G$ for which $\lambda_2(G) < \sigma$, $\lambda_2(G) = \sigma$ and $\lambda_2(G) > \sigma$ will be called $\sigma^-$-*graphs*, $\sigma^0$-*graphs* and $\sigma^+$-*graphs*, respectively. Note that any $\sigma^-$-graph belongs to $\mathcal{C}$, but not vice versa. The class $\mathcal{C}$ was introduced in [Sim7], where each graph $G$ from $\mathcal{C}$ is represented by a weighted rooted tree $T_G$ (called an *expression tree* for $G$), defined recursively as follows:

any subgraph $H = (((H_1 \triangledown H_2) \triangledown \cdots) \triangledown H_m) \dot{\cup} nK_1$ ($m \geq 0$, $n > 0$) of $G$ is represented by a subtree $T_H$ with a root $v$ of weight $n$ whose neighbours in $T_H$ are the roots $v_1, v_2, \ldots, v_m$ of the subtrees representing $H_1, H_2, \ldots, H_m$ respectively.

**Example 3.11.6.** If $G = (((((K_1 \triangledown K_1) \dot{\cup} K_1) \triangledown K_1) \triangledown K_1) \triangledown K_1) \dot{\cup} 3K_1$, then the corresponding expression tree is depicted in Fig. 3.7(a). In Fig. 3.7(b) we represent the same graph by a diagram in which a line between two circled sets of vertices denotes that each vertex inside one set is adjacent to every vertex inside the other set. □

It turns out that these weighted trees can be used to categorize $\sigma^-$-graphs: the weighted tree of any such graph is of one of the nine types illustrated in Fig. 3.8. This result was used in [Sim6] to prove the existence of a finite family of minimal forbidden (induced) subgraphs for the $\sigma^-$-property; except for $2K_2$ and $P_4$, they belong to the class $\mathcal{C}$. Details of such a family may be found in [Sim6]. Some of these minimal forbidden subgraphs have a huge number of vertices (see [Sim6, CvSi4]), and to date the divisor technique and a computer search have proved insufficient to identify all those of the last type illustrated

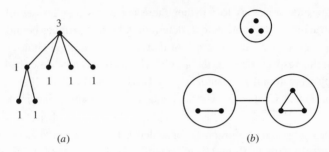

Figure 3.7 An expression tree and associated diagram.

in Fig. 3.8. On the other hand, some arbitrarily large families of $\sigma^-$-graphs are described in [Sim8].

Now we turn our attention to $\sigma$-graphs. It was first observed in [CvSi2] that, apart from complete multipartite graphs and subgraphs of $C_5$, the structure of such a graph $G$ can be specified in relation to a triangle of $G$ (see [CvRS2, Chapter 9] or [PetRa, Chapter 3]). There exists a finite set of minimal forbidden subgraphs for the $\sigma$-property, but such a set remains to be constructed. The following result provides some additional information:

**Theorem 3.11.7.** *If $H$ is a minimal forbidden subgraph for the $\sigma$-property, then either*

(a)  *$H$ is one of the graphs $2K_2$, $J_1$, $J_2$, $J_3$, $J_4$ (see Fig. 3.9), or*
(b)  *$H$ belongs to the class $\mathcal{C}$.*

The problem of finding the graphs $G$ with $\lambda_2(G) \leq 1$ is attributed to Hoffman. Cvetković [Cve8] showed that if $G$ is such a graph then either $G$ has girth at most 6, or $G$ is a tree of diameter at most 4. Petrović [Pet1] showed that the bipartite graphs which arise fall into seven classes, three of them infinite:

**Theorem 3.11.8.** *Let $G$ be a connected bipartite graph. Then $\lambda_2(G) \leq 1$ if and only if $G$ is an induced subgraph of a graph illustrated in Fig. 3.10.*

Fig. 3.10 depicts three infinite families of graphs and four individual graphs. In all cases, encircled vertices form a co-clique, and a full line between co-cliques indicates a complete bipartite subgraph. Parallel broken lines between the vertices of two co-cliques indicate a graph obtained from some $K_{n,n}$ by deleting $n$ independent edges. Parallel full lines between the vertices of two co-cliques indicate a graph of the form $nK_2$.

As a consequence of Theorem 3.11.8, the bipartite graphs with $\lambda_2 \leq 1$ can be characterized by a family of 12 forbidden subgraphs (see [Pet1] or [PetRa, Chapter 3]).

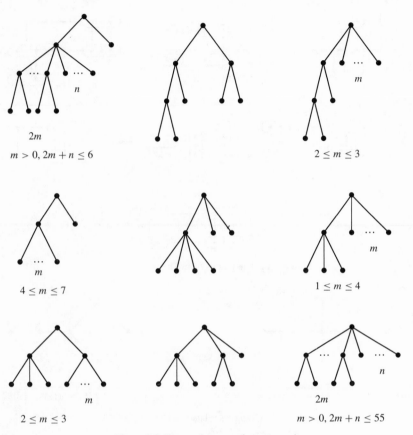

$2m$

$m > 0, 2m + n \leq 6$

$2 \leq m \leq 3$

$4 \leq m \leq 7$

$1 \leq m \leq 4$

$2 \leq m \leq 3$

$m > 0, 2m + n \leq 55$

$2m$

Figure 3.8  Expression trees for $\sigma^-$-graphs.

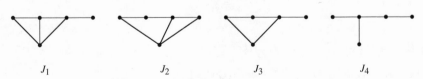

$J_1$            $J_2$            $J_3$            $J_4$

Figure 3.9  Some graphs from Theorem 3.11.7.

By Theorem 1.3.13, we have $\lambda_2(G) + \lambda_{n-1}(\overline{G}) \geq -1$ and $\lambda_2(G) + \lambda_n(\overline{G}) \leq -1$. These inequalities provide a natural link between the graphs with $\lambda_2 \leq 1$ and the graphs with least eigenvalue $\geq -2$:

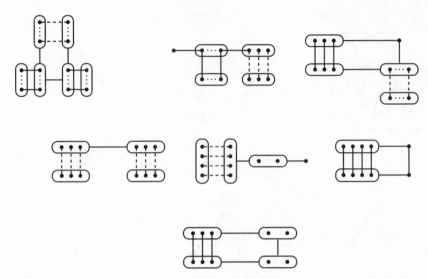

Figure 3.10 The graphs from Theorem 3.11.8.

(a)  $\lambda_n(\overline{G}) = -2.074,\ \lambda_2(G) = 0.753$         (b)  $\lambda_n(\overline{G}) = -2.136,\ \lambda_2(G) = 1.082$

Figure 3.11 Examples related to Theorem 3.11.9.

**Theorem 3.11.9** [Cve8].   *Let $G$ be a graph on $n$ vertices with $\lambda_2(G) \leq 1$. Then either*

(a) $\lambda_n(\overline{G}) \geq -2$, *or*
(b) $\lambda_n(\overline{G}) < -2$   *and*   $\lambda_{n-1}(\overline{G}) \geq -2$.

*Conversely, if $\lambda_n(G) \geq -2$ then $\lambda_2(G) \leq 1$.*

A graph which satisfies condition (b) may or may not have $\lambda_2(G) \leq 1$: see the graphs in Fig. 3.11, where in each case $\lambda_n(\overline{G})$ and $\lambda_2(G)$ are as shown. Note that if a graph has $\lambda_n(G) = -2$ with multiplicity at least 2, then $\lambda_n(G) = \lambda_{n-1}(G)$, and so necessarily $\lambda_2(\overline{G}) = 1$.

Since $\mathcal{E}_G(\lambda) \cap \mathbf{j}^\perp = \mathcal{E}_{\overline{G}}(-1 - \lambda) \cap \mathbf{j}^\perp$ for any eigenvalue $\lambda$ of $G$, we can say a little more by way of a converse in Theorem 3.11.9:

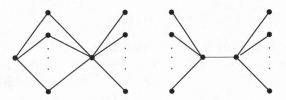

Figure 3.12 The graphs from Theorem 3.11.12.

**Theorem 3.11.10** [Cve8]. *Let G be a graph on n vertices. Then:*

(i) *if G is a graph with $\lambda_n(G) > -2$ then $\lambda_2(\overline{G}) < 1$;*
(ii) *if G is a graph with $\lambda_n(G) = -2$ then $\lambda_2(\overline{G}) \leq 1$, with strict inequality if and only if $-2$ is a simple main eigenvalue of G.*

We conclude this subsection with two general results. The first applies to trees, and is due to Neumaier:

**Theorem 3.11.11** [Neu]. *If T is a tree with $\lambda_2(T) \leq \lambda$ then either*

(a) *there exists a vertex v of T such that $\lambda_1(T - v) \leq \lambda$, or*
(b) *there exists an edge uv of T such that $T - uv = T_1 \mathbin{\dot{\cup}} T_2$ where $\lambda_1(T_1 - u) < \lambda < \lambda_1(T)$ and $\lambda_1(T_2 - v) < \lambda < \lambda_1(T)$.*

Taking $\lambda = 2$ in Theorem 3.11.11, we see that if $T$ is a tree with $\lambda_2(T) \leq 2$ then either an edge or a vertex may be deleted to obtain subtrees that have index at most 2 and are therefore of the very restricted type described in Theorem 3.11.1. A similar approach is used by Neumaier and Seidel [NeuSe] to investigate arbitrary graphs $G$ with $\lambda_2(G) \leq 2$: such graphs are called *reflexive* graphs because of their relation to automorphism groups of certain lattices generated by reflections. More on reflexive graphs can be found in [PetRa, Chapter 3].

The second general result is due to Howes:

**Theorem 3.11.12** [How]. *For an infinite set of graphs $\mathcal{G}$, the following statements are equivalent.*

(i) *There exists a real number $\alpha$ such that $\lambda_2(G) \leq \alpha$ for every $G \in \mathcal{G}$.*
(ii) *There exists a positive integer s such that for each $G \in \mathcal{G}$, the graphs $(K_s \mathbin{\dot{\cup}} K_1) \triangledown K_s$, $(sK_1 \mathbin{\dot{\cup}} K_{1,s}) \triangledown K_1$, $(K_{s-1} \mathbin{\dot{\cup}} sK_1) \triangledown K_1$, $K_s \mathbin{\dot{\cup}} K_{1,s}$, $2K_{1,s}$, $2K_s$ and the graphs in Fig. 3.12 (each obtained from two copies of $K_{1,s}$ by adding extra edges) are not induced subgraphs of G.*

## Exercises

**3.1** Derive the formula given in Theorem 3.1.5.

**3.2** Provide the details required to complete the proof of Theorem 3.1.6.

**3.3** Let $G$ be an $r$-regular graph, and let $c(i)$ be the number of cycles in $G$ of length $i$ ($i = 3, 4, 5$). Show that, in the notation of Corollary 2.3.3,

$$c(3) = -\frac{1}{2}c_3, \quad c(4) = \frac{1}{4}(c_2^2 + 2rc_2 - c_2 - 2c_4),$$

$$c(5) = \frac{1}{2}(c_2c_3 + 3rc_3 - 3c_3 - c_5).$$

**3.4** Show that the eigenvalues and angles of a graph $G$ determine whether or not $G$ is a tree.

**3.5** Let $G$ be a graph with index $\lambda_1$ and adjacency matrix $A$. Show that $G$ is connected if and only if for any $\lambda > \lambda_1$, each entry of $(I - \frac{1}{\lambda}A)^{-1}$ is positive.

**3.6** Prove Theorem 3.4.4.

**3.7** Prove Theorem 3.4.5.

**3.8** Show that the Petersen graph is not a generalized line graph. Find a representation of this graph in $E_6$.

**3.9** Prove Lemma 3.4.7.

**3.10** Let $T = Y_6$ (the corona $K_2 \circ 2K_1$). Show that if the cubic graph $G$ is the edge-disjoint union of subgraphs isomorphic to $T$, then 0 is an eigenvalue of $G$.

**3.11** Show that if $G$ is a strongly regular graph with parameters $(n, r, e, f)$ then its complement $\overline{G}$ is strongly regular with parameters $(\bar{n}, \bar{r}, \bar{e}, \bar{f})$, where

$$\bar{n} = n, \quad \bar{r} = n - r - 1, \quad \bar{e} = n - 2 - 2r + f, \quad \bar{f} = n - 2r + e.$$

**3.12** Show that a strongly regular graph $G$ is imprimitive if and only if $G$ or $\overline{G}$ is a complete multipartite graph of the form $K_{m,m,\ldots,m}$.

**3.13** Show that if $n \geq 2m$ then the distinct eigenvalues of the Johnson graph $J(n, m)$ are $(m - i)(n - m - i) - i$, with multiplicity $\binom{n}{i} - \binom{n}{i-1}$ ($i = 0, 1, \ldots, m$).

**3.14** Verify Equations (3.21).

**3.15** Verify the properties claimed for the graph constructed in Example 3.7.1(i).

**3.16** Show that a connected graph is distance-regular if and only if for each positive integer $k$, the number of $i$-$j$ walks of length $k$ depends only on $d(i, j)$.

**3.17** By finding an appropriate eigenvector, show that there is no distance-regular graph with intersection array $\{3, 2, 1; \ 1, 1, 3\}$.

**3.18** Show that the orbits of any group of automorphisms of a graph $G$ form an equitable partition of $G$.

**3.19** Prove that if the vertices $u$, $v$ are similar then the angles $\alpha_{iu}$, $\alpha_{iv}$ coincide for each $i \in \{1, \ldots, m\}$.

**3.20** Let $\pi$ be an automorphism of a graph $G$. Show that if $\pi$ has $s$ cycles of odd length and $t$ cycles of even length (when written as a product of disjoint cycles) then the number of simple eigenvalues of $G$ is at most $s + 2t$.

**3.21** Let $G$ be a graph whose characteristic polynomial is irreducible over the rationals. Show that $G$ has no non-trivial automorphisms.

**3.22** Prove that $K_2$ is the only non-trivial vertex-transitive graph without multiple eigenvalues.

**3.23** Show that if a graph $G$ has mean degree $\overline{d}$ and just two main eigenvalues, $\mu_1$ and $\mu_j$ $(j > 1)$, then [Row16]

$$\frac{1}{n} \sum_{i=1}^{n} (d_i - \overline{d})^2 = (\mu_1 - \overline{d})(\overline{d} - \mu_j).$$

**3.24** Let $G$ be a non-trivial connected graph with index $\mu$. Show that $G$ is a semi-regular bipartite graph if and only if the main eigenvalues of $G$ are $\mu$ and $-\mu$ [Plo, Row16].

**3.25** Let $G$ be a graph with neither $P_4$ nor $2K_2$ as an induced subgraph. Show that either $G$ has an isolated vertex or $\overline{G}$ is not connected.

**3.26** Let $G$ be a graph with spectrum $\lambda_1 \geq \lambda_2 \geq \cdots \geq \lambda_n$. Show that if $G$ is not complete then [Hof6]:

$$\chi(\overline{G}) \geq \frac{n + \lambda_2 - \lambda_1}{1 + \lambda_2}.$$

**3.27** (a) Let $\lambda_1^* \leq \lambda_2^* \leq \cdots \leq \lambda_n^*$ be the eigenvalues of the graph $G$, and let $k = \chi(G)$. Show that $\lambda_1^* + \lambda_2^* + \cdots + \lambda_{k-1}^* + \lambda_n^* \leq 0$.
(b) Show that if further $G$ is $k$-colourable in such a way that two vertices are adjacent if and only if they have different colours, then equality holds in (a).

**3.28** Let $G$ be a graph with eigenvalues $\lambda_1^* \leq \lambda_2^* \leq \cdots \leq \lambda_n^*$. Show that if $G$ has independence number $\alpha(G)$ and clique number $\omega(G)$, then

$$\lambda_{\alpha(G)+1}^* \geq 0, \ \ \lambda_{n-\alpha(G)}^* \leq 0, \ \ \lambda_{n-\omega(G)-1}^* \leq -1, \ \ \lambda_{\omega(G)+1}^* \geq -1.$$

# Notes

Analogues of Theorem 3.5.3, concerning similar measures of expansion for arbitrary graphs, appear in Chapter 7 in the context of the Laplacian spectrum. A proof of the relation (3.17) is given in the monograph [DavSV], along with a self-contained treatment of the Ramanujan graphs of Lubotzky, Phillips and Sarnak [LuPS]. Cubic Ramanujan graphs are discussed in [Chi], and the girth of Ramanujan graphs is investigated in [BigBo].

All rank 3 permutation groups (and by implication all rank 3 graphs) are known as a consequence of the classification of finite simple groups (see [Cam2]). Among strongly regular graphs, the rank 3 graphs are relatively rare; a compilation of strongly regular graphs may be found in [Hub]. Those with strongly regular subconstituents are investigated in [CamGS]. Some further feasibility conditions for the existence of strongly regular graphs with pre-scribed parameters can be found in [BroLi]. The absolute bound and the Krein inequalities for strongly regular graphs are special cases of general inequalities for association schemes (see [BroCN, Chapter 2]).

Distance-regular graphs arose in a paper of Biggs [Big1] and a good intro-duction to the topic may be found in his monograph [Big2], a secondary source for Example 3.7.1(iii). The monograph [BroCN] is the standard reference for a comprehensive treatment, with 800 references. For graphs in the follow-ing categories, it lists all the arrays which pass all known feasibility tests for distance-regularity : (i) graphs with diameter $\geq 5$ and at most 4096 vertices, (ii) non-bipartite graphs with diameter 4 and at most 4096 vertices, (iii) primi-tive graphs with diameter 3 and at most 1024 vertices. (In this context, a graph with diameter $d$ is *primitive* if each of $A_1, A_2, \ldots, A_d$ is the adjacency matrix of a connected graph.) For a survey of distance-transitive graphs, see [Coh].

For a survey of graph automorphisms, see [Cam3]. The proofs of The-orem 3.9.9 and its corollaries are taken from [Row16], a survey of main eigenvalues which includes a discussion of the cases $M_G(x) = x^2 - \mu^2$, $M_G(x) = x(x^2 - \mu^2)$. Theorem 3.9.9 was first established in [Cve6] by means of walk-generating functions. The concept of a divisor has been exploited in coding theory; see [CvLi].

Graphs whose spectra conform to prescribed conditions (such as those inves-tigated in Section 3.11) are said to be *spectrally constrained*; such graphs are the subject of the monograph [PetRa]. The second largest eigenvalue of line graphs and generalized line graphs is discussed in [PetMi1] and [PetMi2] respectively. A survey of results on $\lambda_2$ may be found in [CvSi3]. In view of the relation $\lambda_2(G) + \lambda_n(\overline{G}) \leq -1$, there is a natural link between lower bounds for $\lambda_n$ and upper bounds for $\lambda_2$. An analogue of Theorem 3.11.12 for graphs whose least eigenvalue is bounded below may be found in [Hof8]. Graphs with least eigenvalue $\geq -\sqrt{3}$ are determined in [CvSt]; those with least eigenvalue $\geq -1 - \sqrt{2}$ are discussed in [Hof7] and [WoNe1]. The graphs with maximal least eigenvalue, among the connected non-complete graphs with a prescribed number of vertices, are determined in [Hon3]. The graphs with minimal least eigenvalue, among the connected graphs with prescribed numbers of vertices and edges, are discussed in Section 8.2. The corresponding problem concerning the maximal index is investigated in Section 8.1. The graphs with maximal index, among the graphs with a prescribed number of edges, are determined in [Row4]; see also [CvRS2, Chapter 3], where graphs with maximal index in various classes of graphs are described. A survey of results on the index of a graph may be found in [CvRo3].

Non-regular graphs with just three eigenvalues are discussed in [BriMe], [Dam2], [MuKl], and regular graphs with just four eigenvalues are investigated in [Dam1], [DamSp].

For a discussion of NP-completeness in a combinatorial context, see [BruRy, pp. 245–8].

# 4

# Characterizations by spectra

In this chapter we discuss several instances of the following problem:

*Given the spectrum, or some spectral characteristics of a graph, determine all graphs from a given class of graphs having the given spectrum, or the given spectral characteristics.*

In some cases, the solution of such a problem can provide a characterization of a graph up to isomorphism (see Section 4.1). In other cases we can deduce structural details (see also Chapter 3). Non-isomorphic graphs with the same spectrum can arise as sporadic exceptions to characterization theorems or from general constructions. Accordingly, Section 4.2 is devoted to cospectral graphs; we include comments on their relation to the graph isomorphism problem, together with various examples and statistics. We also discuss the use of other graph invariants to strengthen distinguishing properties. In particular, in Section 4.3, we consider characterizations of graphs by eigenvalues and angles.

## 4.1 Spectral characterizations of certain classes of graphs

In this section we investigate graphs that are determined by their spectra. The three subsections are devoted to (i) elementary characterizations, (ii) characterizations of graphs with least eigenvalue $-2$, and (iii) characterizations of special types. In the case of (i), a graph is uniquely reconstructed from its spectrum, while in cases (ii) and (iii) various exceptions occur due to the existence of cospectral graphs.

### 4.1.1 Elementary spectral characterizations

We say that a graph $G$ is characterized by its spectrum if the only graphs cospectral with $G$ are those isomorphic to $G$.

Note first that this condition is satisfied by graphs which are characterized by invariants (such as the number of vertices and edges) which can be determined from the spectrum. Examples include the complete graphs and graphs with one edge, together with their complements. Given the spectrum spec($G$) of a graph $G$ we can always establish whether or not $G$ is regular (see Corollary 3.2.2); moreover, if $G$ is regular, the largest eigenvalue is the degree of regularity (see Proposition 1.1.2). It follows that if $G$ or $\overline{G}$ is regular of degree 1 then $G$ is characterized by its spectrum.

Regular graphs of degree 2 are unions of cycles. As we saw in Example 1.1.4, the eigenvalues of the cycle $C_n$ are the real parts of the $n$-th roots of $2^n$, i.e. the numbers

$$2\cos\frac{2\pi}{n}j \quad (j = 0, 1, \ldots, n - 1).$$

The largest eigenvalue is $\lambda_1 = 2$ (which arises when $j = 0$) and the second largest is two-fold: $\lambda_2 = \lambda_3 = 2\cos\frac{2\pi}{n}$ (which arises when $j = 1$ and $j = n - 1$). Suppose now that $G = C_{n_1} \dot\cup \cdots \dot\cup C_{n_k}$. Then the eigenvalues of $G$ are the numbers

$$2\cos\frac{2\pi}{n_i}j \quad (j = 0, 1, \ldots, n_i - 1; \ i = 1, \ldots, k).$$

Given spec($G$), we can first establish (as above) that $G$ is regular of degree 2. From the second largest eigenvalue in spec($G$), we can determine the length $m$ of the largest cycle in $G$. Now we eliminate the eigenvalues of $C_m$ and (if eigenvalues remain) repeat the process. Proceeding in this way, we can identify the lengths of all cycles of $G$, and thereby determine $G$ up to isomorphism. Accordingly, we have the following theorem:

**Theorem 4.1.1** [Cve1]. *Any regular graph of degree 2 is characterized by its spectrum.*

**Remark 4.1.2.** From the spectrum of a regular graph $G$ we can find the spectrum of $\overline{G}$ (see Theorem 2.1.2), and so it follows from Theorem 4.1.1 that any $n$-vertex graph which is regular of degree $n - 3$ is characterized by its spectrum. This result was proved for connected multigraphs by Finck [Fin]. □

It is also straightforward to show that a graph of the form $mK_n$ is characterized by its spectrum, a fact established in complementary form by Finck:

**Theorem 4.1.3** [Fin]. *For each positive integer $n$, the complete multipartite graph $K_{n,n,\ldots,n}$ is characterized by its spectrum.*

The next result, however, does not admit a transition to the complement.

**Theorem 4.1.4** [Cve1]. *The spectrum of the graph $G$ is $n_1 - 1, \ldots, n_k - 1, 0^s, (-1)^{n_1 + \cdots + n_k - k}$ if and only if $G = K_{n_1} \dot{\cup} \cdots \dot{\cup} K_{n_k} \dot{\cup} s K_1$.*

It follows from Theorem 3.1.11 that the path $P_n$ is characterized by its spectrum (Exercise 4.6); the same is true of $\overline{P_n}$, but the proof cannot be described as elementary (see [DooHa]). A *lollipop graph* is obtained from $P_n$ ($n \geq 4$) by adding an edge joining non-adjacent vertices of degrees 1 and 2. It has recently been shown, in a long proof, that every lollipop graph is characterized by its spectrum [HaeLZ, BouJo].

We continue with two examples where the Interlacing Theorem (Corollary 1.3.12) is sufficient to obtain a spectral characterization. In each case the pre-scribed spectrum lies in the interval in $[-2, 4]$, and so any graph with an eigenvalue less than $-2$ (or greater than 4) is forbidden as an induced sub-graph. In the next subsection we discuss more general results obtained using our knowledge of the regular graphs with least eigenvalue $\geq -2$.

**Proposition 4.1.5.** *$L(\overline{C_6})$ is characterized by its spectrum.*

**Proof.** Suppose that $G$ is a graph with the spectrum of $L(\overline{C_6})$, namely $4, 2, 1^2, (-1)^2, (-2)^3$. By Corollary 3.2.2, $G$ is 4-regular. For $u \in V(G)$, let $G(u)$ denote the subgraph of $G$ induced by the neighbours of $u$. By Theorem 3.1.1, the average number of edges in the subgraphs $G(u)$ is less than three. Thus there exists a vertex $v$ of $G$ with $|E(G(v))| \leq 2$. Consider the vertex $v$ along with its four neighbours. In order to avoid an induced subgraph on five or six vertices with least eigenvalue less than $-2$, it must be the case that $G(v)$ consists of two independent edges.

Let $G(v)^*$ denote the subgraph induced by $v$ and its neighbours, and let $H$ be the subgraph induced by the remaining four vertices of $G$. Note that (i) there are eight edges between $G(v)^*$ and $H$, (ii) no vertex of $H$ is adjacent to three vertices of $G(v)^*$, again because of forbidden subgraphs. It follows that $H$ is a 4-cycle. The remaining edges can be added in only two ways to avoid forbidden subgraphs: one yields $L(\overline{C_6})$ and the other yields $L(K_{3,3})$ (whose spectrum is $4, 1^4, (-2)^4$).

This completes the proof.                                                    □

**Proposition 4.1.6.** *Let $H_8$ denote the cubic graph on eight vertices formed by taking two copies of the graph on four vertices with five edges and adding two appropriate edges. Then $L(H_8)$ is characterized by its spectrum.*

**Proof.** Suppose that $G$ is a graph with the spectrum of $L(H_8)$, namely $4, 1 + \sqrt{5}, 2, 0^4, 1 - \sqrt{5}, (-2)^4$. As before, let $G(u)$ denote the subgraph of $G$ induced by the neighbours of $u$ ($u \in V(G)$). By Theorem 3.1.1 the average number of edges in the subgraphs $G(u)$ is three. Let us suppose first that every subgraph $G(u)$ has three edges. Then each $G(u)$ is $K_{1,3}$, $K_3 \mathbin{\dot\cup} K_1$ or $P_4$. Suppose that $v$ is a vertex for which $G(v)$ is $K_{1,3}$, and let $w$ be one of the vertices of degree 1 in $G(v)$. Then $G(w)$ has fewer than three edges, contrary to assumption. If $G(v)$ is $K_3 \mathbin{\dot\cup} K_1$, then $G(u)$ is $K_3 \mathbin{\dot\cup} K_1$ for every vertex $u$. In this case, each vertex lies in exactly one complete graph with four vertices, and $G$ consists of three disjoint copies of $K_4$ together with six edges. There is only one regular graph with this property, and it is the line graph of a semi-regular bipartite graph, with spectrum 4, $(1 + \sqrt{2})^2$, $0^3$, $(1 - \sqrt{2})^2$, $(-2)^4$. If $G(u)$ is $P_3$ for every $u$, then there is only one way to complete the graph avoiding forbidden subgraphs, and the spectrum is 4, $(1+\sqrt{3})^2$, $0^3$, $(1-\sqrt{3})^2$, $(-2)^4$.

These contradictions show that there is a vertex $v$ such that $G(v)$ has fewer than three edges, and as in Proposition 4.1.5, $G(v)$ has two independent edges.

Now Fig. 4.1 illustrates all the possible ways of adding further vertices adjacent to neighbours of $v$. In each case, it is straightforward to complete the graph. Among the graphs obtained in this way, only $L(H_8)$ has the given spectrum. $\square$

In the last proof, details of the completions are left to the reader. To prove that a completed graph does not have the given spectrum, it suffices here to count the numbers of triangles, quadrilaterals and pentagons (see Chapter 3).[1]

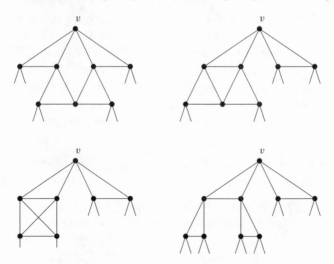

Figure 4.1 The graphs from Proposition 4.1.6.

[1] The graphs can be completed by hand or by the use of graph editing computer packages such as 'newGRAPH' (www.mi.sanu.ac.yu/newgraph/), where interactive facilities enable the spectrum of each extension to be calculated.

### 4.1.2  Graphs with least eigenvalue –2

Most spectral characterization theorems are related to graphs with least eigenvalue $-2$, reflecting the fact that such graphs are well understood: we saw in Section 3.4 that there are only finitely many such graphs which are connected but not generalized line graphs, and these are called exceptional graphs (see also Chapter 5). We have also seen that the properties of regularity, and then connectedness, can be established from the spectrum. In this section we show how knowledge of the regular exceptional graphs leads to spectral characterizations of connected regular graphs with least eigenvalue $-2$. The regular exceptional graphs were determined in 1976 with the aid of a computer; they are not listed here but can be found in [CvRS7, Table A3]. However, the first general results, which we state without proof, were obtained by Hoffman [Hof1, Hof2] in the early 1960s.

**Theorem 4.1.7.** *If $n \neq 8$ then $L(K_n)$ is characterized by its spectrum.*

Note that $L(K_n)$ is a regular graph of degree $2n - 4$ on $n(n - 1)/2$ vertices, with spectrum $2n-4, (n-4)^{n-1}, (-2)^{n(n-3)/2}$.

**Theorem 4.1.8.** *If $n \neq 4$ then $L(K_{n,n})$ is characterized by its spectrum.*

See Example 2.4.5 for the spectrum of $L(K_{n,n})$.

In Theorem 4.1.7 the exceptions which arise when $n = 8$ are the three Chang graphs described in Example 1.2.6. Similarly, the only exception in Theorem 4.1.8 when $n = 4$ is the Shrikhande graph, introduced in Example 1.2.4. The Shrikhande graph and the Chang graphs are exceptional graphs. Since they are obtained by Seidel switching (as noted in Chapter 1), one might think that many other exceptional graphs cospectral with regular line graphs can be constructed in the same way. In fact, the possibilities are severely restricted by the following theorem, which relates the divisor concept to switching in graphs.

**Theorem 4.1.9** [Cve6]. *If a regular graph $G$ of degree $r$ with $n$ vertices can be switched into a regular graph of degree $r^*$, then $r^* - n/2$ is an eigenvalue of $G$.*

**Proof.** If $G$ has the stated property in respect of a switching set $S$ of size $t$ ($0 < t < n$) then $S$ and its complement determine a divisor with adjacency matrix

$$\begin{pmatrix} r - \frac{1}{2}(n - t - r^* + r) & \frac{1}{2}(n - t - r^* + r) \\ \frac{1}{2}(t - r^* + r) & r - \frac{1}{2}(t - r^* + r) \end{pmatrix}.$$

The eigenvalues of this matrix are $r$ and $r^* - n/2$, and by Theorem 3.9.5, they are eigenvalues of $G$. The result follows. $\qquad\square$

Since a rational eigenvalue of a graph is an integer, we have:

**Corollary 4.1.10.** *If $n$ is odd then $G$ cannot be switched into another regular graph.*

**Corollary 4.1.11.** *If the $r$-regular graph $G$ can be switched into a regular graph of the same degree and if $q$ is the least eigenvalue of $G$, then $r - n/2 \geq q$, i.e. $n \leq 2r - 2q$. Since $q \geq -r$, we have $r - n/2 \geq -r$, i.e. $r \geq n/4$.*

**Example 4.1.12.** There is no cospectral pair of non-isomorphic cubic graphs with fewer than 14 vertices. Accordingly it follows from Corollary 4.1.11 that the existence of (non-isomorphic) cospectral cubic graphs cannot be explained by switching. $\qquad\square$

**Example 4.1.13.** If $L(K_s)$ ($s > 1$) can be switched to another regular graph of the same degree then by Corollary 4.1.11, $2s - 4 - s(s-1)/4 \geq -2$, whence $s \leq 8$. (The three Chang graphs arise when $s = 8$.) $\qquad\square$

We extend the argument of Example 4.1.13 to any regular line graph $L(G)$ where $G$ is connected and non-trivial; there are two cases.

(1) If $G$ is regular of degree $r$ with $n$ vertices then $L(G)$ is of degree $2r - 2$ and has $nr/2$ vertices. If $L(G)$ can be switched into another regular graph of the same degree then $2r - 2 - nr/4$ is an eigenvalue of $L(G)$. Clearly, $2r - 2 - nr/4 \geq -2$, which implies $n \leq 8$.

(2) Let $G$ be semi-regular bipartite with parameters $(n_1, n_2, d_1, d_2)$. Then $L(G)$ has $n_1 d_1$ ($= n_2 d_2$) vertices and degree $d_1 + d_2 - 2$. Therefore, we have

$$d_1 + d_2 - 2 - n_1 d_1/2 \geq -2, \qquad n_1 d_1 \leq 2(d_1 + d_2),$$

$$n_1 \leq 2(1 + d_2/d_1) = 2(1 + n_1/n_2), \qquad 1/n_1 + 1/n_2 \geq 1/2.$$

Without loss of generality, $n_1 \leq n_2$. If $n_1 = 1$, then $L(G) = K_{n_2}$, a graph characterized by its spectrum. Doob [Doo3] proved that also $L(K_{2,n_2})$ is characterized by its spectrum; for $n_2 > 16$, this follows from Theorem 4.1.18 below. Accordingly, we suppose that $n_1 \geq 3$. The possibilities for $(n_1, n_2)$ are then $(3, 3)$ (ruled out by Theorem 4.1.2) and $(3, 4)$, $(3, 5)$, $(3, 6)$, $(4, 4)$. In particular, $n_1 + n_2 \leq 9$ and $G$ has at most 18 edges.

In view of the bounds on the number of vertices established in cases (1) and (2), it is straightforward to identify the graphs which arise. We shall see that all graphs cospectral with a connected regular line graph can be constructed from line graphs by switching. For future reference we illustrate four examples in

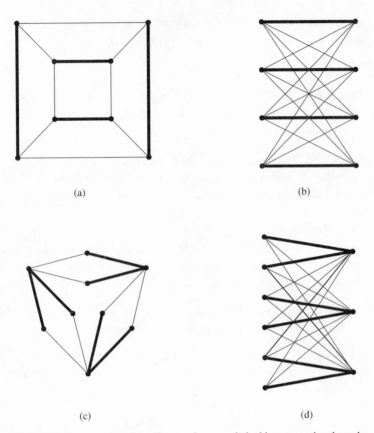

(a)                                    (b)

(c)                                    (d)

Figure 4.2  Some graphs whose line graphs are switched into exceptional graphs.

Fig, 4.2; here, each of (a), (b), (c), (d) is a graph $H$ in which the bold edges denote vertices in the switching set for $L(H)$. The graphs obtained by switching are denoted by $G6$, $G9$, $G69$, $G70$ respectively (the numbers chosen for consistency with [CvRS7, Table A4].)

It can also happen that two non-isomorphic regular line graphs have the same spectrum, and the following theorem specifies the possibilities.

**Theorem 4.1.14** [BuCS1, BuCS2]. *Let $L(G_1)$, $L(G_2)$ denote cospectral, connected, regular line graphs of the connected graphs $G_1$, $G_2$. Then one of the following holds:*

(a) $G_1$ *and* $G_2$ *are cospectral regular graphs with the same degree,*

(b) $G_1$ *and* $G_2$ *are cospectral semi-regular bipartite graphs with the same parameters,*

(c) $\{G_1, G_2\} = \{H_1, H_2\}$, *where $H_1$ is regular and $H_2$ is semi-regular bipartite; in addition there exist positive integers $s > 1$ and $t < \frac{1}{2}s$, and real numbers $\lambda_i$, $0 \le \lambda_i < t\sqrt{s^2 - 1}$, $i = 2, 3, \ldots, \frac{1}{2}s(s - 1)$, such that $H_1$ has $s^2 - 1$ vertices, degree $st$, and the eigenvalues*

$$st, \quad \pm\sqrt{\lambda_i^2 + t^2}, \quad -t \quad (\text{of multiplicity } s);$$

*$H_2$ has $s^2$ vertices, parameters $n_1 = \frac{1}{2}s(s + 1)$, $n_2 = \frac{1}{2}s(s - 1)$, $r_1 = t(s - 1)$, $r_2 = t(s + 1)$, and the eigenvalues $\pm t\sqrt{s^2 - 1}$, $\pm\lambda_i$, $0$ (of multiplicity $s$).*

**Proof.** We know that if the graph $G$ is connected and the line graph $L(G)$ is regular then either $G$ is regular or $G$ is a semi-regular bipartite graph. If $G_1$ and $G_2$ from the theorem are both regular or both semi-regular bipartite we have cases (a) and (b) of the theorem; this follows readily from Theorems 2.4.1, 2.4.2 and Corollary 2.4.3.

Suppose therefore that $\{G_1, G_2\} = \{H_1, H_2\}$ where $H_1$ is regular non-bipartite of degree r with n vertices, $H_2$ is semi-regular bipartite with parameters $(n_1, n_2, r_1, r_2)$, and $n_1 > n_2$. Since $L(H_1)$ and $L(H_2)$ are cospectral they must have the same degree, the same number of vertices and the same multiplicity of the eigenvalue $-2$. This yields the following relations

$$r_1 + r_2 - 2 = 2r - 2, \quad n_1 r_1 = \frac{nr}{2} (= n_2 r_2),$$

$$n_1 r_1 - n_1 - n_2 + 1 = \frac{nr}{2} - n,$$

which may be rewritten as follows:

$$r = \frac{r_1 + r_2}{2}, \tag{4.1}$$

$$nr = 2n_1 r_1 = 2n_2 r_2, \tag{4.2}$$

$$n = n_1 + n_2 - 1. \tag{4.3}$$

If we use (4.1) and (4.3) to substitute for $r$ and $n$ in (4.2), we obtain:

$$n_1 - n_2 = \frac{r_1 + r_2}{r_2 - r_1}. \tag{4.4}$$

Let $\lambda_1, \lambda_2, \ldots, \lambda_{n_2}$ be the first $n_2$ largest eigenvalues of $H_2$. From the proof of Corollary 2.4.3, we know that $H_2$ has also the eigenvalues

$-\lambda_1, -\lambda_2, \ldots, -\lambda_{n_2}$ and $n_1 - n_2$ eigenvalues equal to 0, where $\lambda_1 = \sqrt{r_1 r_2}$. Since the sum of squares of eigenvalues is twice the number of edges we have

$$2r_1 r_2 + 2 \sum_{i=2}^{n_2} \lambda_i^2 = 2n_1 r_1,$$

equivalently,

$$\sum_{i=2}^{n_2} \lambda_i^2 = n_1 r_1 - r_1 r_2. \tag{4.5}$$

Now, by Theorem 2.4.1 and Corollary 2.4.3, the eigenvalues of $H_1$ are $\frac{1}{2}(r_1 + r_2)$ with multiplicity 1 (largest eigenvalue), $\frac{1}{2}(r_1 - r_2)$ with multiplicity $n_1 - n_2$ and $\pm\sqrt{\lambda_i^2 + \frac{1}{4}(r_1 - r_2)^2}$ $(i = 2, 3, \ldots, n_2)$. The sum of eigenvalues must be 0 and this yields again the relation (4.4). Considering the sum of squares we have

$$\left(\frac{r_1 + r_2}{2}\right)^2 + (n_1 - n_2)\left(\frac{r_1 - r_2}{2}\right)^2 + 2\sum_{i=2}^{n_2}\left(\lambda_i^2 + \frac{(r_1 - r_2)^2}{4}\right) = 2n_1 r_1.$$

Using (4.5), we obtain:

$$n_1 + n_2 = \left(\frac{r_1 + r_2}{r_2 - r_2}\right)^2. \tag{4.6}$$

Let $s = \dfrac{r_1 + r_2}{r_2 - r_1}$. Then $s$ is an integer greater than 1, and relations (4.4) and (4.6) yield

$$n_1 = \frac{s^2 + s}{2} \quad \text{and} \quad n_2 = \frac{s^2 - s}{2}.$$

By Equation (4.1), $r_1$ and $r_2$ are of the same parity, and since $r_2 > r_1$ we can take $r_2 = r_1 + 2t$, where $t$ is a positive integer. Then

$$r_1 = t(s - 1) \quad \text{and} \quad r_2 = t(s + 1).$$

Since $r_1 \leq n_2$ and $r_2 \leq n_1$ we see that $t \leq s/2$. If we now express the spectra of $H_1$ and $H_2$ in terms of $s$, $t$ and the $\lambda_i$, the proof of the theorem is complete. $\square$

**Remark 4.1.15.** When $s = 2$ we have $H_1 = K_3$ and $H_2 = K_{1,3}$, but then $L(H_1)$ and $L(H_2)$ are not only cospectral, but also isomorphic. (By a theorem of Whitney [Whi], $\{K_3, K_{1,3}\}$ is the only pair of non-isomorphic connected graphs having isomorphic line graphs.) When $s = 3$, $H_2$ is the graph shown in Fig. 4.2(c); but then $H_1$ does not exist. For $s = 4$ and $t = 2$ we have

$H_2 = K_{10,6}$ and $H_1 = L(K_6)$ and, of course, $L(K_{10,6})$ and $L(L(K_6))$ are cospectral but not isomorphic. In the case $s = 4, t = 1$, $H_2$ belongs to the design with the parameters $v = 6, b = 10, r = 5, k = 3, \lambda = 2$, and $H_1$ is the Petersen graph. For higher values of $s$, in the known examples $H_2$ is the graph of a 2-design. It would be interesting to know whether there exists a pair of graphs $H_1, H_2$ such that (i) $H_2$ is not the graph of a 2-design, and (ii) $H_1, H_2$ satisfy the conditions of Theorem 4.1.14(iii) with $s > 4$. $\qquad\square$

Now we turn again to exceptional graphs. We start with the following definition.

**Definition 4.1.16.** $\mathcal{G}$ is the set of all connected regular graphs, whose adjacency matrix has least eigenvalue $-2$, and which are neither line graphs nor cocktail-party graphs.

Note that a regular generalized line graph is either a line graph or a cocktail-party graph (Exercise 1.11), and so $\mathcal{G}$ is just the set of exceptional regular graphs. Hoffman [Hof5] posed the problem of determining $\mathcal{G}$, and he and Ray-Chaudhuri [HofRa3] showed that graphs in $\mathcal{G}$ cannot have degree $\geq 17$. As exceptional graphs, the graphs in $\mathcal{G}$ have a representation in the root system $E_8$ (see Chapter 3), and we use this fact to prove the following:

**Theorem 4.1.17** [CamGSS, Theorem 4.4]. *Any graph in $\mathcal{G}$ has at most 28 vertices, and degree at most 16.*

**Proof.** If $A$ is the adjacency matrix of an $r$-regular graph $G$ in $\mathcal{G}$ then the matrix $I + \frac{1}{2}A$ has rank at most 8 since $G$ has a representation in $E_8$. Hence the positive semi-definite matrix

$$I + \frac{1}{2}A - \frac{r+2}{2n}J$$

has rank at most 7, and is therefore expressible in the form $Q^{\top}Q$, where $Q = (\mathbf{q}_1 | \cdots | \mathbf{q}_n)$, of size $7 \times n$. Let $Q_i = \frac{1}{\alpha}\mathbf{q}_i\mathbf{q}_i^{\top}$ $(i = 1, \ldots, n)$, where $\alpha = 1 - \frac{r+2}{2n}$. Thus $Q_i$ represents the orthogonal projection of $\mathbb{R}^7$ onto the line spanned by the vector $\mathbf{q}_i$ $(i = 1, \ldots, n)$. These $n$ projections lie in the space of symmetric linear maps $\mathbb{R}^7 \to \mathbb{R}^7$, and with respect to the inner product

$$\langle Q_i, Q_j \rangle = \alpha^2 \mathrm{tr}(Q_i Q_j) = (\mathbf{q_i}^{\top}\mathbf{q_j})^2,$$

their Gram matrix is

$$\left(1 - \frac{r+2}{2n}\right)^2 I + \left(\frac{1}{2} - \frac{r+2}{2n}\right)^2 A + \left(\frac{r+2}{2n}\right)^2 (J - I - A).$$

Since this matrix must be positive semi-definite, each eigenvalue $\lambda$ of $A$ other than $r$ satisfies:

$$\left(1 - \frac{r+2}{2n}\right)^2 + \left(\frac{1}{2} - \frac{r+2}{2n}\right)^2 \lambda + \left(\frac{r+2}{2n}\right)^2 (-1 - \lambda) \geq 0,$$

equivalently:

$$\lambda(n - 2r - 4) \geq -2(2n - 2r - 4). \tag{4.7}$$

We distinguish two cases: (a) $n - 2r - 4 \geq 0$, (b) $n - 2r - 4 < 0$.

In case (a), (4.7) is a strict inequality because $\lambda \geq -2$ ; then the vectors $\mathbf{q}_1, \ldots, \mathbf{q}_n$ are linearly independent, and so $n \leq 28$, $r \leq 12$.

In case (b), (4.7) becomes

$$\lambda \leq \frac{2(2n - 2r - 4)}{2r + 4 - n}.$$

Now $A$ has $-2$ as an eigenvalue of multiplicity at least $n - 8$. If the remaining eigenvalues are $r > \lambda_2 \geq \cdots \geq \lambda_8$ then we have:

$$0 = \text{tr}(A) = r + (n - 8)(-2) + \sum_{i=2}^{8} \lambda_i \leq r - 2n + 16 + \frac{14(2n - 2r - 4)}{2r + 4 - n}.$$

In particular, we have $r \leq 17$ when $n = 28$ and $r \leq 16$ when $n \leq 27$ (Exercise 4.7). To see that $n \leq 28$, consider the positive semi-definite matrix $I + \frac{2}{3}A - \frac{1}{3}(J - I)$, in which the non-diagonal entries are $\pm 1/3$. This is the Gram matrix of $n$ vectors in $\mathbb{R}^8$ which determine $n$ equiangular lines: the angle between any two of them is $\cos^{-1}(1/3)$. However, the maximal number of equiangular lines in $\mathbb{R}^8$ is 28 (see Section 6.6), and so $n \leq 28$; moreover, any set of 28 equiangular lines in $\mathbb{R}^8$ span a 7-dimensional subspace. Hence if $n = 28$ then $I + \frac{1}{2}A$ has rank at most 7, and so $\lambda_8 = -2$. In this case,

$$0 = \text{tr}(A) = r + 21(-2) + \sum_{i=2}^{7} \lambda_i \leq r - 42 + \frac{12(52 - 2r)}{2r - 24},$$

whence $r \neq 17$.                                                                 $\square$

We can now extend Theorem 4.1.8 as follows:

**Theorem 4.1.18** [Cve1, Doo1]. *If $m + n \geq 19$ and if $\{m, n\} \neq \{2s^2 + s, 2s^2 - s\}$, where $s$ is a positive integer, then $L(K_{m,n})$ is characterized by its spectrum.*

**Proof.** As before, we may assume that $m > 1$ and $n > 1$; then the eigenvalues of $L(K_{m,n})$ are $m + n - 2$, $m - 2$, $n - 2$, 2 with multiplicities 1, $n - 1$, $m - 1$, $mn - m - n + 1$ respectively. (This follows from Theorems 2.1.8 and 2.4.1.)

Now let $G$ be a graph with the same spectrum as $L(K_{m,n})$. We know from Corollary 3.2.2 and Theorem 1.3.6 that $G$ is a regular connected graph. Moreover, it has degree $\geq 17$ and least eigenvalue $-2$. We conclude that $G$ is a line graph, say $G = L(H)$, where $H$ has no isolated vertices. Since $G$ is regular, $H$ is either a regular graph or a semi-regular bipartite graph.

Suppose first that $H$ is regular of degree $r$. Then $2(r - 1) = m + n - 2$, whence $m + n$ is even and $r = \frac{1}{2}(m + n)$. The number $q$ of edges in $H$ is the number of vertices in $G$, namely $mn$. Now the number of vertices of $H$ is $2q/r$, or $4mn/m + n$. By considering the multiplicity of $-2$ as a root of $P_{L(H)}(x)$ as given by Theorem 2.4.1 we find that $-\frac{1}{2}(m + n)$ is an eigenvalue of $H$ with multiplicity

$$mn - m - n + 1 - \left( mn - 4\frac{mn}{m+n} \right) = 1 - \frac{(m-n)^2}{m+n}.$$

We deduce that $m = n$, for otherwise $\{m, n\} = \{2s^2 + s, 2s^2 - s\}$, contrary to assumption. Accordingly, the result in this case follows from Theorem 4.1.8.

Secondly, let $H$ be a semi-regular bipartite graph with parameters $(n_1, n_2, r_1, r_2)$, where $n_1 > n_2$. Then $n_1 r_1 = n_2 r_2 = mn$ and $r_1 + r_2 = m + n$. By Corollary 2.4.3, $r_1 - 2$ is an eigenvalue of $L(H)$, and a comparison with the eigenvalues of $G$ yields three possibilities: (1) $r_1 = m$, (2) $r_1 = n$, (3) $r_1 = m + n$. The third cannot arise because $r_2 \neq 0$, while in cases (1) and (2) we have $H = K_{m,n}$, as required. $\qquad\square$

Note that if $m > 2$ and $n > 2$, only case (a) in the proof of Theorem 4.1.17 is pertinent, and so then $L(K_{m,n})$ is characterized by its spectrum when $m + n \geq 15$. We shall see shortly how knowledge of the graphs in $\mathcal{G}$ enables Theorem 4.1.18 to be extended to deal with all the cases in which $m + n \leq 18$. The graphs in $\mathcal{G}$ were determined by Bussemaker, Cvetković and Seidel [BuCS2], partly by means of a computer search for representations in $E_8$ (see [CvRS7, Section 4.4]). The report [BuCS1] contains a table of all 187 graphs from $\mathcal{G}$; this table is reproduced in a slightly different form in the monograph [CvRS7, Table A3].

In view of our earlier remarks, we have the following result.

**Theorem 4.1.19.** *Any regular connected graph with least eigenvalue $-2$ is a line graph, or a cocktail party graph, or one of the of the 187 graphs in $\mathcal{G}$.*

We can now make our characterization theorems more precise by inspecting the graphs in $\mathcal{G}$. We find that (i) there are exactly 17 regular connected line graphs $L(G)$ for which there exists an exceptional graph cospectral with $L(G)$, (ii) there are exactly 68 graphs which are not line graphs but which are cospectral with a regular connected line graph, (iii) each of these 68 graphs is obtained from a regular connected line graph by switching. The 68 graphs are listed in [CvRS7, Table A4]; in the paper [CvRa] they are constructed in such a way that these results can be verified without recourse to a computer.

The following is a refinement of Theorem 4.1.18.

**Theorem 4.1.20** [BuCS1, BuCS2]. $L(K_{m,n})$ *is characterized by its spectrum unless*

(a) $m = n = 4$, *where the graph G69 provides the only exception,*
(b) $m = 6, n = 3$, *where the graph G70 provides the only exception,*
(c) $m = 2t^2 + t, n = 2t^2 - t$, *and there exists a symmetric Hadamard matrix of order $4t^2$ with constant diagonal.*

**Proof.** Graphs cospectral with $L(K_{m,n})$ may or may not be line graphs. If they are not line graphs, then they can be identified immediately from the list of graphs in $\mathcal{G}$, and we have cases (a) and (b) of the theorem. The exceptions which are line graphs are described by Theorem 4.1.14: from $n_1 = r_1 = m$ and $n_2 = r_2 = n$ we have $t = \frac{1}{2}s$ and $n_1 = 2t^2 + t, n_2 = 2t^2 - t$. Since the eigenvalues of $K_{m,n}$ are $\pm\sqrt{mn}$ and 0, the spectrum of the graph $H_1$ in Theorem 4.1.14 consists of eigenvalues $2t^2, \pm t$, and its adjacency matrix $A$ satisfies $A^2 = t^2(I + J)$. Replacing the zeros of $A$ by $(-1)$s, and bordering the matrix with $(-1)$s, we obtain a symmetric Hadamard matrix with diagonal $-I$. This completes the proof.                    □

We can extend this characterization to general 2-designs. If $\mathcal{D}$ is a design with incidence graph $H(\mathcal{D})$ then we refer to $L(H(\mathcal{D}))$ as the *line graph of $\mathcal{D}$*.

**Theorem 4.1.21** [BuCS2]. *Let $G_1$ be the line graph of a 2-design with parameters $v, k, b, r, \lambda$. Let $G_2$ be a graph with the same spectrum as $G_1$. Then one of the following holds:*

(a) $G_2$ *is the line graph of a 2-design having the same parameters;*
(b) $(v, k, b, r, \lambda) = (3, 2, 6, 4, 2)$ *and $G_2$ is the graph G6;*
(c) $(v, k, b, r, \lambda) = (4, 3, 4, 3, 2)$ *and $G_2$ is the graph G9;*
(d) $(v, k, b, r, \lambda) = (4, 4, 4, 4, 4)$ *and $G_2$ is the graph G69;*
(e) $(v, k, b, r, \lambda) = (3, 3, 6, 6, 6)$ *and $G_2$ is the graph G70;*
(f) $v = \frac{1}{2}s(s-1), k = t(s-1), b = \frac{1}{2}s(s+1), r = t(s+1), \lambda = \frac{2t(st-t-1)}{s-2}$, *where $s$ and $t$ are integers with $st$ even, $t \leq \frac{1}{2}s$, $(s-2)|2t(t-1)$, and $G_2 = L(H)$ where $H$ is a regular graph on $s^2 - 1$ vertices with the eigenvalues $st, \pm\sqrt{ts(s-1-t)(s-2)^{-1}}, -t$ of multiplicities $1, \frac{1}{2}(s-2)(s+1)$, $\frac{1}{2}(s-2)(s+1), s$, respectively.*

The following theorem summarizes many of the previous results, and it can be proved without the aid of a computer (see [CvDo1]).

**Theorem 4.1.22** [CvDo2]. *The spectrum of a graph G determines whether or not it is a regular, connected line graph except for* 17 *cases. In these cases G has the spectrum of* $L(H)$ *where H is one of the 3-connected regular graphs on* 8 *vertices or H is a connected, semi-regular bipartite graph on* 6 + 3 *vertices.*

### 4.1.3 Characterizations according to type

We can identify two further sorts of characterization theorems involving graph spectra.

**1.** There are certain families of graphs, defined in terms of graph structure, which have the property that different graphs from the same family have different spectra. In view of Theorem 4.1.1, the regular graphs of degree 2 constitute such a family. Further examples include (i) vertex-transitive graphs with a prime number of vertices [Tur1], (ii) starlike trees (obtained from stars by subdividing edges) [LepGu], (iii) the family $\mathcal{H}$ of all bicyclic Hamiltonian graphs (cycles with one chord). Indeed, different graphs in $\mathcal{H}$ which have the same number of vertices are distinguished by their indices (see [SimKo]).

**2.** A family $\mathcal{G}$ of graphs may be spectrally determined in the following (weaker) sense: if $G \in \mathcal{G}$ and $H$ is cospectral with $G$ then $H \in \mathcal{G}$. We describe without proof three such families in terms of their structural properties.

**Theorem 4.1.23** (cf. [Hof4]). *Let G be the line graph of a projective plane of order n. If the graph H is cospectral with G then it is the line graph of a projective plane of order n.*

**Theorem 4.1.24** (cf. [HofRa1]). *Let G be the line graph of an affine plane of order n. If the graph H is cospectral with G then H is the line graph of an affine plane of order n.*

**Theorem 4.1.25** (cf. [HofRa2]). *Let G be the line graph of a symmetric design with parameters* $(v, k, \lambda) \neq (4, 3, 2)$. *If the graph H is cospectral with G then H is the line graph of a design with the same parameters.*

Further examples of such spectral characterizations may be found in [Cve13] and [Doo2].

## 4.2  Cospectral graphs and the graph isomorphism problem

Cospectral graphs are often called *isospectral* graphs in the literature, and the term '(unordered) pair of isospectral non-isomorphic graphs' is denoted by PING. More generally, the term 'set of isospectral non-isomorphic graphs' is denoted by SING. We say that a SING is *trivial* if it consists of just one graph, and that different members of a SING are *cospectral mates*. Example 1.2.4 includes a PING on 16 vertices, and Example 1.2.6 gives a SING on 28 vertices. Further examples arise in the context of characterization theorems in Section 4.1.

In this section we review what is known about cospectral graphs. Subsection 4.2.1 surveys examples of cospectral graphs, and some constructions of PINGs are discussed in Subsection 4.2.2. Enumeration results for cospectral graphs are described in Subsection 4.2.3, where (together with the spectrum of the adjacency matrix) the spectra of other graph matrices are treated. Subsection 4.2.4 contains a comparison of the characterizing properties of various graph invariants.

### 4.2.1  Examples of cospectral graphs

The literature contains various examples of PINGs (and, more generally, of SINGs). Their importance lies in the following observations:

(1) For every pair of non-isomorphic graphs one can find a set of characteristic properties that are different for the two graphs. Therefore, every PING points to properties of graphs that are not uniquely determined by the spectrum.

(2) The existence of a PING rules out various possibilities in the search for families of graphs with the property that different graphs from the same family have different spectra.

In [Har1], Harary states that his conjecture, that isospectrality implies the isomorphism of graphs, was disproved by Bose, who described a PING with 16 vertices. According to [Har1], Bruck and Hoffman also found PINGs with 16 vertices. In [ColSi], Collatz and Sinogowitz had already noted that the spectrum of a graph does not determine the graph up to isomorphism. They gave an example of two isospectral trees with eight vertices and different sets of vertex degrees. Turner [Tur2] gives a PING consisting of 12-vertex trees which have the same vertex degree sequence; the author expresses his pessimism concerning the possibility of distinguishing even graphs of restricted type by means of their spectra.

Given two graphs $G$ and $H$, we say that $G$ is *smaller* than $H$ if $|V(G)| < |V(H)|$, and in the case $|V(G)| = |V(H)|$, if $|E(G)| < |E(H)|$. Any set of graphs has one or several *smallest* graphs in the above order of graphs. Since graphs in any SING have the same number of vertices and the same number of edges, we can compare SINGs as well in the above sense. For example, Fig. 1.2 shows the smallest PING (with five vertices) and the smallest PING consisting of connected graphs (with six vertices). The smallest PING consisting of regular graphs (with ten vertices) is illustrated in Fig. 4.8. From the first of these examples we see that in general we cannot determine from the spectrum whether or not a graph is connected. This example has been generalized in [Cve1] as follows. The graph having as components $s$ isolated vertices and one complete bipartite graph $K_{n_1,n_2}$ has eigenvalues $\sqrt{n_1 n_2}$, $-\sqrt{n_1 n_2}$ and $n_1 + n_2 - 2 + s$ numbers equal to 0. Now consider a graph with spectrum $\sqrt{m}$, $-\sqrt{m}$ and $n - 2$ numbers equal to 0 ($m$ a natural number). This spectrum belongs to each graph of the above type whose parameters $n_1, n_2, s$ satisfy the equations $n_1 + n_2 + s = n$, $n_1 n_2 = m$.

Among other things, the paper [HarKMR] gives the smallest triplet of connected cospectral graphs (Fig. 4.3), while in [GoHMK] we find the smallest cospectral graphs with cospectral complements (Fig. 4.4) and the smallest cospectral forests ($K_{1,3} \dot\cup K_2$ and $P_5 \dot\cup K_1$).

The paper [DAGT] includes a discussion of some cospectral graphs relevant to chemistry, methods for recognizing cospectrality and certain properties of eigenvectors in cospectral graphs. If the eigenvalues of a graph (with multiplicities) appear among the eigenvalues of another graph then these graphs are said to be *subspectral*. Several cases of subspectral graphs are reviewed, with an observation that in many cases the smaller graph appears as a fragment of the larger one.

Among PINGs the least eigenvalue cannot exceed the smallest root of $x^2 - x - 4$ (approximately $-1.5616$), and the unique smallest PING for which this value is attained is shown in Fig. 4.5. This follows from Theorem 3.4.14 (see [CvLe5]).

Figure 4.3 Three cospectral graphs.

Figure 4.4 Cospectral graphs with cospectral complements.

Figure 4.5 The smallest PING with largest least eigenvalue.

Fisher, who encountered cospectral graphs when investigating the vibration of membranes [Fis] (see Section 9.1), considered connected planar graphs with no vertex of degree 1. He constructed an infinite sequence of PINGs with $5n$ vertices ($n = 3, 4, \ldots$) consisting of such graphs. An infinite sequence of sets of mutually non-isomorphic isospectral graphs was also given by Bruck [Bruc].

A construction for cospectral graphs with cospectral complements will be described in the next subsection (see Theorem 4.2.1). We shall also discuss a well-known theorem of Schwenk [Sch1], which states that almost all trees have a cospectral mate.

### 4.2.2 Constructions of cospectral graphs

Many methods for constructing cospectral graphs are described in the literature, and we have already seen in Proposition 1.1.8 how one can produce cospectral regular graphs using Seidel switching. From a PING consisting of regular graphs of degree greater than 2, we can construct another PING with more vertices by taking the line graphs of the graphs in question (see Theorem 2.4.1).

Several other graph operations and modifications, as described in Chapter 2, can also be used to produce SINGs. One of the simplest ways is to use Theorem 2.1.1: if a SING with $n$ vertices is known, then a SING with $m$ vertices ($m > n$) can easily be constructed by adding an arbitrary graph with $m - n$ vertices as a new component in each of the two graphs.

More generally, for two SINGs $\mathcal{S}$ and $\mathcal{P}$ we define the *composition* $\mathcal{S} \diamond \mathcal{P}$ by $\mathcal{S} \diamond \mathcal{P} = \{G \mathbin{\dot\cup} H \ : \ G \in \mathcal{S}, \ H \in \mathcal{P}\}$. Then $\mathcal{S} \diamond \mathcal{P}$ is a SING.

A SING $S$ is called *reducible* if each graph in $S$ contains a component isomorphic to a fixed graph; otherwise, $S$ is called *irreducible*. A reducible non-trivial SING can be *reduced* to an irreducible one by extracting components common to each graph in the SING. Accordingly, reducible SINGs are not normally recorded in tables such as those found in [CvLe1, CvLe3]. However, reducible SINGs are not without interest, as the following examples demonstrate. The reducible PING $\{K_{1,4} \dot\cup K_2, C_4 \dot\cup K_1 \dot\cup K_2\}$ extends to the irreducible SING $\{K_{1,4} \dot\cup K_2, C_4 \dot\cup K_1 \dot\cup K_2, Y_6 \dot\cup K_1\}$, where $Y_6$ is the tree on six vertices with index 2 (see Fig. 3.5). Another interesting irreducible SING from [CvLe3, CvLe5] is the quadruple shown in Fig. 4.6: this is the union of two reducible PINGs (the first and second graph, and the third and fourth graph).

The procedures described above have been formalized in [CvLe4] to describe an algebra of SINGs using formal linear combinations of graphs and of their spectra. This generalizes a technique used in [CvGu1] to characterize the SINGs in the set $S$ of all graphs whose largest eigenvalue does not exceed 2. The main result of [CvGu1] is that any bipartite graph in $S$ is cospectral with a union of paths and 4-cycles. Examples include two PINGs already mentioned, namely $\{K_{1,4}, C_4 \cup K_1\}$ and $\{K_{1,3} \cup K_2, P_5 \cup K_1\}$.

In addition, the results from [CvGu1] enable us to decide whether a finite family of reals from the interval $[-2, 2]$ is the spectrum of a graph, and an algorithm is given which constructs all graphs having this spectrum. This result is significant since, in general, we do not know any reasonable algorithm (that is, an algorithm essentially different from an exhaustive search) for deciding whether there is a graph with a given spectrum.

Next, we prove a theorem which provides a construction for cospectral trees with cospectral complements.

**Theorem 4.2.1** [GoMK1]. *Let $G$ be an arbitrary rooted graph. Let $S$ and $T$ be rooted trees as shown in Fig. 4.7. Then the coalescences $G \cdot S$ and $G \cdot T$ are not isomorphic (unless the root of $G$ is isolated) but are cospectral and have cospectral complements.*

**Proof.** Consider $S, T$ as the tree $H$ rooted at $u, v$ respectively. The graphs $H - u$ and $H - v$ are isomorphic, and so $G \cdot S$ and $G \cdot T$ are cospectral

Figure 4.6 An irreducible SING.

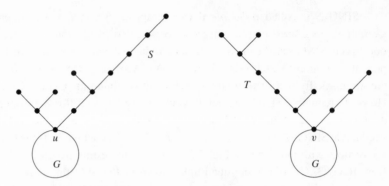

Figure 4.7 The construction for Theorem 4.2.1.

by Theorem 2.2.3. Since $P_{H-u}(x) = P_{H-v}(x)$, the angles of $H$ at $u$ coincide with those at $v$; that is, $\alpha_{iu} = \alpha_{iv}$ ($i = 1, \ldots, m$) in the notation of Proposition 2.2.6. For $k \in I\!N$, let $n_k(u)$ be the number of $u$-$u$ walks of length $k$ in $H$. By Proposition 3.1.2, $n_k(u) = n_k(v)$, and it follows that $G \cdot S$ and $G \cdot T$ have the same walk-generating function. Now Equation (2.14) shows that $\overline{G \cdot S}$ and $\overline{G \cdot T}$ are cospectral.                                                                    □

Similar techniques are used to prove Scwhenk's important result [Sch1] that almost all trees have a cospectral mate. We describe this result in more detail.

**Definition 4.2.2.** A branch of a tree at a vertex $v$ is a maximal subtree containing $v$ as an endvertex. The union of one or more branches at $v$ is called a limb at $v$.

Considered in its own right, a limb at the vertex $v$ is a rooted tree, with $v$ as its root. Schwenk proved that the proportion of trees on $n$ vertices which avoid a specified limb tends to zero as $n$ tends to infinity. Moreover, the number of trees on $n$ vertices which do not contain a specified limb depends only on the number of edges of the limb.

**Definition 4.2.3.** Vertices $u$ and $v$ in cospectral (not necessarily non-isomorphic) graphs $G$ and $H$ are said to be cospectral if $P_{G-u}(x) = P_{H-v}(x)$.

Schwenk observed that vertices $u$ and $v$ in the tree $T$ of Fig. 4.8 are cospectral but lie in different orbits of the automorphism group of $T$. Using Theorem 2.2.3 again, we see that the graphs $G_1$ and $G_2$ of Fig. 4.8 are cospectral, whatever the rooted graph $G$. Now, Schwenk's argument was that almost all trees are of the form $G_1$ and hence have a (non-isomorphic) cospectral mate $G_2$.

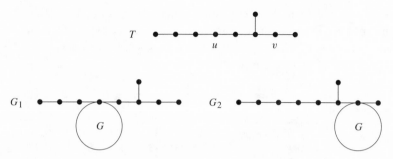

Figure 4.8 Schwenk's construction.

**Definition 4.2.4.** If $v$ is a vertex in a graph $G$, then the pair $(d, e)$, where $d$ is the degree of $v$ and $e$ is the sum of degrees of all neighbours of $v$ in $G$, is called the *degree pair* of $v$.

Note that in a tree $T$, the sequence of vertex degree pairs $(d_i, e_i)$ is determined by the eigenvalues and angles of $T$, because $n_2(i) = d_i$ and $n_4(i) = d_i^2 - d_i + e_i$. Now the graphs $G_1$ and $G_2$ in Fig. 4.8 have different sequences of vertex degree pairs, and hence different angles. The significance of this observation is that all of the cospectral graphs constructed by Schwenk can be distinguished by angles. In Subsection 4.3.2 we shall see to what extent trees are better characterized if not only the eigenvalues but also the angles are known. The results include an algorithm for constructing all the trees with prescribed eigenvalues and angles.

### 4.2.3 Statistics of cospectral graphs

It seems that PINGs with a large number of vertices are a common occurrence: this was suggested by Baker on the basis of statistical evidence presented in [Bak2]. The table in [CvLe3] of cospectral graphs with least eigenvalue $-2$ contains 201 irreducible SINGs with at most 8 vertices; this number includes 178 pairs, 20 triplets and 3 quadruples of cospectral graphs. The paper [GoMK1] presents the results of a computational study of graph spectra: the characteristic polynomials of all graphs with at most 9 vertices are computed, and the cospectral graphs identified. Statistics are given for cospectral graphs in various classes of graphs. The data is extended to cospectral graphs on 10 vertices in [Lep1]. Before we give the results for graphs on 11 vertices we need a definition.

Let $S$ be a finite set of graphs, and let $S'$ be the set of graphs in $S$ which have a cospectral mate in $S$. The ratio $r_S = |S'|/|S|$ is called the *spectral uncertainty* of $S$ (with respect to the adjacency matrix). The papers [DamHa4], [HaeSp] provide spectral uncertainties $r_n, s_n, t_n$ of the sets of all graphs on $n$ vertices for $n \leq 11$ with respect to the adjacency matrix, the Laplacian and the signless Laplacian, respectively:

| $n$ | 4 | 5 | 6 | 7 | 8 | 9 | 10 | 11 |
|-----|-----|-----|-----|-----|-----|-----|-----|-----|
| $r_n$ | 0 | 0.059 | 0.064 | 0.105 | 0.139 | 0.186 | 0.213 | 0.211 |
| $s_n$ | 0 | 0 | 0.026 | 0.125 | 0.143 | 0.155 | 0.118 | 0.090 |
| $t_n$ | 0.182 | 0.118 | 0.103 | 0.098 | 0.097 | 0.069 | 0.053 | 0.038 |

For the Seidel matrix of a graph $G$, the corresponding ratios are 1 for all $n > 1$, because we can use Seidel switching to construct from $G$ a graph with the same Seidel spectrum but with a different number of edges.

We see that for $n \geq 7$ we have $t_n < r_n$. In addition, the sequence $t_n$ is decreasing for $n \leq 11$ while the sequence $r_n$ is increasing for $n \leq 10$. This is a basis for believing that the $Q$-spectrum (discussed in Chapter 7) provides a means of studying graphs that is more effective than the adjacency spectrum. Given the direct relation between the $Q$-spectrum of a graph and the spectrum of its line graph (see Chapter 1), this in turn indicates that the theory of graphs with least eigenvalue $-2$ is important for the whole theory of graph spectra.

### 4.2.4 A comparison of various graph invariants

We have now encountered many counterexamples to the early conjecture that a graph is determined, to within isomorphism, by its spectrum. Had the conjecture been valid, it would have provided a polynomial algorithm for the solution of the *graph isomorphism problem*, that is, the problem of deciding whether two graphs are isomorphic. As things stand, the algorithmic complexity of this problem is not known. The problem belongs to the class NP but it is not known whether it is NP-complete or belongs to the class P (see, for example, [Cve10] or [BruRy, pp. 245–8]).

A set of graph invariants (which might consist of numbers, vectors, matrices, etc.) is called *complete* if it determines any graph to within isomorphism. Although the spectrum of a graph does not, in general, constitute a complete set of invariants, complete sets of invariants do exist. For instance, it is clear that a graph $G$ is determined up to isomorphism by the largest (or least) binary number obtained by concatenation of the rows (or the rows of the upper triangle) of an adjacency matrix of $G$. However, the known algorithms for computing such an invariant are exponential.

Although it would be useful if a complete set of invariants were computable in polynomial time, no such set has been identified to date, and pessimism has been expressed in the literature concerning this question [ReCo]. Optimists point to the fact (see Theorem 1.3.1) that a graph is determined by its eigenvalues and eigenspaces, both of which can be found in polynomial time, but this is to ignore the non-invariant nature of eigenspaces: the components of eigenvectors are ordered according to a labelling of vertices. Nevertheless the study of eigenspaces has enabled us to extend spectral techniques in graph theory; some of the results (such as those concerning graph angles and star complements) are included in this book, and others may be found in the monographs [CvRS2] and [CvRS7].

Let us consider the extent to which various graph invariants determine graph structure.

1. *The vertex degrees*. The family of vertex degrees can be calculated readily from the adjacency matrix or from other common graph representations. In general this can be regarded as a set of local invariants which says little about the graph structure. In the particular case that all degrees are 1, the graph is determined uniquely; if all degrees are 1 or 2 then several non-isomorphic graphs may arise, each graph being a union of paths and cycles. For larger degrees, there are few general conclusions. Hakimi [Hak] provided a polynomial algorithm for determining whether or not a family of integers is a family of vertex degrees for a graph; the algorithm may be adapted to determine in exponential time all the graphs which arise.

2. *The spectrum*. In general, the eigenvalues depend on structural details beyond the vertex degrees. For example, consider again the spectral characterization of a regular graph $G$ of degree 2 (Theorem 4.1.1). Given the spectrum of $G$, we first establish that $G$ is regular of degree 2, and so we know the family of vertex degrees. But the spectrum tells us more: from the second largest eigenvalue we can determine the length of the largest cycle in $G$. Gradually, by analysing the whole spectrum we can find the lengths of all cycles of $G$, and thereby determine $G$ up to isomorphism, in contrast to the case where only the degrees are given. The importance of this result has been demonstrated in [CvCK3] in relation to the Travelling Salesperson Problem (see Section 9.4).

It seems that those graph-theoretical invariants which contain significant structural information (and are therefore useful for the graph isomorphism problem) can be obtained by solving some kind of optimization problem: graph eigenvalues can be obtained by considering extremal values of the Rayleigh quotient of the adjacency matrix, while angles can be obtained as extrema of the scalar product of vectors of a standard othonormal basis of $I\!R^n$ with unit eigenvectors. See [CvRS2, Chapter 8] and [CvCK3] for other examples of such invariants (called 'highly informative' in [CvCK3]).

3. *A binary number.* The ordering of vertices which yields a characterizing binary number (as described above) can be considered as a canonical vertex ordering. One can consider several variations of this idea but it turns out that the known algorithms for determining the invariant that characterizes the graph are exponential (cf. [ReCo], [Bab2]). Here a high price has been paid: we have an invariant which tells us everything about the graph but is time consuming to compute. Nevertheless the extremal binary number has been used repeatedly and successfully to recognize graphs.

From the point of view of practical computation it is usually not necessary to know whether the graph isomorphism problem is NP-complete or belongs to P. Experience has shown that any reasonable algorithm for testing graph isomorphism performs well on average; however, the problem has great theoretical significance. Leaving aside the implications for complexity theory, one can say that to understand the difficulties arising in the graph isomorphism problem is to understand the difficulties that emerge in treating graph theory problems in general.

Having acquainted ourselves with these three examples we might be inclined to believe that spectral invariants provide a good balance between the opposing reqirements of graph invariants, and to conclude that this accounts for the appeal of spectral graph theory as an area of research.

## 4.3  Characterizations by eigenvalues and angles

In this section we treat the problem of constructing all graphs with prescribed eigenvalues and angles. Although graphs cannot, in general, be characterized by eigenvalues and angles, for certain classes of graphs (for example, trees, unicyclic graphs, bicyclic graphs, tree-like cubic graphs) it is possible to construct all the graphs in a given class with prescribed eigenvalues and angles. Details may be found in [CvRS2, Chapter 5]. Here we first discuss cospectral graphs with the same angles (Subsection 4.3.1). In Subsection 4.3.2 we describe an algorithm for constructing all the trees with prescribed eigenvalues and angles. In Subsection 4.3.3 we discuss some instances of characterization by eigenvalues and angles.

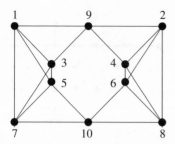 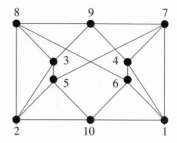

Figure 4.9 A pair of cospectral regular graphs.

### 4.3.1 Cospectral graphs with the same angles

The following example shows that a graph may not be determined by its angles, main angles and spectrum.

**Example 4.3.1.** The two graphs depicted in Fig. 4.9 are non-isomorphic, but they are both 4-regular and have the same eigenvalues, the same angles and the same main angles. The ten vertices are labelled so that the angle sequences $(\alpha_{1j}, \alpha_{2j}, \ldots, \alpha_{mj})$ coincide for $j = 1, 2, \ldots, 10$; equivalently, for each $j$, the graphs obtained by deleting the vertex $j$ are cospectral. □

It has been shown by a computer search (see [CvLe1]) that graphs with fewer than 10 vertices are characterized by their eigenvalues and angles. However, there are 58 pairs of cospectral graphs on 10 vertices with the property that the graphs within each pair have the same angles. Moreover, they also have the same main angles (a fact for which we do not have an explanation), and no multiple eigenvalue is a main eigenvalue. By Proposition 2.1.3, the characteristic polynomial of a complementary graph $\overline{G}$ is determined by the characteristic polynomial and the main angles of $G$, and this explains why 29 of the 58 pairs are the complements of those from the other 29 pairs.

A construction described in the next subsection shows that there is an infinite series of cospectral trees with the same angles. The trees in the smallest example given there have 35 vertices, but an exhaustive computer search has revealed that the following example is the sole example among trees with at most 20 vertices (see [CvLe1]). The trees from this pair have 19 vertices, and it is surprising that there are no examples with 20 vertices.

**Example 4.3.2.** Fig. 4.10 shows the smallest pair of cospectral trees $T_1$ and $T_2$ with the same angles. The subtree $T$, identified by the bold lines in Fig. 4.10, is well known in constructions of cospectral graphs, mainly because the graphs $T - 4$ and $T - 7$ are cospectral (cf. Fig. 4.8).

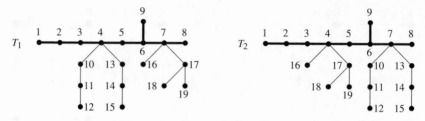

Figure 4.10 Cospectral trees with the same angles.

The vertices in $T_1$ and $T_2$ are labelled so that $T_1 - i$ is cospectral with $T_2 - i$, for $i = 1, 2, \ldots, 19$. Note that although $T_1 - 5$ and $T_2 - 5$ both have two components with 10 and 8 vertices, the components are not cospectral. In $T_1 - 5$ the components have the following spectra (where non-integer eigenvalues are given to three places of decimals):

$$\pm 2.074, \ \pm 1.414^2, \ \pm 0.835, 0^2 \text{ and } \pm 2.222, \ \pm 1.240, \ \pm 0.726, 0^2.$$

On the other hand, in $T_2 - 5$ the two components have spectra

$$\pm 2.222, \ \pm 1.414, \ \pm 1.240, \ \pm 0.726, 0^2 \text{ and } \pm 2.074, \ \pm 1.414, \ \pm 0.835, 0^2.$$

If we delete vertex 6, the components even have different numbers of vertices, yet $T_1 - 6$ and $T_2 - 6$ are still cospectral.                                □

If we try to generalize this example, we encounter difficulties. Suppose that we form the graph $H_1$ by attaching any two rooted graphs $K$ and $L$ at vertices 4 and 7 of $T$, and then form $H_2$ by interchanging $L$ and $K$. The formula (2.18) shows that the following pairs are cospectral: $H_1$ and $H_2$, $H_1 - i$ and $H_2 - i$ for $i = 4, 7$ or any vertex $i$ in $K$ or $L$. However, for other values of $i$, the pairs are not cospectral, except in the special case illustrated in Fig. 4.10.

An exhaustive search for cospectral graphs on 10 vertices [Lep1] shows that there exists a set $Q$ of 21 cospectral graphs with 10 vertices and 20 edges. The complements of these graphs are also cospectral (and they have 25 edges). Computations show also that, in both cases, the graphs are distinguished by their angles (see [Cve11]). We reproduce here some data concerning the graphs in $Q$.

Spectrum:

4.3803  1.6861  1.1620  0.5423  0  0  −1.2950  −1.5260  −2.2864  −2.6631.

Coefficients of the characteristic polynomial:

1   0   −20   −18   84   76   −119   −72   56   0   0.

Main angles:

0.9563  0.0248  0.0659  0.1505  0.2070  0.0436  0.1086  0.0185  0.0323.

These huge sets of cospectral graphs should perhaps be exploited in experiments to order graphs by their angles, for the following reason. Experience shows that it is appropriate to order graphs first by their eigenvalues or spectral moments; see the Appendix for examples, and [CvPe2] for an explanation. Then cospectral graphs remain to be ordered, and it is natural to use angles for this purpose because they determine the vertex degrees.

### 4.3.2 Constructing trees

As we noted in Chapter 3, the number of vertices and the number of edges in a graph $G$ are determined by the spectrum of $G$. It now follows from Theorem 3.3.3 that given the eigenvalues and angles of $G$ we can tell whether or not $G$ is a tree. Here we present an algorithm for constructing all trees with given eigenvalues and angles. The algorithm is based on the following result, known as the *Reconstruction Lemma*.

**Lemma 4.3.3** [Cve9]. *Given a limb $R$ of a tree $T$ at a vertex $i$ which is adjacent to a unique vertex of $T$ not in $R$, that vertex is among the vertices $j$ for which $P_{T-j}(x) = g_i^R(x)$, where*

$$g_i^R(x) = \frac{P_R(x)}{P_{R-i}(x)^2}\{P_R(x)P_{T-i}(x) - P_{R-i}(x)P_T(x)\}. \qquad (4.8)$$

**Proof.** Let $S$ denote the maximal limb of $T$ at $j$ not containing $i$, as shown in Fig. 4.11. From Theorem 2.2.4 we have

$$P_T(x) = P_R(x)P_S(x) - P_{R-i}(x)P_{S-j}(x). \qquad (4.9)$$

Clearly, $P_{T-i}(x) = P_{R-i}(x)P_S(x)$ and $P_{T-j} = P_R(x)P_{S-j}(x)$. By eliminating $P_S(x)$ and $P_{S-j}(x)$ we obtain (4.8). $\qquad \square$

By specifying that $R$ consists only of vertex $i$, so that $P_R(x) = x$ and $P_{R-i}(x) = 1$, we obtain the following result.

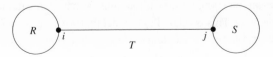

Figure 4.11 Construction of a tree.

**Proposition 4.3.4.** *If $i$ is a vertex of degree 1 in a tree $T$, then the neighbour of the vertex $i$ is among those vertices $j$ such that $P_{T-j}(x) = f_i(x)$, where*

$$f_i(x) = x^2 P_{T-i}(x) - x P_T(x).$$

Now we describe the reconstruction algorithm (let us call it Algorithm EA), which provides a means of constructing all trees with prescribed eigenvalues and angles. Note that examples from Subsection 4.3.1 show that, in general, trees are not EA-reconstructible (as defined in Section 3.1). Indeed we shall see that almost all trees have non-isomorphic mates with the same eigenvalues and angles.

*Algorithm EA* [Cve9]. Let $T$ be a tree with prescribed eigenvalues and angles. First we use Theorem 3.1.3 to find the degrees of vertices in $T$, and then we begin to construct possible edges as follows. For each vertex $i$ of degree 1 we choose a neighbour $j$ from the set $A_i = \{j \in V(T) : P_{T-j}(x) = f_i(x)\}$ (cf. Proposition 4.3.4). The number of times an individual vertex $j$ is chosen as a neighbour of an endvertex is bounded above by the degree of $j$. Now let $T'$ be the graph obtained from $T$ by deleting all endvertices. A vertex of degree 1 in $T'$ is necessarily one of the vertices $j$ chosen above and in this case we may apply Lemma 4.3.3 to the limb $R$ at $j$ consisting of all pendant edges at $j$. The neighbour of $j$ in $T'$ lies in the set $B_j^R = \{k \in V(T') : P_{T-k}(x) = g_j^R(x)\}$ (see (4.8)), and for each such $j$ we choose a neighbour $k \in B_j^R$. Continuing in this way we may construct a tree by successive construction of limbs provided that (i) at each stage there are vertices $j$ of degree 1 in the subtree $T''$ which remains to be constructed, and (ii) the corresponding sets $B_j^R$ are non-empty. If $T''$ is non-trivial and one or other of these requirements is not met, then the algorithm proceeds with a different choice of neighbours at the previous stage. If $T''$ is trivial then a tree $T$ has been constructed and the algorithm is repeated with a new choice of neighbour. Using such a backtracking algorithm one constructs a collection of trees which includes all those with the given eigenvalues and angles. Finally one excludes those which do not have the specified eigenvalues. $\qquad\square$

Let us consider how big a step has been made in determining the structure of trees by the introduction of angles. The difference is that now we can construct all of the trees in question, while without angles that seems not to be possible in a reasonable way. This is related to the fact that we know exactly which features are responsible for the existence of non-isomorphic trees with the same eigenvalues and angles. Indeed, in the notation of Fig. 4.11, non-isomorphic trees can arise as follows.

(1) The limb $R$ may be replaced with a cospectral limb $R'$ such that $P_{R-i}(x) = P_{R'-i}(x)$.

(2) The choice of neighbours $j$ with given $P_{T-j}(x)$ may not be unique.

In view of (1), we may use four copies of the tree $T$ from Fig. 4.8 to construct the trees $T_1, T_2$ shown in Fig. 4.12, where $H$ denotes a rooted tree. For any choice of $H$, the trees $T_1$ and $T_2$ are non-isomorphic and have the same eigenvalues and angles. Corresponding vertices (i.e. vertices for which the vertex-deleted subgraphs in $T_1$ and $T_2$ are cospectral) are denoted by the same numbers for some specific vertices.

The construction illustrated in Fig. 4.12 also shows that almost all trees are not characterized by eigenvalues and angles (cf. p. 122). It also shows (e.g. by reference to the vertices labelled 2 in $T_1$ and $T_2$) that eigenvalues and angles do not determine degree sequences of vertices. (The *degree sequence* of a vertex $v$ consists of the degrees of the neighbours of $v$, in non-increasing order.)

If, in applying the reconstruction algorithm, we know the degree sequences in $T$ (in particular, if we know the vertex-deleted subgraphs of $T$) then the choice in (2) above is limited to the extent that the trees in question are determined up to cospectral limbs with a constant degree sequence of the root. We do not know of an example of non-isomorphic cospectral trees $G_1, G_2$ for which there exists a bijection $\theta : V(G_1) \to V(G_2)$ such that for each $v \in V(G_1)$, the vertices $v$ and $\theta(v)$ are cospectral with the same degree sequence.

It is well known that a tree is in fact (uniquely) reconstructible from its vertex-deleted subgraphs. Also, the characteristic polynomial of a tree is reconstructible from the characteristic polynomials of vertex-deleted subgraphs (see Section 8.3). The reconstruction algorithm can be used to construct all trees for which only the characteristic polynomials of vertex-deleted subgraphs are specified.

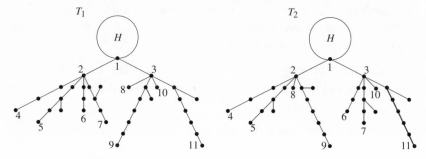

Figure 4.12 More cospectral trees with the same angles.

### 4.3.3 Some characterization theorems

Graphs which appear in the Hückel molecular orbital theory from quantum chemistry have vertex degrees at most 3 (see Section 9.2). In this subsection we show that many of these chemically interesting graphs are characterized by eigenvalues and angles. Excluded from our considerations are the cubic graphs corresponding to fullerenes. An algorithm for constructing all graphs with the eigenvalues and angles of a prescribed fullerene is given in [CvFRS].

One of the basic tools is the algorithm EA for reconstructing trees from eigenvalues and angles described in previous subsection. The essence of this algorithm is that we can reconstruct an edge of a tree if we know the structure of the tree on one side of the edge. More generally, we can reconstruct a bridge of a graph if we know the structure of the graph on one side of the bridge.

We start with a result related to trees obtained from three paths, each rooted at one of its endvertices, by identifying the roots.

**Proposition 4.3.5.** *A forest in which each component has exactly one vertex of degree* 3, *others being of smaller degree, can be reconstructed uniquely from eigenvalues and angles.*

**Proof.** We apply Algorithm EA repeatedly, starting from a vertex of degree 1 and traversing the path connecting this vertex to the vertex of degree 3. We reconstruct this path uniquely, thereby identifying its length and the terminal vertex (of degree 3). □

Next we consider the set $S$ of graphs with largest eigenvalue $\lambda_1 \leq 2$. The connected graphs in $S$ with $\lambda_1 = 2$ are shown in Fig. 3.5. Note that $Y_5 (= K_{1,4})$ is the only example with a vertex of degree 4. Further, $Y_n$ $(n > 5)$ has two vertices of degree 3, while all other connected graphs in $S$ have at most one vertex of degree 3 (see Theorem 3.11.1).

Cospectral graphs in the set $S$ are very frequent, and an algorithm to find all graphs cospectral with a given graph from $S$ is described in [CvGu1] (see also Subsection 4.2.2). However, if we know angles as well as eigenvalues, the situation is much improved; indeed, graphs in $S$ are $EA$-reconstructible. To prove this, we need the following observation:

**Lemma 4.3.6.** *Let the eigenvalues and angles of a graph $G$ be given. Given also the vertex-set of $\tilde{G}$, where $\tilde{G}$ is a union of components of $G$, we can find the eigenvalues and angles of $\tilde{G}$.*

**Proof.** For any vertex $j$ of $\tilde{G}$, the number of $j$-$j$ walks of length $k$ in $\tilde{G}$ is the same as the number of $j$-$j$ walks of length $k$ in $G$, namely $a_{jj}^{(k)} = \sum_{i=1}^{m} \alpha_{ij}^2 \mu_i^k$. Now the spectral moments of $\tilde{G}$ are $\sum_{j \in V(\tilde{G})} a_{jj}^{(k)}$ $(k = 0, 1, 2, \ldots)$, and these determine the spectrum of $\tilde{G}$. Moreover, $a_{jj}^{(k)} = \sum_{i=1}^{t} \tilde{\alpha}_{ij} \tilde{\mu}_i^k$ $(k = 0, 1, 2, \ldots)$, where $\tilde{\mu}_1, \ldots, \tilde{\mu}_t$ are the distinct eigenvalues of $\tilde{G}$ and $\tilde{\alpha}_{ij}$ is the angle of $\tilde{G}$ corresponding to $\tilde{\mu}_i$ and $j$. These equations now determine $\tilde{\alpha}_{1j}, \ldots, \tilde{\alpha}_{tj}$ $(j \in V(\tilde{G}))$. $\qquad\square$

**Theorem 4.3.7** [Cve11]. *A graph whose largest eigenvalue does not exceed 2 is characterized by its eigenvalues and angles.*

**Sketch proof.** As in the proof of Theorem 4.1.1, we first identify some components of the graph $G$ in question, then extract them and consider what remains. Suppose first that $\lambda_1(G) = 2$. The non-zero angles in the angle sequence $(\alpha_{1j})$ determine the vertices $j$ belonging to components whose index is equal to 2. By Lemma 4.3.6 we can calculate the characteristic polynomial and angles of the subgraph consisting of these components. Components isomorphic to $K_{1,4}$ can be recognized by vertices of degree 4 and then Algorithm EA shows which vertices of degree 1 are adjacent to vertices of degree 4. In a similar way we can treat components with two vertices of degree 3. Simple calculations show that if $j$ is a vertex of degree 3 in $Y_k$ $(k > 5)$ then $\alpha_{1j} = 1/\sqrt{k-3}$ and this enables us to find the length of the path connecting two such vertices of degree 3. Components isomorphic to $F_7$, $F_8$ or $F_9$ (see Fig. 3.5.) can also be easily identified using Algorithm EA as in the proof of Proposition 4.3.5. The remaining vertices are of degree 2 and they belong to cycles; for a vertex $j$ of an $s$-cycle we have $\alpha_{1j} = 1/\sqrt{s}$.

Finally, if components with index 2 are extracted or are not present, then we consider the graph $\tilde{G}$ that remains. By Lemma 4.3.6, we can determine the eigenvalues and angles of $\tilde{G}$. Each non-trivial component of $\tilde{G}$ is either a path or a tree of the type described in Proposition 4.3.5. Isolated vertices are recognized directly, while Algorithm EA identifies the paths. For what is left of the graph we apply Proposition 4.3.5. $\qquad\square$

It is possible to prove that graphs from some other classes of graphs, which are of interest in the Hückel theory, are characterized up to isomorphism by eigenvalues and angles (cf. [CvFRS]).

# Exercises

**4.1** Show that graphs with four vertices are characterized by their spectra.

**4.2** Show that a connected graph with just two distinct eigenvalues is complete.

**4.3** Show that the Petersen graph is characterized by its spectrum.

**4.4** Show that the complement of the Clebsch graph is the unique graph with spectrum $5, 1^{10}, (-3)^5$.

**4.5** Prove Theorem 4.1.3.

**4.6** Use Theorem 3.1.11 and Corollary 2.3.3 to prove that the path $P_n$ is characterized by its spectrum.

**4.7** Deduce the following from Equation (4.7): $r \leq 17$ when $n = 28$ and $r \leq 16$ when $n \leq 27$.

**4.8** A *partial graph* of a graph $G$ is the union of some components of $G$. A SING $S$ is *weakly reducible* if there is a graph $H$ such that any graph in $S$ contains a partial graph cospectral with $H$. If $G$ is not weakly reducible it is called *strongly irreducible*. Prove that a SING is irreducible and strongly irreducible if it contains a connected graph.

**4.9** We say that a SING $\mathcal{P}$ is *relevant* to the SING $\mathcal{S}$, denoted by $\mathcal{P} \vdash \mathcal{S}$, if some graph $G$ in $\mathcal{P}$ is cospectral with a partial graph of a graph in $\mathcal{S}$. (Any such graph $G$ is called a *basis* of $(\mathcal{P}, \mathcal{S})$.) Prove that If $\mathcal{P} \vdash \mathcal{S}$ and $\mathcal{S} \vdash \mathcal{P}$, then $\mathcal{P}$ and $\mathcal{S}$ are cospectral.

**4.10** If $\mathcal{P} \vdash \mathcal{S}$, then for any basis $G$ of $(\mathcal{S}, \mathcal{P})$ (see Exercise 4.9) we can define the *expansion* $E(\mathcal{S}, \mathcal{P}, G)$ of $\mathcal{S}$ by $\mathcal{P}$ through the basis $G$ as follows. The graphs in $E(\mathcal{S}, \mathcal{P}, G)$ are those obtained from $\mathcal{S}$ by replacing a partial graph isomorphic to $G$ with a graph from $\mathcal{P}$. Prove that the set $E(\mathcal{S}, \mathcal{P}, G)$ is a SING.

# Notes

The problem of characterizing graphs with least eigenvalue $-2$ was one of the earliest problems in the theory of graph spectra. As we saw in Section 3.4, the problem was essentially settled by Cameron, Goethals, Seidel and Shult [CamGSS], who established a link between such graphs and the theory of root systems. Every exceptional graph is an induced subgraph of one of 473 maximal exceptional graphs initially found with the aid of a computer [CvLRS2]; the underlying theory is described in Section 5.4, and full details are given in the monograph [CvRS7]. A refinement of Theorem 4.1.17, also with a computer-free proof, was given by Brouwer, Cohen and Neumaier [BroCN, Theorem 3.12.2]; their result appears as Theorem 4.1.5 of [CvRS7].

Few spectral characterizations emerged in the 1980s and 1990s. Early results on cospectral graphs were surveyed in the thesis [Cve1] of 1971. Another review of cospectral graphs appeared in the same year, written by Harary, King, Mowshowitz and Read [HarKMR]. A third review of cospectral graphs in 1971 appeared in the paper [BalHa], which gives a PING consisting of trees on 12 vertices with the same degrees, the maximum degree being 4. Since these trees are relevant to chemistry the authors justify in this way the main message of the paper, expressed by its title: *the characteristic polynomial does not uniquely determine the topology of a molecule.*

The expository article [GoHMK] contains a list of smallest PINGs in various classes of graphs. We have restricted our attention to undirected graphs without loops or multiple edges. It is relatively easy to construct PINGs for other kinds of graphs. For example, all digraphs without cycles have a spectrum containing only numbers equal to zero [Sed].

Concerning the result of Schwenk [Sch1] that almost all trees have a cospectral mate, his construction of cospectral graphs uses not only cospectral vertices but also the notion of *unrestricted vertices*: these are vertices at which arbitrary graphs may be attached without destroying vertex-cospectrality. Both concepts feature in general procedures for constructing PINGs described in [HeEl2]. This paper describes methods for constructing graphs with such vertices, and discusses cospectral graphs with cospectral complements.

Graphs with cospectral vertices are called *endospectral* graphs [Ran]. From Section 4.2 we see that the study of endospectral graphs is closely related to the study of cospectral graphs. Some constructions of endospectral trees are given in [RanKl], while the endospectral trees with at most 16 vertices have been found by a computer search [KMSTKR].

Other references concerning cospectral graphs are [Ach], [Bab1], [Bab3],[Bak1], [Bens], [Cha1], [Cha2], [Chao], [Con1], [Cou], [CvGu1], [DAGT], [DinKZ], [Doo5], [Doo8], [FaGr], [GoMK2],[Hei], [Herm], [Hern1], [Hern2], [HeEl1], [Hof1],[Jia], [KoSu], [KrPa1], [KrPa2], [LiWZ], [Mey], [RanTŽ], [Sch4], [Sei1], [SimmMe], [StewMa] and [ZiTR]. Graphs cospectral with respect to the generalized adjacency matrix $yJ - A$ are discussed in [DamHK].

More spectral characterizations of line graphs appear in [Doo3], [Doo4], [Doo6] and [RaoRa]. Some spectral characterizations of distance-regular graphs may be found in [DamHa3]. For some investigations concerning complete sets of invariants, see [BalaPa], [BosMe], [Kri], [Mas], [RiMW] and [Tur2]. Characterizations of certain trees by their Laplacian spectrum may be found in [OmTa] and [WaXu].

# 5

# Structure and one eigenvalue

In Chapters 3 and 4 we have concentrated on the relation between the structure and spectrum of a graph. Here we discuss the connection between structure and a single eigenvalue, and for this the central notion is that of a star complement. In Section 5.1 we define star complements both geometrically and algebraically, and note their basic properties. In Section 5.2 we illustrate a technique for constructing and characterizing graphs by star complements. In Section 5.3 we use star complements to obtain sharp upper bounds on the multiplicity of an eigenvalue different from $-1$ or $0$ in an arbitrary graph, and in a regular graph. In Section 5.4 we describe how star complements can be used to determine the graphs with least eigenvalue $-2$, and in Section 5.5 we investigate the role of certain star complements in generalized line graphs.

## 5.1 Star complements

Let $G$ be a graph with vertex set $V(G) = \{1, \ldots, n\}$ and adjacency matrix $A$. Let $\{\mathbf{e}_1, \ldots, \mathbf{e}_n\}$ be the standard orthonormal basis of $I\!R^n$ and let $P$ be the matrix which represents the orthogonal projection of $I\!R^n$ onto the eigenspace $\mathcal{E}(\mu)$ of $A$ with respect to $\{\mathbf{e}_1, \ldots, \mathbf{e}_n\}$. Since $\mathcal{E}(\mu)$ is spanned by the vectors $P\mathbf{e}_j$ $(j = 1, \ldots, n)$ there exists $X \subseteq V(G)$ such that the vectors $P\mathbf{e}_j$ $(j \in X)$ form a basis for $\mathcal{E}(\mu)$. Such a subset $X$ of $V(G)$ is called a *star set* for $\mu$ in $G$. The terminology reflects the fact that the vectors $P\mathbf{e}_1, \ldots, P\mathbf{e}_n$ form a *eutactic star*: in general, such a star consists of vectors which are an orthogonal projection of pairwise orthogonal vectors of the same length.

**Proposition 5.1.1.** *Let $G$ be a graph with $\mu$ as an eigenvalue of multiplicity $k > 0$. The following conditions on a subset $X$ of $V(G)$ are equivalent:*

(i) $X$ *is a star set for* $\mu$;
(ii) $\mathbb{R}^n = \mathcal{E}(\mu) \oplus \mathcal{V}$, *where* $\mathcal{V} = \langle \mathbf{e}_i : i \notin X \rangle$;
(iii) $|X| = k$ *and* $\mu$ *is not an eigenvalue of* $G - X$.

**Proof.** ((i) $\Rightarrow$ (ii)) Since $\dim \mathcal{E}(\mu) = k$ and $\dim \mathcal{V} = n - k$, it suffices to show that $\mathcal{E}(\mu) \cap \mathcal{V} = \{\mathbf{0}\}$. Accordingly, let $\mathbf{x} \in \mathcal{E}(\mu) \cap \mathcal{V}$. Then $\mathbf{x} = P\mathbf{x}$ and $\mathbf{x}^\top \mathbf{e}_j = 0$ for all $j \in X$. Hence $\mathbf{x}^\top (P\mathbf{e}_j) = \mathbf{x}^\top (P^\top \mathbf{e}_j) = (P\mathbf{x})^\top \mathbf{e}_j = 0$ for all $j \in X$. Thus $\mathbf{x} \in \langle P\mathbf{e}_j : j \in X \rangle^\perp = \mathcal{E}(\mu)^\perp$ and so $\mathbf{x} = \mathbf{0}$.

((ii) $\Rightarrow$ (iii)) Suppose that $\mathbb{R}^n = \mathcal{E}(\mu) \oplus \mathcal{V}$. We consider an adjacency matrix $A$ of $G$ in the form $\begin{pmatrix} * & * \\ * & A' \end{pmatrix}$, where $A'$ is the adjacency matrix of $G - X$. Suppose that $A'\mathbf{x}' = \mu \mathbf{x}'$. If $\mathbf{y} = \begin{pmatrix} \mathbf{0} \\ \mathbf{x}' \end{pmatrix}$, then

$$A\mathbf{y} = \begin{pmatrix} * & * \\ * & A' \end{pmatrix} \begin{pmatrix} \mathbf{0} \\ \mathbf{x}' \end{pmatrix} = \begin{pmatrix} * \\ \mu\mathbf{x}' \end{pmatrix}.$$

Now let $\mathbf{x} \in \mathcal{V}$. Then $\mathbf{x}^\top$ has the form $(\mathbf{0}^\top | \mathbf{z}^\top)$, and $\mathbf{x}^\top A\mathbf{y} = \mu \mathbf{z}^\top \mathbf{x}' = \mu \mathbf{x}^\top \mathbf{y}$. Hence $(A - \mu I)\mathbf{y} \in \mathcal{V}^\perp$. On the other hand, if $\mathbf{x} \in \mathcal{E}(\mu)$, then $\mathbf{x}^\top A\mathbf{y} = \mathbf{x}^\top A^\top \mathbf{y} = (A\mathbf{x})^\top \mathbf{y} = (\mu\mathbf{x})^\top \mathbf{y} = \mu \mathbf{x}^\top \mathbf{y}$ and so $(A - \mu I)\mathbf{y} \in \mathcal{E}(\mu)^\perp$. Hence $(A - \mu I)\mathbf{y} \in \mathcal{V}^\perp \cap \mathcal{E}(\mu)^\perp = (\mathcal{E}(\mu) + \mathcal{V})^\perp$, which is the zero subspace. Therefore, $\mathbf{y} \in \mathcal{E}(\mu)$. But $\mathbf{y} \in \mathcal{V}$, and since $\mathcal{E}(\mu) \cap \mathcal{V} = \{\mathbf{0}\}$ we have $\mathbf{y} = \mathbf{0}$. Hence $\mathbf{x}' = \mathbf{0}$ and $\mu$ is not an eigenvalue of $G - X$.

((iii) $\Rightarrow$ (i)) Here, it suffices to prove that $\langle P\mathbf{e}_j : j \in X \rangle = \mathcal{E}(\mu)$. Suppose, by way of contradiction, that $\langle P\mathbf{e}_j : j \in X \rangle \subset \mathcal{E}(\mu)$. Then there is a non-zero vector $\mathbf{x} \in \mathcal{E}(\mu) \cap \langle P\mathbf{e}_j : j \in X \rangle^\perp$. Thus $\mathbf{x}^\top P\mathbf{e}_j = 0$ for all $j \in X$. Hence $(P\mathbf{x})^\top \mathbf{e}_j = (\mathbf{x}^\top P)\mathbf{e}_j = 0$ for all $j \in X$. Consequently $P\mathbf{x} \in \langle \mathbf{e}_j : j \in X \rangle^\perp = \langle \mathbf{e}_s : s \notin X \rangle = \mathcal{V}$. But $\mathbf{x} = P\mathbf{x}$ and so we have a non-zero vector $\mathbf{x} \in \mathcal{E}(\mu) \cap \mathcal{V}$. Since $\mathbf{x} = \begin{pmatrix} \mathbf{0} \\ \mathbf{x}' \end{pmatrix}$ with $\mathbf{x}' \neq \mathbf{0}$ it follows that $\mathbf{x}'$ is an eigenvector of $G - X$, a contradiction. $\qquad\square$

Here $G - X$ is the subgraph of $G$ induced by the complement of $X$; it is called the *star complement* for $\mu$ corresponding to $X$. (Star complements for $\mu$ are sometimes called $\mu$-*basic* subgraphs, as in [Ell].) It is clear from the definitions that star sets and star complements exist for any eigenvalue of any graph. Statement (iii) of Proposition 5.1.1 provides a characterization of star sets and star complements which is often the most useful in practice. For instance, the claims in the following example are easily verified in this way.

**Example 5.1.2.** In Fig. 5.1, the vertices of the Petersen graph are labelled with eigenvalues in such a way that the vertices labelled $\mu$ form a star set for $\mu$.

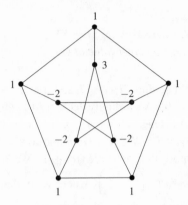

Figure 5.1 The Petersen graph (Example 5.1.2).

For example, $-2$ is an eigenvalue of multiplicity 4, and if we delete the four vertices labelled $-2$ we obtain a subgraph $H$ consisting of a 5-cycle with a single pendant edge attached. Since $H$ does not have $-2$ as an eigenvalue, this subgraph is a star complement for $-2$.                                    □

**Remark 5.1.3.** It can be shown (see [CvRS2, Chapter 7]) that if $G$ is a graph with $\mu_1, \ldots, \mu_m$ as its distinct eigenvalues then $V(G)$ has a partition $X_1 \dot{\cup} \cdots \dot{\cup} X_m$ such that $X_i$ is a star set for $\mu_i$ ($i = 1, \ldots, m$). Such a partition is called a *star partition*, and in this context the star sets $X_i$ are called *star cells*. Each star partition determines a basis for $I\!R^n$ consisting of eigenvectors of an adjacency matrix. The finite number of bases obtained in this way may be ordered lexicographically, and an extremal basis is determined uniquely by $G$. These ideas were introduced as a means of investigating the complexity of the graph isomorphism problem.                                    □

**Proposition 5.1.4.** *Let $X$ be a star set for $\mu$ in $G$, and let $\overline{X} = V(G) \setminus X$.*
*(i) If $\mu \neq 0$ then $\overline{X}$ is a dominating set for $G$;*
*(ii) If $\mu \neq -1$ or 0 then $\overline{X}$ is a location-dominating set for $G$ – that is, the $\overline{X}$-neighbourhoods of distinct vertices in $X$ are distinct and non-empty.*

**Proof.** The matrix $P$, which represents the orthogonal projection of $I\!R^n$ onto $\mathcal{E}(\mu)$, is a polynomial function of the adjacency matrix $A$ (see Section 1.1), and so $AP = PA$. For each vertex $u$ of $G$ we have

$$\mu P\mathbf{e}_u = AP\mathbf{e}_u = PA\mathbf{e}_u = P(\sum_{i \sim u} \mathbf{e}_i) = \sum_{i \sim u} P\mathbf{e}_i. \tag{5.1}$$

For part (i), we have to show that any vertex $u$ in $X$ is adjacent to a vertex in $\overline{X}$. Since $\mu \neq 0$, we know from Equation (5.1) that the vectors in $\{P\mathbf{e}_u\} \cup \{P\mathbf{e}_i : i \sim u\}$ are linearly dependent. Since the vectors $P\mathbf{e}_j$ ($j \in X$) are linearly independent, it follows that there is a vertex adjacent to $u$ which lies outside $X$.

For part (ii), let $\Gamma(u)$ be the set of neighbours of $u$ in $X$. Suppose by way of contradiction that $u$ and $v$ are vertices in $X$ with the same neighbourhoods in $\overline{X}$. From equation (5.1) and its counterpart for $v$ we have

$$\mu P\mathbf{e}_u - \mu P\mathbf{e}_v - \sum_{j \in \Gamma(u)} P\mathbf{e}_j + \sum_{j \in \Gamma(v)} P\mathbf{e}_j = \mathbf{0}.$$

This is a relation on vectors in $\{P\mathbf{e}_j : j \in X\}$. Since these vectors are linearly independent, it follows that either (a) $\mu = 0$, $u \not\sim v$ and $u, v$ have the same neighbourhoods in $G$, or (b) or $\mu = -1$, $u \sim v$ and $u, v$ have the same closed neighbourhood in $G$, contrary to assumption. $\qquad\square$

In case (a) above, $u$ and $v$ are called *duplicate vertices*, and in case (b), *co-duplicate vertices*.

It follows from Proposition 5.1.4(ii) that there are only finitely many graphs with a prescribed star complement for an eigenvalue $\mu \neq 0$ or $-1$, for if $|\overline{X}| = t$ then $|X| < 2^t$. This exponential bound will be improved to a quadratic bound in the next section. If $\mu = 0$ or $-1$ then $|X|$ cannot be bounded by a function of $t$: this can be seen by considering $K_2 \,\dot\cup\, (n-2)K_1$ (with $\mu = 0$) or $K_n$ (with $\mu = -1$). Alternatively, when $\mu = 0$ or $-1$ we can add arbitrarily many duplicate or co-duplicate vertices: this corresponds to repeating rows (and corresponding columns) of $A - \mu I$ without increasing the rank of $A - \mu I$. It can be shown that if $\mu \in \{-1, 0\}$ and $G$ has no duplicate or co-duplicate vertices then $n$ is at most $O(2^{t/2})$ (see [KotLo]).

It follows from Proposition 5.1.4(i) that if $\mu \neq 0$ and $G - X$ is connected then $G$ is connected. In the reverse direction, a connected graph always has a connected star complement for each eigenvalue. In fact we can establish a stronger result (Theorem 5.1.6), and to prove it, we require the following observation.

**Lemma 5.1.5.** *If the column space of the symmetric matrix* $\begin{pmatrix} C & D^\top \\ D & E \end{pmatrix}$ *has the columns of* $\begin{pmatrix} C \\ D \end{pmatrix}$ *as a basis, then the columns of $C$ are linearly independent.*

**Proof.** Since each column of $\begin{pmatrix} D^\top \\ E \end{pmatrix}$ is a linear combination of the columns of $\begin{pmatrix} C \\ D \end{pmatrix}$, there exists a matrix $L$ such that $D^\top = CL$, equivalently $D = L^\top C$. Thus if $C\mathbf{x} = \mathbf{0}$ then $\begin{pmatrix} C \\ D \end{pmatrix} \mathbf{x} = \mathbf{0}$, whence $\mathbf{x} = \mathbf{0}$ as required. $\square$

**Theorem 5.1.6.** *Let $\mu$ be an eigenvalue of the connected graph $G$, and let $K$ be a connected induced subgraph of $G$ not having $\mu$ as an eigenvalue. Then $G$ has a connected star complement for $\mu$ containing $K$.*

**Proof.** Let $|V(K)| = r$. Since $G$ is connected we may label its vertices $1, \ldots, n$ so that each vertex after the first is adjacent to a predecessor. Since $K$ is connected we may take $1, \ldots, r$ to be the vertices of $K$. Let $A$ be the adjacency matrix of $G$, with columns $\mathbf{c}_1, \ldots, \mathbf{c}_n$, and let $\{\mathbf{c}_k : k \in Y\}$ be the basis of the column space of $\mu I - A$ obtained by deleting each column which is a linear combination of its predecessors. Note that $\{1, \ldots, r\} \subseteq Y$ because $\mu$ is not an eigenvalue of $K$. By Lemma 5.1.5, the principal submatrix of $\mu I - A$ determined by $Y$ is invertible. Since $|Y| = \operatorname{codim} \mathcal{E}(\mu)$, $\overline{Y}$ is a star set for $\mu$ and the subgraph $H$ induced by $Y$ is a star complement for $\mu$.

We prove that $H$ is connected by showing that each vertex $y$ of $Y$ with $y > 1$ is adjacent to a previous vertex $j$ of $Y$. We take $j$ to be the least element of $\{1, \ldots, n\}$ such that $j$ is adjacent to $y$ in $G$. Then $j < y$ and the $y$-th entry of $\mathbf{c}_j$ is $-1$. On the other hand, the $y$-th entry of each $\mathbf{c}_i$ ($i < j$) is 0, and so $\mathbf{c}_j$ is not a linear combination of its predecessors. Thus $j \in Y$ as required. $\square$

The next result, which establishes the basic property of star complements, is known as the Reconstruction Theorem and its converse.

**Theorem 5.1.7.** *Let $X$ be a set of $k$ vertices in the graph $G$, and suppose that $G$ has adjacency matrix $\begin{pmatrix} A_X & B^\top \\ B & C \end{pmatrix}$, where $A_X$ is the adjacency matrix of the subgraph induced by $X$. Then $X$ is a star set for $\mu$ in $G$ if and only if $\mu$ is not an eigenvalue of $C$ and*

$$\mu I - A_X = B^\top (\mu I - C)^{-1} B. \tag{5.2}$$

*In this situation, the eigenspace of $\mu$ consists of the vectors $\begin{pmatrix} \mathbf{x} \\ (\mu I - C)^{-1} B\mathbf{x} \end{pmatrix}$, where $\mathbf{x} \in \mathbb{R}^k$.*

**Proof.** Suppose first that $X$ is a star set for $\mu$. Then $\mu$ is not an eigenvalue of $C$, and we have

$$\mu I - A = \begin{pmatrix} \mu I - A_X & -B^\top \\ -B & \mu I - C \end{pmatrix},$$

where $\mu I - C$ is invertible. In particular, if $|V(G)| = n$ then the matrix $(-B \mid \mu I - C)$ has rank $n - k$; but $\mu I - A$ also has rank $n - k$ and so the rows of $(-B \mid \mu I - C)$ form a basis for the row space of $\mu I - A$. Hence there exists a $k \times (n - k)$ matrix $L$ such that $(\mu I - A_X \mid -B^\top) = L(-B \mid \mu I - C)$. Now $\mu I - A_X = -LB$, $-B^\top = L(\mu I - C)$ and Equation (5.2) follows by eliminating $L$.

Conversely, if $\mu$ is not an eigenvalue of $C$ and Equation (5.2) holds, then it is straightforward to verify that the vectors specified lie in $\mathcal{E}(\mu)$. They form a $k$-dimensional space, and, by interlacing, the multiplicity of $\mu$ is exactly $k$. Hence $X$ is a star set for $\mu$. □

Note that if $X$ is a star set for $\mu$ then the corresponding star complement $H(= G - X)$ has adjacency matrix $C$, and Equation (5.2) tells us that $G$ is determined by $\mu$, $H$ and the $H$-neighbourhoods of vertices in $X$. If $\mu \neq -1$ or $0$ then by Proposition 5.1.4(ii), there is a one-one correspondence between the vertices in $X$ and their $H$-neighbourhoods. To find all the graphs with a prescribed star complement for $\mu$, we have to find all solutions $A_X$, $B$ of Equation (5.2), given $\mu$ and $C$. In this situation, let $|V(H)| = t$ and define a bilinear form on $\mathbb{R}^t$ by

$$\langle\!\langle \mathbf{x}, \mathbf{y} \rangle\!\rangle = \mathbf{x}^\top (\mu I - C)^{-1} \mathbf{y} \quad (\mathbf{x}, \mathbf{y} \in \mathbb{R}^t).$$

If we denote the columns of $B$ by $\mathbf{b}_u$ $(u \in X)$ and equate matrix entries in Equation (5.2), we obtain the following consequence of Theorem 5.1.7.

**Corollary 5.1.8.** *Suppose that $\mu$ is not an eigenvalue of the graph $H$, where $|V(H)| = t$. There exists a graph $G$ with a star set $X$ for $\mu$ such that $G - X = H$ if and only if there exist $(0, 1)$-vectors $\mathbf{b}_u$ $(u \in X)$ in $\mathbb{R}^t$ which satisfy*

(i) $\langle\!\langle \mathbf{b}_u, \mathbf{b}_u \rangle\!\rangle = \mu$ *for all* $u \in X$, *and*
(ii) $\langle\!\langle \mathbf{b}_u, \mathbf{b}_v \rangle\!\rangle \in \{-1, 0\}$ *for all pairs* $u, v$ *in* $X$.

In this situation, $u \sim v$ when $\langle\!\langle \mathbf{b}_u, \mathbf{b}_v \rangle\!\rangle = -1$ and $u \not\sim v$ when $\langle\!\langle \mathbf{b}_u, \mathbf{b}_v \rangle\!\rangle = 0$.

## 5.2 Construction and characterization

In this section we give four basic examples to illustrate the use of Corollary 5.1.8 in constructing graphs with a prescribed star complement $H$ for a prescribed eigenvalue $\mu$. Note that if only $H$ is prescribed then there are only finitely many possibilities for $\mu$: they can be identified from all possible one-vertex extensions of $H$, as in Example 5.2.8 below. The examples serve to illustrate how star complements have been used in the literature to characterize certain graphs. In practice it is often convenient to write Equation (5.2) in the form

$$m(\mu)(\mu I - A_X) = B^\top m(\mu)(\mu I - C)^{-1} B \tag{5.3}$$

where $m(x)$ is the minimal polynomial of $C$. This is because $m(\mu)(\mu I - C)^{-1}$ is given explicitly as follows. The proof is left to the reader.

**Proposition 5.2.1.** *Let $C$ be a square matrix with minimal polynomial*

$$m(x) = x^{d+1} + c_d x^d + c_{d-1} x^{d-1} + \cdots + c_1 x + c_0.$$

*If $\mu$ is not an eigenvalue of $C$ then*

$$m(\mu)(\mu I - C)^{-1} = a_d C^d + a_{d-1} C^{d-1} + \cdots + a_1 C + a_0 I$$

*where $a_d = 1$ and for $0 < i \le d$,*

$$a_{d-i} = \mu^i + c_d \mu^{i-1} + c_{d-1} \mu^{i-2} + \cdots + c_{d-i+1}.$$

If $G$ has $H$ as a star complement for $\mu$, with a corresponding star set $X$ of size $k$, then the deletion of any $r$ vertices in $X$ results in a graph with $\mu$ as an eigenvalue of multiplicity $k - r$. The reason is that the multiplicity of an eigenvalue changes by 1 at most when any vertex is deleted (see Corollary 1.3.12). It follows that each induced subgraph $G - Y$ ($Y \subset X$) also has $H$ as a star complement for $\mu$. Moreover any graph with $H$ as a star complement for $\mu$ is an induced subgraph of such a graph $G$ for which $X$ is maximal, because $H$-neighbourhoods determine adjacencies among vertices in a star set. Accordingly, in determining all the graphs with $H$ as a star complement for $\mu$, it suffices to describe those for which a star set $X$ is maximal. By Proposition 5.1.4(ii), such maximal graphs always exist when $\mu \ne -1$ or $0$.

**Example 5.2.2.** We begin with the simple problem of finding the graphs that have a 5-cycle 123451 as a star complement $H$ for $-2$. In the notation of Proposition 5.2.1, $C$ is the circulant matrix with first row 01001, $\mu = -2$ and $m(x) = (x - 2)(x^2 + x - 1)$. Here $m(\mu) = -4$ and the proposition yields

$$4(2I + C)^{-1} = C^2 - 3C + 3I = \begin{pmatrix} 5 & -3 & 1 & 1 & -3 \\ -3 & 5 & -3 & 1 & 1 \\ 1 & -3 & 5 & -3 & 1 \\ 1 & 1 & -3 & 5 & -3 \\ -3 & 1 & 1 & -3 & 5 \end{pmatrix}.$$

Now we apply Corollary 5.1.8(i). From Equation (5.3) we know that $\langle\!\langle \mathbf{b}_u, \mathbf{b}_u \rangle\!\rangle = -2$ if and only if $\mathbf{b}_u^\top (C^2 - 3C + 3I)\mathbf{b}_u = 8$. In this situation the neighbours of $u$ in $H$ constitute a set $S$ such that the $i$-th entry of $\mathbf{b}_u$ is 1 if $i \in S$, 0 if $i \notin S$. Accordingly we have to find the subsets $S$ of $\{1, 2, 3, 4, 5\}$ such that the sum of entries in the principal submatrix of $C^2 - 3C + 3I$

determined by $S$ is equal to 8. It is straightforward to verify that this occurs precisely when $|S| = 4$. All five possiblities for $S$ occur simultaneously in $L(K_5)$, which is therefore the unique maximal graph that arises. The graphs with a 5-cycle as a star complement for $-2$ are therefore the induced subgraphs of $L(K_5)$ containing $C_5$. Since $C_5 = L(C_5)$, these graphs are just the graphs $L(G)$, where $G$ is a Hamiltonian graph on five vertices. $\square$

The arguments of Example 5.2.2 can be generalized to show that for any odd $t > 3$, $L(K_t)$ is the unique maximal graph with a $t$-cycle as a star complement for $-2$. Determination of the possible subsets $S$ requires substantial effort in the general case. An inspection of $L(K_t)$ reveals easily that such sets include those consisting of two pairs of consecutive vertices on the $t$-cycle, and the work lies in proving that there are no other possibilities for $S$. The graphs in which the path $P_t$ is a star complement for $-2$ have also been determined: when $t \geq 3$ and $t \neq 7, 8$, such graphs are precisely the line graphs of bipartite graphs with $t + 1$ vertices (other than $P_{t+1}$) which have a Hamiltonian path.

In Example 5.2.2, there was no need to apply part (ii) of Corollary 5.1.8 because we had prior knowledge of a graph in which all possible vertices were added to the prescribed star complement. We cannot expect that a unique maximal graph always exists, and in the general case, where a graph $H$ occurs as a star complement for an eigenvalue $\mu$, it is useful to consider a *compatibility graph* defined as follows. The vertices are those $\mathbf{b}_u$ for which $\langle\!\langle \mathbf{b}_u, \mathbf{b}_u \rangle\!\rangle = \mu$, and $\mathbf{b}_u$ is adjacent to $\mathbf{b}_v$ if and only if $\langle\!\langle \mathbf{b}_u, \mathbf{b}_v \rangle\!\rangle \in \{-1, 0\}$. It is convenient to represent the edge $\mathbf{b}_u \mathbf{b}_v$ by a full line if $\langle\!\langle \mathbf{b}_u, \mathbf{b}_v \rangle\!\rangle = -1$, and by a broken line if $\langle\!\langle \mathbf{b}_u, \mathbf{b}_v \rangle\!\rangle = 0$. If each vertex $\mathbf{b}_u$ is labelled instead with the $H$-neighbourhood of $u$, then this same graph is called the *extendability graph* $\Gamma(H, \mu)$. Note that when $\mu \neq -1$ or $0$ there is a one-one correspondence between cliques in $\Gamma(H, \mu)$ and graphs with $H$ as a star complement for $\mu$; moreover, the full lines in a clique determine the subgraph induced by the corresponding star set. In particular, if we use a computer to find the maximal graphs with $H$ as a star complement for $\mu$, we can invoke an algorithm for finding the maximal cliques in a graph. The next example illustrates the procedure in a small case.

**Example 5.2.3.** Here we find the graphs having a 5-cycle 123451 as a star complement $H$ for 1. In this case, Proposition 5.1.11 yields

$$(I - C)^{-1} = 3I - C^2 = \begin{pmatrix} 1 & 0 & -1 & -1 & 0 \\ 0 & 1 & 0 & -1 & -1 \\ -1 & 0 & 1 & 0 & -1 \\ -1 & -1 & 0 & 1 & 0 \\ -0 & -1 & -1 & 0 & 1 \end{pmatrix}.$$

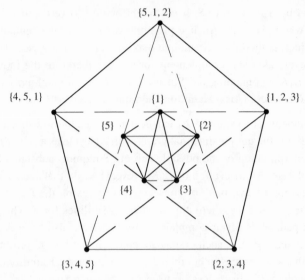

Figure 5.2 The extendability graph $\Gamma(C_5, 1)$.

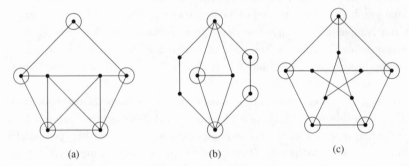

Figure 5.3 The maximal graphs with $C_5$ as a star complement for 1.

First, we apply Corollary 5.1.8(i). From Equation (5.2) we know that $\langle\langle \mathbf{b}_u, \mathbf{b}_u \rangle\rangle = 1$ if and only if $\mathbf{b}_u^\top(3I - C^2)\mathbf{b_u} = 1$. Now we have to find the subsets $S$ of $\{1, 2, 3, 4, 5\}$ such that the sum of entries in the principal submatrix of $3I - C^2$ determined by $S$ is equal to 1. It is straightforward to verify that this occurs if and only if $S$ consists of a single vertex or three consecutive vertices of the 5-cycle. Next we apply part (ii) of Corollary 5.1.8 to construct the extendability graph $\Gamma(C_5, 1)$ shown in Fig. 5.2. The automorphism group of $\Gamma(C_5, 1)$ has three orbits of maximal cliques (with 2, 3 and 5 vertices). These determine the three maximal graphs illustrated in Fig. 5.3, where the vertices of $H$ are circled. The Petersen graph has already featured in

Example 5.2.2. Alternatively, its occurrence here could have been predicted from Example 5.1.2, where $C_5$ is a star complement for $-2$ in $L(K_5)$: since $-2$ is not a main eigenvalue of $L(K_5)$, we deduce from Proposition 2.1.3 that $\overline{C_5}$ is a star complement for $-(-2) - 1$ in $\overline{L(K_5)}$ – that is, $C_5$ is a star complement for 1 in the Petersen graph.  □

For characterizations among regular graphs, the following result is very useful in restricting the vectors $\mathbf{b}_u$ that need to be considered.

**Proposition 5.2.4** [CvRS3]. *If $\mu$ is a non-main eigenvalue of G then, in the notation of Corollary 5.1.8,*

$$\langle\!\langle \mathbf{b}_u, \mathbf{j} \rangle\!\rangle = -1 \quad \text{for all} \quad u \in X.$$

**Proof.** Here the all-1 vector in $\mathbb{R}^n$ is orthogonal to $\mathcal{E}(\mu)$. From the specification of $\mathcal{E}(\mu)$ in Theorem 5.1.7 we deduce that $\langle\!\langle \mathbf{b}_u, \mathbf{j} \rangle\!\rangle = -1$ for all $u \in X$.  □

**Example 5.2.5.** Proposition 5.2.4 tells us that if $G$ is $r$-regular and $\mu \neq r$ then, for each $u \in X$, $-1$ is the sum of entries in the columns of $(\mu I - C)^{-1}$ indexed by the $H$-neighbourhood of $u$. Thus if we restrict $G$ to be regular in Example 5.2.3, the only candidates for an $H$-neighbourhood are the singletons of $V(H)$. It follows that the Petersen graph is characterized among regular graphs by a 5-cycle as a star complement for 1.  □

The procedures illustrated above are known collectively as the *star complement technique* for constructing and characterizing graphs with a prescribed star complement for a prescribed eigenvalue. We give a further example of the technique which illustrates the purely combinatorial nature of the arguments when the star complement $H$ is essentially devoid of structure. Here the $H$-neighbourhood of a vertex $u$ is denoted by $\Delta_H(u)$.

**Example 5.2.6.** Suppose that $K_8$ is a star complement $H$ for $-2$. In this situation we have $C = J - I$, $m(x) = (x + 1)(x - 7)$ and Equation (5.3) becomes

$$9(2I + A_X) = B^\top(9I - J)B.$$

Equating $(u, u)$-entries here, we have $18 = 9h - h^2$, where $h = |\Delta_H(u)|$. Hence $h = 3$ or $6$. Equating non-diagonal entries, we find that the following conditions on $H$-neighbourhoods are necessary and sufficient for the simultaneous addition of two vertices $u$ and $v$:

if $|\Delta_H(u)| = |\Delta_H(v)| = 3$ then $|\Delta_H(u) \cap \Delta_H(v)| \in \{1, 2\}$;

if $|\Delta_H(u)| = 3$ and $|\Delta_H(v)| = 6$ then $|\Delta_H(u) \cap \Delta_H(v)| \in \{2, 3\}$;

if $|\Delta_H(u)| = |\Delta_H(v)| = 6$ then $|\Delta_H(u) \cap \Delta_H(v)| \in \{4, 5\}$.

Note that the third condition is satisfied automatically because there $\Delta_H(u)$ and $\Delta_H(v)$ are 6-subsets of an 8-set. For the maximal graphs $G$ with $K_8$ as a star complement for $-2$, we need to find the maximal families of 3-sets and 6-sets satisfying the other two conditions. We give just three of many examples of such a family $\mathcal{F}$.

(a) $\mathcal{F}$ consists of all 28 subsets of $V(H)$ of size 6; in this case, the maximal graph $G$ is the graph obtained from $L(K_9)$ by switching with respect to $K_8$.

(b) $\mathcal{F}$ consists of all 21 subsets of size 3 containing a fixed vertex of $H$; in this case, $G$ is the cone over $L(K_8)$.

(c) $\mathcal{F}$ consists of all 7 subsets of size 6 not containing a fixed vertex $w$ of $H$, together with 7 subsets of size 3 which form the lines of a geometry $PG(3, 2)$ on $V(H) \setminus \{w\}$; in this case, $G$ is the unique smallest maximal graph that arises. $\qquad \square$

In order to describe the general form of a maximal family of neighbourhoods in Example 5.2.6, we give some further definitons. Suppose that $\mathcal{F}$ is a family of 3-subsets of $\{1, 2, \ldots, 8\}$, and let $\mathcal{F}^{(2)}$ be the family of 2-sets which are contained in some 3-set of $\mathcal{F}$. We say that $\mathcal{F}$ is an *intersecting* family if $U \cap V \neq \emptyset$ for all $U, V \in \mathcal{F}$; and such a family $\mathcal{F}$ is *complete* if there does not exist an intersecting family of 3-sets $\mathcal{F}_0$ such that $\mathcal{F} \subset \mathcal{F}_0$ and $\mathcal{F}^{(2)} = \mathcal{F}_0^{(2)}$. (For example, if $\mathcal{F} = \{138, 157, 568\}$ then $\mathcal{F}$ is not complete because we can take $\mathcal{F}_0 = \mathcal{F} \cup \{158\}$.) The final result of this section shows that a maximal exceptional graph with $K_8$ as a star complement for $-2$ is determined by a complete intersecting family of 3-subsets of $\{1, 2, \ldots, 8\}$, and *vice versa*. Here we take $V(H) = \{1, 2, \ldots, 8\}$ and write $\overline{ij}$ for the complement of $\{i, j\}$ in $V(H)$.

**Theorem 5.2.7** [Row14]. *Let $G$ be a graph with $K_8$ as a star complement for $-2$, say $H = G - X \cong K_8$. Then $G$ is a maximal exceptional graph if and only if the family of $H$-neighbourhoods $\Delta_H(u)$ ($u \in X$) has the form $\mathcal{F}_3 \cup \mathcal{F}_6$ where $\mathcal{F}_3$ is a complete intersecting family of 3-sets and $\mathcal{F}_6 = \{\overline{ij} : ij \notin \mathcal{F}_3^{(2)}\}$.*

**Proof.** First suppose that $G$ is a maximal exceptional graph, and let $\mathcal{F}_3$ be the family of $H$-neighbourhoods of size 3. From the remarks above we know that $\mathcal{F}_3$ is an intersecting family. If $ij \notin \mathcal{F}_3^{(2)}$ then the 6-set $\overline{ij}$ intersects each member of $\mathcal{F}_3$ in at least two elements. Now the maximality of $X$ ensures first that the $H$-neighbourhoods include every such 6-set, and secondly that $\mathcal{F}_3$ is complete.

Conversely, if the family of $H$-neighbourhoods has the form given then $X$, and hence $G$, is maximal. Moreover $G$ is exceptional because a graph obtained from $K_8$ by adding a vertex of degree 3 or 6 is itself exceptional. $\qquad \square$

In Example 5.2.6(a), $\mathcal{F}_3 = \emptyset$, and in Example 5.2.6(b), $\mathcal{F}_6 = \emptyset$. It has been shown by computer that there are exactly 363 maximal graphs with $K_8$ as a star complement for $-2$ [Lep2]; all are maximal exceptional graphs as defined in Section 5.4.

We complete this section with two examples which illustrate the situation in which a star complement is prescribed but an associated eigenvalue is not. Recall that a graph $H$ can be a star complement for only finitely many values of $\mu$, since then $\mu$ is an eigenvalue of a graph obtained from $H$ by adding a vertex. In our illustrations (and in Exercise 5.7) the star complement is a complete bipartite graph, and so we begin with some remarks on the general case $H \cong K_{r,s}$ $(r + s > 2)$.

If $V(H)$ has the bipartition $R \cup S$, where $|R| = r$ and $|S| = s$, then we say that a vertex $u$ added to $H$ is of *type* $(a, b)$ if the $H$-neighbourhood $\Delta_H(u)$ of $u$ consists of $a$ vertices in $R$ and $b$ vertices in $S$. If $H$ has adjacency matrix $C$ then $C$ has minimal polynomial $x(x^2 - rs)$, and $\mu(\mu^2 - rs)(\mu I - C)^{-1} = (\mu^2 - rs)I + \mu C + C^2$ by Proposition 5.2.1. Thus $\mu(\mu^2 - rs) \neq 0$ and we may write Equation (5.2) in the form

$$\mu(\mu^2 - rs)(\mu I - A_X) = B^\top \{(\mu^2 - rs)I + \mu C + C^2\}B. \qquad (5.4)$$

Now suppose that $u, v$ are distinct vertices in $X$ of types $(a, b), (c, d)$ respectively. If we let $A_X = (a_{ij})$ and equate $(u, v)$-entries in Equation (5.4) we obtain

$$-\mu(\mu^2 - rs)a_{uv} = (\mu^2 - rs)\rho_{uv} + \mu(ad + bc) + acs + bdr, \qquad (5.5)$$

where $\rho_{uv} = |\Delta_H(u) \cap \Delta_H(v)|$.

**Theorem 5.2.8.** *If $G$ is a graph with $K_{1,5}$ as a star complement for some multiple eigenvalue $\mu \neq -1$ then $\mu = 1$ and $\overline{G}$ is an induced subgraph of the Clebsch graph.*

**Proof.** We write $H + u$ for the subgraph induced by $\overline{X} \cup \{u\}$. Since $\mu \neq 0$, $H + u$ is connected by Proposition 5.1.4(i). From the spectra [CvDGT] of the 11 connected graphs $H + u$ (corresponding to the 11 possible types $(a, b) \neq (0, 0)$) we find that the only non-zero eigenvalue common to non-isomorphic graphs is $-1$. Thus $\mu$ arises as a multiple eigenvalue only when all vertices in $X$ are of the same type, $(a, b)$ say. In this situation, if we equate diagonal and non-diagonal entries in Equation (5.4) we obtain

$$\mu^2(\mu^2 - 5) = (\mu^2 - 5)(a + b) + 2\mu ab + 5a^2 + b^2 \qquad (5.6)$$

and

$$-\mu(\mu^2 - 5)a_{uv} = (\mu^2 - 5)\rho_{uv} + 2\mu ab + 5a^2 + b^2. \qquad (5.7)$$

On subtracting Equation (5.7) from Equation (5.6), and dividing by $\mu^2 - 5$, we obtain $\mu^2 + \mu a_{uv} = a + b - \rho_{uv}$. Note that $a + b - \rho_{uv} \in \{1, 2\}$. Suppose by way of contradiction that $\mu$ is not an integer. If $\mu^2 \notin \mathbb{Z}$ then $a_{uv} = 1$, $a + b - \rho_{uv} = 1$ and $\mu = \frac{1}{2}(-1 \pm \sqrt{5})$; if $\mu^2 \in \mathbb{Z}$ then $a_{uv} = 0$ and $\mu = \pm\sqrt{2}$. But none of the 11 graphs $H + u$ has $\pm\sqrt{2}$ or $\frac{1}{2}(-1 \pm \sqrt{5})$ as an eigenvalue. Accordingly $\mu \in \mathbb{Z}$. The only integer other than $-1$ or $0$ to be found among the eleven spectra is 1, which arises as an eigenvalue of $H + u$ precisely when $u$ is of type $(0, 2)$. Now the solutions of Equation (5.7) are given by $(a_{uv}, \rho_{uv}) \in \{(0, 1), (1, 0)\}$. Thus *all* $\binom{1}{0}\binom{5}{2} = 10$ possible vertices may be included in $X$: in this case $X$ induces a Petersen graph because $u \sim v$ if and only if the 2-element subsets $\Delta_H(u)$, $\Delta_H(v)$ of the 5-element set of endvertices in $H$ are disjoint. The 16-vertex graph so obtained is necessarily the complement of the Clebsch graph, since a strongly regular graph with parameters $(16, 5, 0, 2)$ has $K_{1,5}$ as a star complement for 1. Note that foreknowledge of this example shows that $\Gamma(K_{1,5}, 1) \cong K_{10}$ and obviates the need to solve Equation (5.2). We conclude that the complement of the Clebsch graph is the unique maximal graph with $K_{1,5}$ as a star complement for a multiple eigenvalue different from $-1$. $\qquad\square$

The last theorem of this section illustrates the use of Proposition 5.2.4.

**Theorem 5.2.9.** *Let $G$ be an $r$-regular graph with $n$ vertices. If $G$ has $K_{1,s}$ ($s > 1$) as a star complement for $\mu$ then one of the following holds:*

(a) $\mu = \pm 2$, $r = s = 2$ *and $H$ is a 4-cycle;*
(b) $\mu = \frac{1}{2}(-1 \pm \sqrt{5})$, $r = s = 2$ *and $H$ is a 5-cycle;*
(c) $\mu \in \mathbb{N}$, $r = s$ *and $G$ is strongly regular with parameters $((\mu^2 + 3\mu)^2, \mu(\mu^2 + 3\mu + 1), 0, \mu(\mu + 1))$.*

**Proof.** By Proposition 5.1.4(i), $G$ is connected since $\mu \neq 0$. If $\mu = r$ then $n = 4$ by Corollary 1.3.8, and we have $r = s = 2$, $G = C_4$. Accordingly we suppose that $\mu \neq r$ and consider a vertex in $X$ of type $(a, b) \neq (0, 0)$; note that $a^2 = a$. From Proposition 5.2.4 we have

$$-\mu(\mu^2 - s) = a\mu^2 + a\mu s + b\mu^2 + b\mu, \tag{5.8}$$

and from Corollary (5.1.8) we have

$$\mu^2(\mu^2 - s) = a\mu^2 + 2\mu ab + b^2 + (\mu^2 - s)b. \tag{5.9}$$

Equations (5.8) and (5.9) yield just two possibilities:

$$a = 0, \ b = \mu^2 + \mu \neq 0, \ s = \mu(\mu^2 + 3\mu + 1) \text{ or } a = 1, \ \mu = -1, \ b \in \{1, s\}.$$

Thus if $\mu = -1$ then the central vertex of $H$ is adjacent to all other vertices, and this contradicts the regularity of $G$ since other vertices of $H$ have degree less than $n-1$. It follows that $\mu \neq -1$ and the central vertex of $H$ is adjacent to no vertices in $X$; in particular, $r = s = \mu(\mu^2 + 3\mu + 1)$. All vertices in $X$ are of type $(0, \mu^2 + \mu)$, and counting in two ways the edges between $X$ and $H$ we have

$$|X|(\mu^2 + \mu) = \mu(\mu^2 + 3\mu + 1)(\mu^3 + 3\mu^2 + \mu - 1),$$

whence $|X| = (\mu^2 + 3\mu + 1)(\mu^2 + 2\mu - 1)$ and $n = |X| + s + 1 = (\mu^2 + 3\mu)^2$. From Equation (5.7) we have

$$\rho_{uv} = \begin{cases} 0 \text{ if } u \sim v \\ \mu \text{ if } u \not\sim v \end{cases}. \tag{5.10}$$

If $X$ induces a clique then $|X| - 1 = r - \mu^2 - \mu$, whence

$$(\mu + 1)(\mu + 2)(\mu^2 + \mu - 1) = 0.$$

Therefore, either $\mu = -2$ and we have case (a), or $\mu = \frac{1}{2}(-1 \pm \sqrt{5})$ and we have case (b). If $X$ does not induce a clique then it follows from (5.10) that $\mu \in \mathbb{N}$. In this situation, let $k = |X|$, and let $\theta_1, \ldots, \theta_r$ be the eigenvalues of $G$ other than $\mu$ and $r$. We have

$$\sum_{i=1}^{r} \theta_i + k\mu + r = 0 \quad \text{and} \quad \sum_{i=1}^{r} \theta_i^2 + k\mu^2 + r^2 = nr = (1 + k + r)r.$$

It follows that if $\overline{\theta} = \frac{1}{r} \sum_{i=1}^{r} \theta_i$ then

$$\sum_{i=1}^{r}(\theta_i - \overline{\theta})^2 = \sum_{i=1}^{r} \theta_i^2 - r\overline{\theta}^2 = k(r - \mu^2 - \tfrac{k}{r}\mu^2 - \mu).$$

On expressing $r$ and $k$ in terms of $\mu$, we find that $r - \mu^2 - \frac{k}{r}\mu^2 - \mu = 0$. Hence $\theta_i = \overline{\theta}$ $(i = 1, \ldots, r)$ and $G$ has just three distinct eigenvalues. By Theorem 3.6.4, $G$ is strongly regular, and we have case (c) of the Theorem. This completes the proof. □

In case (c) of Theorem 5.2.9, let $\mathcal{D} = \{\Delta_H(u) : u \in X\}$. If $\mu = 1$ then $\mathcal{D}$ consists of all 2-subsets of $X$, and so the star complement technique yields a unique graph $G$, necessarily the complement of the Clebsch graph. If $\mu = 2$ then $\mathcal{D}$ is a Steiner system $S(3, 6, 22)$: this is a design with 22 points and 77 blocks of size 6, with the property that any 3 points lie in a unique block. By a Theorem of Witt [Witt], there is only one such design, and so again $G$ is unique. Here $G$ is the Higman–Sims graph, the strongly regular graph with parameters $(100, 22, 0, 6)$ first constructed from $S(3, 6, 22)$ in [HiSi]. Accordingly, we have:

**Corollary 5.2.10.** *Let G be a regular graph with $K_{1,s}$ $(s > 1)$ as a star complement for $\mu$. If $\mu = 1$ then $\overline{G}$ is the Clebsch graph. If $\mu = 2$ then G is either a 4-cycle or the Higman–Sims graph.*

Note that conversely, if $d \in \mathbb{N}$ and if $G$ is a strongly regular graph with parameters $((d^2 + 3d)^2, d(d^2 + 3d + 1), 0, d(d + 1))$ then $G$ has, as a star complement for $d$, the star induced by the closed neighbourhood of a vertex. Thus our proofs establish both the existence and uniqueness of strongly regular graphs with parameters $(16, 5, 0, 2)$ and $(100, 22, 0, 6)$. It is shown in [KasÖs] that there is no strongly regular graph with parameters $(324, 57, 0, 12)$ (the case $d = \mu = 3$).

## 5.3 Bounds on multiplicities

We saw in Section 5.1 that if a graph $G$ has a star complement with $t$ vertices, for an eigenvalue $\mu \neq -1$ or $0$, then $|V(G)| < t + 2^t$. Here we first improve this upper bound to one which is a quadratic function of $t$.

**Theorem 5.3.1** [BelRo]. *Let G be a graph with n vertices, and let $\mu$ be an eigenvalue of G, $\mu \notin \{-1, 0\}$. If the eigenspace of $\mu$ has codimension $t$ then either*

(a) $n \leq \frac{1}{2}t(t + 1)$ *or*
(b) $\mu = 1$ *and* $G = K_2$ *or* $2K_2$.

**Proof.** Suppose first that $G$ is connected. Using the notation of Theorem 5.1.7, we let $S = (B \,|\, C - \mu I)$, with columns $\mathbf{s}_u$ $(u = 1, \dots, n)$. Using Equation (5.2), we see that

$$\mu I - A = S^\top(\mu I - C)^{-1}S,$$

and so, for all vertices $u, v$ of $G$,

$$\langle\!\langle \mathbf{s}_u, \mathbf{s}_v \rangle\!\rangle = \begin{cases} \mu & \text{if } u = v \\ -1 & \text{if } u \sim v \\ 0 & \text{otherwise} \end{cases}.$$

We define quadratic functions $F_1, \dots, F_n$ as follows:

$$F_u(\mathbf{x}) = \langle\!\langle \mathbf{s}_u, \mathbf{x} \rangle\!\rangle^2 \quad (\mathbf{x} \in \mathbb{R}^t).$$

It is easily checked that if $k = \dim \mathcal{E}(\mu)$ and $\mathbf{x} = (x_{k+1}, \dots, x_n)^\top$ then $F_u(\mathbf{x}) = x_u^2$ $(u = k + 1, \dots, n)$.

We show that $F_1, \ldots, F_n$ are linearly independent unless $\mu = 1$ and $G = K_2$. If $\mu$ is the index of $G$, then $k = 1$ and $F_u(\mathbf{x}) = x_u^2$ $(u = 2, \ldots, n)$. If $F_1, \ldots, F_n$ are linearly dependent, then, since $F_1$ is the square of a linear function, $F_1$ must be a multiple of one of $F_2, \ldots, F_n$, say of $F_v$. The continuity of the functions $\mathbf{x} \mapsto \langle\!\langle \mathbf{s}_1, \mathbf{x} \rangle\!\rangle$ and $\mathbf{x} \mapsto \langle\!\langle \mathbf{s}_v, \mathbf{x} \rangle\!\rangle$ ensures that $\langle\!\langle \mathbf{s}_1, \mathbf{x} \rangle\!\rangle$ is a constant multiple of $x_v$, and therefore $\mathbf{s}_1$ is a multiple of the $v$-th column of $\mu I - C$. But the entries of $\mathbf{s}_1$ and of $C$ are all either 0 or 1; and since $\mu \neq -1, 0$, we deduce that the vertices 1 and $v$ are adjacent to each other but to no other vertices of $G$. Since $G$ is connected we have $G = K_2$ and $\mu = 1$.

Now let $\mu_1$ be the index of $G$, and consider the case in which $\mu \neq \mu_1$. Let $\mathbf{w}$ be an eigenvector of $G$ corresponding to $\mu_1$, with all entries of $\mathbf{w}$ positive. Let $\mathbf{w} = (w_1, \ldots, w_n)^\top$, and let $\mathbf{w}^* = (w_{k+1}, \ldots, w_n)^\top$. Since $\mathbf{w}$ lies in $\mathcal{E}(\mu)^\perp$, it follows from Theorem 5.1.7 that

$$\langle\!\langle \mathbf{s}_u, \mathbf{w}^* \rangle\!\rangle = -w_u \quad (u = 1, \ldots, n).$$

Suppose that $\sum_u \alpha_u F_u = 0$, that is, $\sum_u \alpha_u \langle\!\langle \mathbf{s}_u, \mathbf{x} \rangle\!\rangle^2 = 0$ for all $\mathbf{x} \in \mathbb{R}^t$. Taking $\mathbf{x} = \mathbf{s}_i$, we obtain $\mu^2 \alpha_i + \sum_{u \sim i} \alpha_u = 0$ $(i = 1, \ldots, n)$. Thus

$$(\mu^2 I + A)\mathbf{a} = \mathbf{0}, \quad \text{where } \mathbf{a} = (\alpha_1, \ldots, \alpha_n)^\top.$$

From $\sum_u \alpha_u \langle\!\langle \mathbf{s}_u, \mathbf{x} + \mathbf{y} \rangle\!\rangle^2 = \mathbf{0}$, we obtain $\sum_u \alpha_u \langle\!\langle \mathbf{s}_u, \mathbf{x} \rangle\!\rangle \langle\!\langle \mathbf{s}_u, \mathbf{y} \rangle\!\rangle = \mathbf{0}$ for all $\mathbf{x}, \mathbf{y} \in \mathbb{R}^t$. Taking $\mathbf{x} = \mathbf{s}_i$ and $\mathbf{y} = \mathbf{w}^*$, we obtain $\mu \alpha_i w_i - \sum_{u \sim i} \alpha_u w_u = 0$ $(i = 1, \ldots, n)$. Thus

$$(\mu I - A)\mathbf{a}' = \mathbf{0}, \quad \text{where } \mathbf{a}' = (\alpha_1 w_1, \ldots, \alpha_n w_n)^\top.$$

Because $\mu \neq -1, 0$, we have $\mu \neq -\mu^2$, and so $\mathbf{a}^\top \mathbf{a}' = 0$, that is, $\alpha_1^2 w_1 + \cdots + \alpha_n^2 w_n = 0$. It follows that $\alpha_u = 0$ for all $u$, and so $F_1, \ldots, F_n$ are linearly independent. Now the functions $F_u$ lie in the space of all homogeneous quadratic functions on $\mathbb{R}^t$, and since this space has dimension $\frac{1}{2}t(t + 1)$, we have $n \leq \frac{1}{2}t(t + 1)$.

Finally, suppose that $G$ is not connected. It is clear that, for any vertex $u$, $F_u(\mathbf{x})$ involves only those entries of $\mathbf{x}$ which correspond to vertices in the same component as $u$. Thus, if in each component the $F_u$ are linearly independent, then all the $F_u$ are linearly independent. It follows that the bound holds except possibly when $G = rK_2$ for some $r$. In this case $n = 2r$, $t = r$, and the inequality holds whenever $r \geq 3$. This completes the proof. $\square$

The bound in Theorem 5.3.1 is attained in the graph obtained from $L(K_9)$ by switching with respect to $K_8$: here $\mu = -2$ and $t = 8$. Apart from a few trivial exceptions, the bound is not attained in any regular graph; in fact, if $G$ is regular and $t > 2$, the bound can be reduced by 1, as we now show.

**Theorem 5.3.2** [BelRo]. *Let $\mu$ be a non-main eigenvalue of a graph with $n$ vertices, and let $t$ be the codimension of $\mathcal{E}(\mu)$. If $\mu \notin \{-1, 0\}$ and $t > 2$ then*

$$n \le \frac{1}{2}t(t+1) - 1 = \frac{1}{2}(t-1)(t+2).$$

**Proof.** Since $\mu$ is non-main, we have $\mathbf{j}_n \in \mathcal{E}(\mu)^\perp$, and it follows from Theorem 5.1.7 that $\langle\!\langle \mathbf{s}_u, \mathbf{j} \rangle\!\rangle = -1$ $(u = 1, \ldots, n)$, where $\mathbf{j}$ denotes the all-1 vector in $\mathbb{R}^t$. Consider the function $F(\mathbf{x}) = \langle\!\langle \mathbf{j}, \mathbf{x} \rangle\!\rangle^2$. We will show that $F$ does not belong to the span of $F_1, \ldots, F_n$. Suppose, by way of contradiction, that $F = \sum_u \beta_u F_u$, i.e. $\langle\!\langle \mathbf{j}, \mathbf{x} \rangle\!\rangle^2 = \sum_u \beta_u \langle\!\langle \mathbf{s}_u, \mathbf{x} \rangle\!\rangle^2$ for all $\mathbf{x} \in \mathbb{R}^t$. By considering $\langle\!\langle \mathbf{j}, \mathbf{x} + \mathbf{y} \rangle\!\rangle^2$, we see that

$$\langle\!\langle \mathbf{j}, \mathbf{x} \rangle\!\rangle \langle\!\langle \mathbf{j}, \mathbf{y} \rangle\!\rangle = \sum_u \beta_u \langle\!\langle \mathbf{s}_u, \mathbf{x} \rangle\!\rangle \langle\!\langle \mathbf{s}_u, \mathbf{y} \rangle\!\rangle$$

for all $\mathbf{x}, \mathbf{y} \in \mathbb{R}^t$. Taking $\mathbf{x} = \mathbf{y} = \mathbf{s}_i$, we have $1 = \mu^2 \beta_i + \sum_{u \sim i} \beta_u$ $(i = 1, \ldots, n)$, that is,

$$\mathbf{j}_n = (\mu^2 I + A)\mathbf{b}, \tag{5.11}$$

where $\mathbf{b} = (\beta_1, \ldots, \beta_n)^\top$. Next, taking $\mathbf{x} = \mathbf{s}_i, \mathbf{y} = \mathbf{j}$, we obtain $-\langle\!\langle \mathbf{j}, \mathbf{j} \rangle\!\rangle = -\mu \beta_i + \sum_{u \sim i} \beta_u$ $(i = 1, \ldots, n)$, that is,

$$\langle\!\langle \mathbf{j}, \mathbf{j} \rangle\!\rangle \mathbf{j}_n = (\mu I - A)\mathbf{b}. \tag{5.12}$$

From (5.11) and (5.12),

$$(\mu + \mu^2)\mathbf{b} = (1 + \langle\!\langle \mathbf{j}, \mathbf{j} \rangle\!\rangle)\mathbf{j}_n$$

Since $\mu + \mu^2 \ne 0$, $\mathbf{b}$ is a scalar multiple of $\mathbf{j}_n$, say $\mathbf{b} = \beta \mathbf{j}_n$, so that $\beta_u = \beta$ $(u = 1, \ldots, n)$. Thus

$$\beta^2 \left( \sum_u \langle\!\langle \mathbf{s}_u, \mathbf{x} \rangle\!\rangle \langle\!\langle \mathbf{s}_u, \mathbf{y} \rangle\!\rangle \right)^2 = \langle\!\langle \mathbf{j}, \mathbf{x} \rangle\!\rangle^2 \langle\!\langle \mathbf{j}, \mathbf{y} \rangle\!\rangle^2 = \beta^2 \sum_u \langle\!\langle \mathbf{s}_u, \mathbf{x} \rangle\!\rangle^2 \sum_u \langle\!\langle \mathbf{s}_u, \mathbf{y} \rangle\!\rangle^2.$$

From (5.11) we know that $\beta \ne 0$, and so a Cauchy–Schwarz bound is attained. It follows that $\langle\!\langle \mathbf{s}_u, \mathbf{x} \rangle\!\rangle = \alpha \langle\!\langle \mathbf{s}_u, \mathbf{y} \rangle\!\rangle$ $(u = 1, \ldots, n)$, for some $\alpha = \alpha(\mathbf{x}, \mathbf{y})$. Then $\langle\!\langle \mathbf{s}_u, \mathbf{x} - \alpha \mathbf{y} \rangle\!\rangle = 0$ for all $u$, that is,

$$\mathbf{s}_u^\top (\mu I - C)^{-1} (\mathbf{x} - \alpha \mathbf{y}) = 0 \quad (u = 1, \ldots, n).$$

It follows that $(C - \mu I)(\mu I - C)^{-1}(\mathbf{x} - \alpha \mathbf{y}) = \mathbf{0}$, whence $\mathbf{x} = \alpha \mathbf{y}$. Since this holds for all $\mathbf{x}, \mathbf{y} \in \mathbb{R}^t$, $t$ must be 1, contrary to assumption. Thus $F$ does not belong to the subspace spanned by $F_1, \ldots, F_n$. Since $t > 2$, it follows from the proof of Theorem 5.3.1 that $F_1, \ldots, F_n$ are linearly independent. Hence $F, F_1, \ldots, F_n$ are linearly independent, and we have $n + 1 \le \frac{1}{2}t(t+1)$, as required. $\qquad\square$

**Theorem 5.3.3** [BelRo]. *The regular graphs attaining the bound of Theorem 5.3.2 are precisely the extremal strongly regular graphs.*

**Proof.** First let $G$ be an extremal strongly regular graph with eigenvalues $r, \mu', \mu$ of multiplicities $1, k', k$, where $1 < k' \le k$. Thus if $G$ has $n$ vertices then $n = \frac{1}{2}k'(k' + 3)$. If $t = n - k$ then $k' = t - 1$ and so $n = \frac{1}{2}(t - 1)(t + 2)$, as required.

For the converse, we give a proof due to B. Tayfeh-Rezaie. If $G$ is a regular graph that attains the bound of Theorem 5.3.2, then every homogeneous quadratic function on $\mathbb{R}^t$ is a linear combination of $F_1, F_2, \dots F_n$ and $F$. In particular,

$$\langle\!\langle \mathbf{x}, \mathbf{x} \rangle\!\rangle = \sum_{u=1}^{n} \epsilon_u F_u(\mathbf{x}) + \gamma F(\mathbf{x}), \tag{5.13}$$

for some scalars $\epsilon_1, \epsilon_2, \dots, \epsilon_n$ and $\gamma$. It follows that

$$\langle\!\langle \mathbf{x}, \mathbf{y} \rangle\!\rangle = \sum_{u=1}^{n} \epsilon_u \langle\!\langle \mathbf{s}_u, \mathbf{x} \rangle\!\rangle \langle\!\langle \mathbf{s}_u, \mathbf{y} \rangle\!\rangle + \gamma \langle\!\langle \mathbf{j}, \mathbf{x} \rangle\!\rangle \langle\!\langle \mathbf{j}, \mathbf{y} \rangle\!\rangle. \tag{5.14}$$

Let $\mathbf{e} = (\epsilon_1, \epsilon_2, \dots, \epsilon_n)^\top$. Taking $\mathbf{x} = \mathbf{s}_i$, $\mathbf{y} = -\mathbf{j}$ $(i = 1, 2, \dots, n)$ in (5.14), we find that

$$(\mu I - A)\,\mathbf{e} = (1 - \gamma \langle\!\langle \mathbf{j}, \mathbf{j} \rangle\!\rangle)\,\mathbf{j}. \tag{5.15}$$

Taking $\mathbf{x} = \mathbf{s}_i$ in (5.13), we find that

$$(\mu^2 I + A)\,\mathbf{e} = (\mu - \gamma)\,\mathbf{j}. \tag{5.16}$$

From (5.15) and (5.16) we see that $(\mu^2 + \mu)\mathbf{e}$ is a scalar multiple of $\mathbf{j}$. Since $\mu^2 + \mu \ne 0$, $\mathbf{e} = \epsilon \mathbf{j}$ for some $\epsilon$. Now, taking $\mathbf{x} = \mathbf{s}_i$, $\mathbf{y} = \mathbf{s}_j$ $(i \ne j)$ in (5.14), we have

$$\langle\!\langle \mathbf{s}_i, \mathbf{s}_j \rangle\!\rangle = \epsilon \sum_{i=1}^{n} \langle\!\langle \mathbf{s}_u, \mathbf{s}_i \rangle\!\rangle \langle\!\langle \mathbf{s}_u, \mathbf{s}_j \rangle\!\rangle + \gamma.$$

It follows that if $i \not\sim j$ then $0 = \epsilon a_{ij}^{(2)} + \gamma$, where $A^2 = (a_{ij}^{(2)})$. Since $G$ is not complete, we deduce that $\epsilon \ne 0$, and $a_{ij}^{(2)} = -\epsilon^{-1}\gamma$ when $i \not\sim j$. Similarly, if $i \sim j$ then $a_{ij}^{(2)} = 2\mu - \epsilon^{-1}(\gamma + 1)$, and the result follows. $\qquad\square$

The five known extremal strongly regular graphs are described in Section 3.6.

## 5.4 Graphs with least eigenvalue $-2$

We denote the least eigenvalue of a graph $G$ by $\lambda(G)$. We noted in Chapter 1 that if $G$ is a generalized line graph then $\lambda(G) \geq -2$. On the other hand, we saw in Chapter 3 that not every graph whose spectrum is contained in $[-2, \infty)$ is a generalized line graph; examples include the Chang graphs (Examples 1.2.6 and 4.1.13), the Clebsch graph (Example 1.2.4), the Petersen graph and the wheel $W_6$. Recall that a graph $G$ is said to be *exceptional* if (i) $G$ is connected, (ii) $\lambda(G) \geq -2$, and (iii) $G$ is not a generalized line graph. Determination of the exceptional graphs was an early problem in spectral graph theory, attributed to A. J. Hoffman in the early 1960s. In 1976, root systems were used to show that an exceptional graph has at most 36 vertices [CamGSS]. In 1979 the exceptional graphs $G$ with $\lambda(G) > -2$ were determined independently of root systems [DooCv]: in Chapter 3 we noted that there are 573 such graphs (20 with 6 vertices, 110 with 7 vertices and 443 with 8 vertices, comprising the families $\mathcal{G}_6$, $\mathcal{G}_7$, and $\mathcal{G}_8$). In 1980, generalized line graphs were characterized by a collection $\mathcal{H}$ of 31 forbidden induced subgraphs; the forbidden graphs with least eigenvalue greater than $-2$ are precisely the graphs in $\mathcal{G}_6$, while the other 11 forbidden graphs have least eigenvalue less than $-2$. In this section we describe briefly how star complements can be used to find all the exceptional graphs from the 443 exceptional graphs in $\mathcal{G}_8$.

**Theorem 5.4.1** [CvRS5]. *Let $G$ be a graph with least eigenvalue $-2$. Then $G$ is exceptional if and only if it has an exceptional star complement for $-2$.*

**Proof.** Suppose that $G$ has an exceptional star complement $H$ for $-2$. Then $G$ is not a generalized line graph. By Proposition 5.1.4(i), $G$ is connected because $H$ is connected, and so $G$ is exceptional. Conversely, suppose that $G$ is exceptional. Then $G$ contains an induced subgraph $F$ from the family $\mathcal{H}$ identified above. Since $\lambda(G) \geq -2$ we know from interlacing that $F$ is necessarily one of the 20 exceptional graphs in $\mathcal{G}_6$. By Theorem 5.1.6, $G$ has a connected star complement $H$ for $-2$ which contains $F$ as an induced subgraph. Thus $H$ is exceptional, and the theorem follows.    $\square$

In Theorem 5.4.1, the candidates for an exceptional star complement are (by interlacing) precisely the 573 exceptional graphs with least eigenvalue greater than $-2$. These graphs have at most 8 vertices, a fact which follows either from Theorem 3.4.6 or from their explicit determination independently of root sytems. In any case, we can now see from Theorem 5.3.1 that an exceptional graph has at most 36 vertices.

If $G$ is a maximal exceptional graphs then $G$ is a maximal graph with some prescribed exceptional star complement $H$ for $-2$. In the reverse direction, it turns out that if $G$ is a maximal graph with a prescribed exceptional star complement $H$ for $-2$ then $G$ is a maximal exceptional graph only if $H$ has 8 vertices. Accordingly, to find the maximal exceptional graphs, it suffices to consider exceptional star complements with 8 vertices, and so there are 443 possibilities. For each of these, Lepović used a computer implementation of the star complement technique to determine the maximal exceptional graphs which arise. There are only 473 such graphs, and the distribution of the number of vertices is as follows:

| number of vertices | 22 | 28 | 29 | 30 | 31 | 32 | 33 | 34 | 36 |
|---|---|---|---|---|---|---|---|---|---|
| number of graphs | 1 | 1 | 432 | 25 | 7 | 3 | 1 | 2 | 1 |

It transpires that 363 of these graphs have $K_8$ as a (non-exceptional) star complement for $-2$; for example, the unique largest graph and the unique smallest graph are the graphs with 36 and 22 vertices which feature in Examples 5.2.5(a) and (c) respectively. The remaining 110 graphs are among the 430 maximal exceptional graphs which are cones over a graph switching-equivalent to $L(K_8)$. In addition to these 430 graphs, there are a further 37 graphs with maximal degree 28, while the remaining 6 examples have maximal degree less than 28 (see [CvRS6]).

## 5.5 Graph foundations

Let $G$ be a generalized line graph, say $G = L(\hat{H})$, where $\hat{H}$ is a $B$-graph. Let $\mu$ be an eigenvalue of $G$, and let $Y$ be a set of edges of $\hat{H}$. We say that $Y$ is a *line star set* for $\mu$ in $\hat{H}$ if it is a star set for $\mu$ in $L(\hat{H})$. In this situation, $\hat{H} - Y$ (the spanning subgraph of $\hat{H}$ obtained by deleting the edges in $Y$) is the corresponding *line star complement* for $\mu$ in $\hat{H}$. A line star complement for $-2$ is called a *foundation* for $\hat{H}$. We first discuss foundations for simple graphs (i.e. $B$-graphs without petals).

**Example 5.5.1.** The graph $L(K_5)$ has spectrum $6, \ 1^4, (-2)^5$, and a star complement for $-2$ has the form $L(F)$ where the foundation $F$ is one of the graphs of Fig. 5.4. Here the graphs are shown in increasing order of index. $\square$

**Theorem 5.5.2.** (i) *Let $H$ be a connected graph. Then the least eigenvalue of $L(H)$ is greater than $-2$ if and only if $H$ is a tree or an odd-unicyclic graph.*

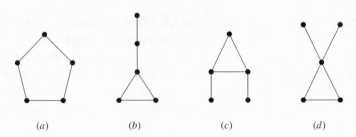

Figure 5.4 The foundations for $K_5$.

(ii) *Let H be a connected bipartite graph such that L(H) has least eigen-value −2. Then the subgraph F of H is a foundation of H if and only if F is a spanning tree for H.*

(iii) *Let H be a connected non-bipartite graph such that L(H) has least eigenvalue −2. Then the subgraph F of H is a foundation of H if and only if F is a spanning subgraph in which each component is an odd-unicyclic graph.*

**Proof.** Part (i) follows from Theorem 3.4.14, and so we suppose that $H$ is a connected graph which is neither a tree nor an odd-unicyclic graph. Suppose that $H$ has $n$ vertices and $m$ edges. Let $E$ be a set of $k$ edges in $H$, and let $F = H - E$. If $F$ is a foundation for $H$ then $k$ is the minimum number of edges whose removal from $H$ results in a graph whose line graph has least eigenvalue greater than $-2$. From Theorem 3.4.9, we see that if $H$ is bipartite then $k = m - n + 1$ and $F$ is a spanning tree, while if $H$ is non-bipartite then $k = m - n$ and each component of $F$ is odd-unicyclic.

To prove that, conversely, a graph of the type specified in (ii) and (iii) is a foundation, we shall identify $k$ linearly independent vectors $\mathbf{v}_e$ ($e \in E$) in $\mathcal{E}_{L(H)}(-2)$. Thus $-2$ has multiplicity at least $k$ in $L(H)$. By interlacing, this multiplicity is precisely $k$, and so $H - E$ is a foundation for $H$.

The vectors $\mathbf{v}_e$ ($e \in E$) are constructed as follows. Here we fix $e$ and let $x_l$ ($l \in E(H)$) be the coordinates of $\mathbf{v}_e$. If $H$ is bipartite then $F + e$ contains a unique cycle $Z$, and $Z$ is of even length. We take $x_l$ to be 1 and $-1$ for alternate edges $l$ of $Z$, with $x_e = 1$, and we define $x_l = 0$ for all $l \notin E(Z)$; see Fig. 5.5.

If $H$ is not bipartite and the addition of $e$ to $F$ creates an even cycle $Z$, then $Z$ is the only even cycle in $F + e$ and we repeat the construction above. Otherwise, the addition of $e$ creates either an odd cycle or a link between two components of $F$. In either case, some component of $F + e$ has just two cycles, say $Z$ and $Z'$; they have odd length and are edge-disjoint. Let $P$ be the unique path of least length (possibly zero) between a vertex of $Z$ and a vertex of $Z'$. If $P$ has non-zero length then we take $x_l$ to be 2 and $-2$ for alternate

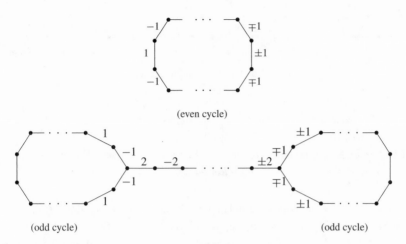

Figure 5.5 A construction for eigenvectors of a line graph.

edges $l$ of $P$. Then we take $x_l = \pm 1$ for $l \in E(Z) \cup E(Z')$ as shown for the dumbbell shape in Fig. 5.5. Finally we define $x_l = 0$ for all remaining edges $l$ of $H$. Reversing all signs if necessary, we may take $x_e > 0$ to determine $\mathbf{v}_e$ uniquely. In all cases, it is straightforward to check that $\mathbf{v}_e$ is an eigenvector of $L(H)$, with corresponding eigenvalue $-2$. These eigenvectors are linearly independent because, for each $e \in E$, the $f$-entry of $\mathbf{v}_e$ is non-zero only for $f = e$. This completes the proof. □

We call the vectors $\mathbf{v}_e$ ($e \in E$) the eigenvectors of $L(H)$ *constructed from* $F$. In case (i) of Theorem 5.5.2, $H$ itself is the unique foundation of $H$. From the proof in cases (ii) and (iii), we have the following:

**Corollary 5.5.3.** *The eigenspace of $-2$ for a line graph $L(H)$ has as a basis the set of eigenvectors constructed from any foundation of $H$.*

We now turn to generalized line graphs that are not line graphs, and in this context the following definitions will be helpful. An *orchid* is a graph which is either odd-unicyclic or a tree with one petal; an *orchid garden* is a graph whose components are orchids.

**Example 5.5.4.** Let $\hat{H}$ be the $B$-graph consisting of a triangle with single petals added at two vertices. The graph $\hat{H}$ and all non-isomorphic foundations of $\hat{H}$ are shown in Fig. 5.6. Note that each foundation is an orchid garden. □

**Theorem 5.5.5.** *Let $\hat{H}$ be a connected $B$-graph with at least one petal. (Thus $L(\hat{H})$ is a generalized line graph which, in general, is not a line graph.)*

Figure 5.6 A *B*-graph and its foundations.

(i) *The graph $L(\hat{H})$ has least eigenvalue greater than $-2$ if and only if $\hat{H}$ is an orchid.*

(ii) *Suppose that the least eigenvalue of $L(\hat{H})$ is $-2$. Then $F$ is a foundation of $\hat{H}$ if and only if $F$ is an orchid garden which spans $\hat{H}$.*

**Proof.** The proof mirrors that of Theorem 5.5.2, and part (i) follows from Theorem 3.4.14. To prove part (ii), let $m = |E(H)|$, where $\hat{H} = H(a_1, \ldots, a_n)$, and let $F = \hat{H} - E$ where $E$ is a set of $k$ edges in $\hat{H}$. If $F$ is a foundation then $k$ is the minimum number of edges whose removal from $\hat{H}$ results in a *B*-graph whose least eigenvalue is greater than $-2$. By Theorem 3.4.13, $k = m - n + \sum_{i=1}^{n} a_i$, and $F$ is an orchid garden.

Conversely, if $F$ is an orchid garden then we can identify $k$ linearly independent vectors $\mathbf{v}_e$ ($e \in E$) in $\mathcal{E}_{L(\hat{H})}(-2)$. By interlacing, $-2$ has multiplicity exactly $k$ in $L(\hat{H})$, and so $F$ is a foundation for $\hat{H}$.

It remains to construct the vectors $\mathbf{v}_e$ ($e \in E$). We fix $e$ and let $x_l$ ($l \in E$) be the coordinates of $\mathbf{v}_e$. We use the term *supercycle* to mean either an odd cycle or a petal. There are $m - n + \sum_{i=1}^{n} a_i$ edges of $\hat{H}$ not in $F$, and three possibilities arise when such an edge $e$ is added to the orchid garden $F$: (1) the edge creates an even cycle, (2) the edge creates a supercycle (that is, it creates an odd cycle or a petal), (3) the edge joins a vertex of one orchid to a vertex of another orchid. We now ascribe weights $x_l$ to the edges of $\hat{H}$ as follows.

In case (1) all weights are 0 except for 1 and $-1$ alternately on edges of the even cycle. In cases (2) and (3), $F + e$ contains a unique shortest path $P$ between vertices of two different supercycles, and we first ascribe weights of 2 and $-2$ alternately to the edges of $P$. To within a choice of sign, weights are ascribed to the edges of the two supercycles as illustrated in Fig. 5.7, and all remaining weights are 0. (In all cases the construction may be seen as ascribing weights $\pm 1$ alternately to the edges in a closed walk, with the assumption that double edges are assigned the same value; in edges traversed twice, the values are added.)

In each case, we choose signs so that $x_e > 0$. The weights $x_l$ of edges in $\hat{H}$ are taken as coordinates of a vector $\mathbf{v}_e$ whose entries are indexed by the corresponding vertices of $L(\hat{H})$. It is straightforward to check that each such vector is an eigenvector of $L(\hat{H})$ corresponding to $-2$. These $m - n + \sum_{i=1}^{n} a_i$

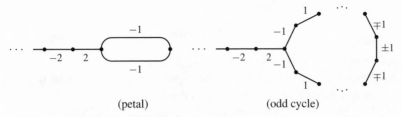

(petal)          (odd cycle)

Figure 5.7 A construction for eigenvectors of a generalized line graph.

vectors are linearly independent because each of the aforementioned closed walks contains an edge not present in any of the others (as in the proof of Theorem 5.5.2). □

Again we call the vectors $\mathbf{v}_e$ ($e \in E$) the eigenvectors *constructed from* $F$. In case (i) of Theorem 5.5.6, $\hat{H}$ itself is the unique foundation for $\hat{H}$. In case (ii) our arguments establish the following result (formulated to subsume Corollary 5.5.3):

**Theorem 5.5.6.** *The eigenspace for the eigenvalue $-2$ of a generalized line graph has as a basis the set of eigenvectors constructed from any foundation of the corresponding root multigraph.*

**Remark 5.5.7.** We can construct a foundation $F$ for the $B$-graph $H(a_1, \ldots, a_n)$ from a foundation $F'$ of $H$ as follows. If $H$ is not bipartite then $F'$ is an orchid garden which spans $H$ and we may take $F$ to consist of $F'$ together with $a_i$ (single) pendant edges attached at vertex $v_i$ ($i = 1, \ldots, n$). If $H$ is bipartite then $F'$ is a tree which spans $H$: here we first modify $F'$ by adding $a_i$ pendant edges at vertex $v_i$ ($i = 1, \ldots, n$) and then obtain $F$ by replacing one of these pendant edges by a double edge. In general, not all foundations for $H(a_1, \ldots, a_n)$ can be constructed in this way. □

Finally, for future reference (see Subsection 9.3.2), we give two simple results that arise as corollaries of the above proofs. Here an *odd dumbbell* is a $B$-graph consisting of two supercycles connected by a path (possibly of zero length).

**Corollary 5.5.8.** *A connected generalized line graph $L(\hat{H})$ has least eigenvalue $-2$ if and only if $\hat{H}$ contains an even cycle or an odd dumbbell.*

**Corollary 5.5.9.** *The edge $e$ of the $B$-graph $\hat{H}$ lies in an even cycle or an odd dumbbell if and only if there exists an eigenvector $\mathbf{x} \in \mathcal{E}_{L(\hat{H})}(-2)$ such that $x_e \neq 0$.*

# Exercises

**5.1** Find a star partition for (i) $K_{m,n}$, (ii) $L(K_n)$.

**5.2** Show that the multiplicity of any eigenvalue of a non-trivial tree $T$ is bounded above by the number of endvertices of $T$.

**5.3** Use Theorem 2.2.4 to show that if $u$, $v$ are adjacent vertices in a star set then $uv$ is not a bridge.

**5.4** Let $\mu$ be a non-zero eigenvalue of the graph $G$, and let $X$ be a star set for $\mu$ in $G$ with corresponding star complement $H$. Show that if $H$ is 2-connected then one of the following holds:

 (a) $G$ is 2-connected;

 (b) $\mu \neq -1$ and $G$ has a pendant edge at a vertex of $H$;

 (c) $\mu = -1$ and $G$ has a cutvertex $v$ in $H$ whose neighbours in $X$ induce a complete subgraph which is a component of $G - v$.

**5.5** Use the Reconstruction Theorem to find, for given $t \in I\!N$, the values of $\mu$ for which $K_t$ arises as a star complement for $\mu$.

**5.6** Consider the graphs with $K_{1,5}$ as a star complement for $-1$ (cf. Theorem 5.2.7). Show that there are two maximal such graphs without co-duplicate vertices. (One has 15 vertices, the other 16 vertices.)

**5.7** Show that the complement of the Schläfli graph is the unique maximal graph with $K_{2,5}$ as a star complement for a multiple eigenvalue other than $-1$ [JaRo].

**5.8** Find in terms of $n$, $\mu$ and $r$ the parameters of the strongly regular graphs which arise in Theorem 5.3.3.

**5.9** Let $X$ be a star set for the eigenvalue $\mu$ in the graph $G$, and let $H = G - X$. The vertex $u$ of $X$ is said to be *amenable to switching* if $\mu$ is an eigenvalue of the graph obtained from $H + u$ by switching with respect to $\{u\}$. Suppose that $\mu$ is non-main and that every vertex in $X$ is amenable to switching. Show that if $G'$ is obtained from $G$ by switching with respect to a subset of $X$ then $\mu$ is a non-main eigenvalue of $G'$ and $X$ is a star set for $\mu$ in $G'$ [RowJa].

**5.10** Let $H$ be a graph with $t$ vertices such that $-2$ is an eigenvalue of $K_1 \triangledown H$ but not of $H$. Let $\Gamma^*(H; -2)$ be the subgraph of $\Gamma(H; -2)$ induced by the $(0, 1)$-vectors $\mathbf{b}$ such that $\langle\!\langle \mathbf{b}, \mathbf{j} \rangle\!\rangle = -1$ (cf. Proposition 5.1.4). Show that $\overline{\Gamma^*(H; -2)}$ has a perfect matching, say $\mathbf{b}_1 \mathbf{c}_1, \ldots, \mathbf{b}_m \mathbf{c}_m$, with $\mathbf{b}_i + \mathbf{c}_i = \mathbf{j}$ $(i = 1, \ldots, m)$. Deduce that any two graphs with $t + m$ vertices having $H$ as a star complement for the non-main eigenvalue $-2$ are switching-equivalent [CvRS9].

**5.11** Let $\hat{H}$ be a connected $B$-graph with at least one petal, such that $L(\hat{H})$ has least eigenvalue $-2$. Show that $-2$ is a main eigenvalue if and only if $\hat{H}$ has an odd cycle or two petals connected by a path of odd length.

**5.12** Prove Corollaries 5.5.8 and 5.5.9.

# Notes

The star complement technique has its origins in the Schur complement of a principal submatrix (see [Pra, p. 17]); its application in a graph-theoretical context was noted independently by Ellingham [Ell] and Rowlinson [Row8] in 1993. Some consequences of the Reconstruction Theorem are discussed in [BelCRS1]. A survey of star complements appears in [Row13], and a database of some 1500 examples is described in [CvLRS1]. Subsequent papers include a characterization of the Hoffman–Singleton graph [HofSi] among regular graphs (see Section 6.4 and [RowSc]), and a reduction to a combinatorial problem in the case of extremal strongly regular graphs with an independent set of maximal size [Row15]. Theorem 5.2.9 is a stronger version of a result which appeared in [Row12], where the degree of regularity was prescribed and strong regularity was not established. Regular graphs with regular star complements are investigated in [Row10].

Further dominating properties of star complements are discussed in [Row9] and [LiuRo]. The relation between star complements and switching is discussed in [CvRS9] and [RowJa]. Several generalized line graphs (and their complements) are characterized by star complements in [CvRS5]. In [BelLMS], the authors investigate the possible star complements for $-2$ in graphs for which $-2$ is the least eigenvalue. Odd cycles and paths as star complements for $-2$ are treated in [Bel3], [Bel4] and [BelSi]. The determination of the maximal exceptional graphs is described in [CvLRS2], and treated comprehensively in the monograph [CvRS7]. The relation of star partitions to the complexity of the graph isomorphism problem is discussed in [CvRS2, Chapter 8] and [CvRS1].

# 6

# Spectral techniques

This chapter is devoted to structural results which do not refer to eigenvalues but which are proved using spectral techiniques. We include classical results such as the Friendship Theorem and constraints on Moore graphs, as well as more recent results concerning graph homomorphisms. We also discuss decompositions of complete graphs, generalized quadrangles and equiangular lines. In the final section, we calculate the number of walks of given length in graphs of a particular type.

## 6.1 Decompositions of complete graphs

An *r-decomposition* of the graph $G$ is a set of $r$ spanning subgraphs $H_1, \ldots, H_r$ such that each edge of $G$ lies in exactly one of the $H_i$. For example, it is easy to see that $K_7$ is the edge-disjoint union of three 7-cycles. Is $K_{10}$ (which has 45 edges) the edge-disjoint union of three copies of the Petersen graph (which has 15 edges)? This question was posed by Schwenk in the *American Mathematical Monthly* (Problem 6434(b) of June-July 1983). The following simple argument involving eigenspaces shows that the answer is 'no'. If $K_{10}$ has such a decomposition then

$$A + B + C + I = J \tag{6.1}$$

where each of $A$, $B$, $C$ is an adjacency matrix of a Petersen graph. Since $\mathcal{E}_A(1)$ and $\mathcal{E}_B(1)$ are 5-dimensional subspaces of the 9-dimensional space $\mathbf{j}^\perp$, there exists a non-zero vector $\mathbf{x} \in \mathcal{E}_A(1) \cap \mathcal{E}_B(1)$ such that $J\mathbf{x} = \mathbf{0}$. From Equation (6.1) we have $A\mathbf{x}+B\mathbf{x}+C\mathbf{x}+\mathbf{x} = 0$, whence $C\mathbf{x} = -3\mathbf{x}$. This is a contradiction because $-3$ is not an eigenvalue of $C$.

We can extend the above argument from the Petersen graph to an arbitrary strongly regular graph $G$, and thereby obtain the following result.

162

**Theorem 6.1.1** [Row2]. *Let $G$ be a connected strongly regular graph with parameters $(n, r, e, f)$. If $K_n$ is the edge-disjoint union of three spanning subgraphs isomorphic to $G$ then there exists a positive integer $k$ such that, with a consistent choice of sign,*

$$n = (3k \pm 1)^2, \ r = 3k^2 \pm 2k, \ e = k^2 - 1 \text{ and } f = k^2 \pm k.$$

**Proof.** By Theorem 3.6.5, $G$ has eigenvalues $r, \mu_2, \mu_3$ with multiplicities $1, k_2, k_3$ where

$$\mu_2, \mu_3 = \frac{1}{2} \left( e - f \pm \sqrt{(e-f)^2 + 4(r-f)} \right)$$

and

$$k_2, k_3 = \frac{1}{2} \left\{ (n-1) \pm \frac{(n-1)(f-e) - 2r}{\sqrt{(e-f)^2 + 4(r-f)}} \right\}.$$

If $K_n$ has a 3-decomposition as described in the statement of the theorem then a consideration of degrees shows that $n - 1 = 3r$. It follows that $k_2 \neq k_3$ for otherwise $f - e = 2r/(n-1) = \frac{2}{3}$. Equation (6.1) holds with each of $A$, $B$, $C$ now an adjacency matrix of $G$. On repeating the original argument with $\mu_2$ in place of the eigenvalue 1 we find that $-2\mu_2 - 1$ is an eigenvalue of $C$. This eigenvalue is different from $r$ because a corresponding eigenvector is orthogonal to $\mathbf{j}$; and different from $\mu_2$ because $\mu_2 \neq -\frac{1}{3}$. Hence $-2\mu_2 - 1 = \mu_3$ and on expressing $k_2$ in the form

$$k_2 = \frac{1}{2} \left\{ 3r - \frac{3r(\mu_2 + \mu_3) + 2r}{\mu_2 - \mu_3} \right\} \tag{6.2}$$

we see that $k_2 = 2r, k_3 = r$. It follows from (6.2) that $3(\mu_2 + \mu_3) + 2 = \mu_3 - \mu_2$. Similarly, if $k_3 > k_2$ then $\mu_2$ and $\mu_3$ are interchanged and we have $3(\mu_2 + \mu_3) + 2 = \mu_2 - \mu_3$. Hence always

$$(3e - 3f + 2)^2 = (e - f)^2 + 4(r - f). \tag{6.3}$$

For any strongly regular graph, we have $r(r - e - 1) = (n - r - 1)f$ by Equation (3.14). Since here $n - 1 = 3r$, we have $r = 2f + e + 1$, and it follows from (6.3) that

$$(f - e + 1)^2 = e + 1.$$

Thus $e$ has the form $k^2 - 1$ and the result follows. □

In the terminology of Mesner [Mes] a strongly regular graph which satisfies the conclusions of Theorem 6.1.1 with $n = (3k + 1)^2$ is a graph of *negative Latin square type* $NL_k(3k + 1)$. If $n = (3k - 1)^2$ then the graph is called by Bose and Shrikhande [BosSh] a *pseudo net graph* of type $L_k(3k - 1)$.

**Remark 6.1.2.** Suppose for definiteness that $k_2 > k_3$ in the proof of Theorem 6.1.1. Our eigenspace argument shows that $\mathcal{E}_A(\mu_2) \cap \mathcal{E}_B(\mu_2) \subseteq \mathcal{E}_C(\mu_3)$. Now $\dim \mathcal{E}_C(\mu_3) = r$ and $\dim(\mathcal{E}_A(\mu_2) \cap \mathcal{E}_B(\mu_2)) = \dim \mathcal{E}_A(\mu_2) + \dim \mathcal{E}_B(\mu_2) - \dim(\mathcal{E}_A(\mu_2) + \mathcal{E}_B(\mu_2)) \geq 2k_2 - (n-1) = r$. Hence $\mathcal{E}_C(\mu_3) = \mathcal{E}_A(\mu_2) \cap \mathcal{E}_B(\mu_2)$. Similarly $\mathcal{E}_B(\mu_3) = \mathcal{E}_C(\mu_2) \cap \mathcal{E}_A(\mu_2)$ and $\mathcal{E}_A(\mu_3) = \mathcal{E}_B(\mu_2) \cap \mathcal{E}_C(\mu_2)$. Since $\mathcal{E}_A(\mu_2) \cap \mathcal{E}_A(\mu_3) = \{\mathbf{0}\}$, we have $\mathcal{E}_A(\mu_3) \cap \mathcal{E}_B(\mu_3) = \{\mathbf{0}\}$ and, on comparing dimensions, $\mathcal{E}_C(\mu_2) = \mathcal{E}_A(\mu_3) \oplus \mathcal{E}_C(\mu_3)$. Therefore,

$$I\!R^n = \langle \mathbf{j} \rangle \oplus \mathcal{E}_A(\mu_3) \oplus \mathcal{E}_B(\mu_3) \oplus \mathcal{E}_C(\mu_3). \tag{6.4}$$

If $k_3 > k_2$ then $\mu_2$ replaces $\mu_3$ in (6.4).

The disposition of the various eigenspaces ensures that the matrices $A$, $B$, $C$ are simultaneously diagonalizable and so they commute: in terms of edge-colourings of $K_n$ this means that if we use three different colours $c_1, c_2, c_3$ for the three subgraphs isomorphic to $G$ then for any two vertices $u, v$ and any two colours $c_i, c_j$ the number of $u$-$v$ walks of length 2 coloured $c_i, c_j$ is the same as the number of $u$-$v$ walks of length 2 coloured $c_j, c_i$. This is not generally the case for a 3-decomposition of $K_n$ into isomorphic regular subgraphs as may be seen from the following decomposition of $K_7$ into three 7-cycles: if the vertices of $K_7$ are labelled 1,2,3,4,5,6,7 and the cycles 12345671, 14275361, 13746251 are coloured blue, red, green respectively then the walk 153 is coloured green-red, but there is no walk from 1 to 3 coloured red-green.                                                                                    □

The following class of examples illustrates Theorem 6.1.1.

**Example 6.1.3.** Let $I\!K$ be a finite field of order $q = p^{2h}$, where $h \in I\!N$ and $p$ is a prime congruent to 2 mod 3. Let $g$ be a generator for the multiplicative group of $I\!K$, and let $H = \langle g^3 \rangle$. The subgroup $H$ has index 3 in $\langle g \rangle$ and consists of all the non-zero cubes in $I\!K$. Since $-1 \in H$ we may define (undirected) graphs $G_i$ ($i = 0, 1, 2$), with vertices the elements of $I\!K$, as follows: vertices $u$ and $v$ are adjacent in $G_i$ if and only if $u - v \in Hg^i$ ($i = 0, 1, 2$). The map $x \longmapsto xg^i$ ($x \in I\!K$) is an isomorphism $G_0 \to G_i$, and it follows that $G_0, G_1, G_2$ constitute a 3-decomposition of $K_q$. Moreover $G_0$ is strongly regular because any pair of adjacent vertices may be mapped to $0, g^3$, and any pair of non-adjacent vertices may be mapped to $0, g$, by an automorphism of $I\!K$. Thus there are infinitely many graphs $G$ which satisfy the hypotheses of Theorem 6.1.1. (The smallest connected example is the complement of the Clebsch graph, which arises as $G_0$ when $q = 16$.)

We can use the relation between parameters given by Theorem 6.1.1 to find the number of solutions of the Fermat equation $x^3 + y^3 = z^3$ in the field $I\!K$. Note that $e = k^2 - 1$ where $k = \frac{1}{3}(p^h - 1)$ if $h$ is even and $k = \frac{1}{3}(p^h + 1)$ if $h$ is odd. Now for given $u \in H$, $e$ is the number of solutions $(v, w)$ of the equation

$u + v = w$ $(v, w \in H)$. It follows that the number of solutions $(u, v, w)$ of the equation $u + v = w$ $(u, v, w \in H)$ is $e|H|$. Each element of $H$ has 3 cube roots in $I\!K$ and so the number of non-trivial solutions $(x, y, z)$ of the equation $x^3 + y^3 = z^3$ $(x, y, z \in I\!K)$ is $f_3(p^{2h})$, where $f_3(p^{2h}) = 27e|H| = (p^{2h} - 1)(p^{2h} - 2(-p)^h - 8)$. $\qquad\square$

If $K_n$ is the edge-disjoint union of subgraphs (not necessarily spanning subgraphs) isomorphic to the graph $G$ then (i) $\frac{1}{2}n(n-1)$ is divisible by the number of edges in $G$ and (ii) $n - 1$ is divisible by the greatest common divisor of the degrees of vertices in $G$. An asymptotic converse was proved by R. M. Wilson [Wils]: given a graph $G$ then for large enough $n$ satisfying conditions (i) and (ii), $K_n$ is the edge-disjoint union of subgraphs isomorphic to $G$. Note that if $G$ is regular of degree $r$ then conditions (i) and (ii) reduce to the single requirement that $r$ divides $n - 1$.

For our next observation on the decomposition of a complete graph, we make use of the following more general result, attributed to H. S. Witsenhausen.

**Lemma 6.1.4.** *Let $G$ be a graph with $n^+$ positive eigenvalues and $n^-$ negative eigenvalues. If $G$ has an $r$-decomposition into complete bipartite graphs then $r \geq \max\{n^+, n^-\}$.*

**Proof.** Here the adjacency matrix $A$ of $G$ is $A_1 + \cdots + A_r$, where $A_i$ is the adjacency matrix of a complete bipartite graph $G_i$ $(i = 1, \ldots, r)$. Thus if $G_i$ is determined by the bipartition $V(G) = U_i \,\dot\cup\, V_i$ then $A_i = \mathbf{u}_i \mathbf{v}_i^\top + \mathbf{v}_i \mathbf{u}_i^\top$, where $\mathbf{u}_i, \mathbf{v}_i$ are the characteristic vectors of $U_i, V_i$ respectively $(i = 1, \ldots, r)$.

Now suppose by way of contradiction that $r < n^+$. Then the eigenvectors of $A$ corresponding to positive eigenvalues span a subspace $V^+$ of dimension greater than $r$. Hence $V^+$ contains a non-zero vector $\mathbf{w}$ orthogonal to $\mathbf{u}_1, \ldots, \mathbf{u}_r$. Now $\mathbf{w}^\top A \mathbf{w} = 0$, a contradiction. We obtain a similar contradiction if $r < n^-$. $\qquad\square$

**Theorem 6.1.5.** *If $K_n$ has an $r$-decomposition into complete bipartite graphs then $r \geq n - 1$.*

**Proof.** The result follows immediately from Lemma 6.1.4 because $K_n$ has $n - 1$ negative eigenvalues. $\qquad\square$

## 6.2 Graph homomorphisms

The topic of graph homomorphisms forms a natural sequel to the previous section, because we can give an alternative solution to the original problem concerning $K_{10}$ in this context. A *homomorphism* from the graph $G$ to the graph $H$ is a function $\sigma : V(G) \to V(H)$ that maps edges to edges; that is, if $i, j$ are adjacent vertices of $G$, then $\sigma(i), \sigma(j)$ are adjacent vertices of $H$. In this situation, let $S$ be the matrix whose rows are indexed by $V(G)$, whose columns are indexed by $V(H)$, and whose $(i, u)$-entry is 1 if $\sigma(i) = u$, and 0 otherwise. Thus if $V(H) = \{1, \ldots, k\}$ then

$$S^\top S = \operatorname{diag}(\, |\sigma^{-1}(1)|, \ldots, |\sigma^{-1}(k)| \,),$$

while the $(u, v)$-entry of $S^\top A(G)S$ is the number $N(u, v)$ of edges between $\sigma^{-1}(u)$ and $\sigma^{-1}(v)$. We say that $\sigma$ is *uniform*, with parameters $p, q$, if each vertex of $H$ is the image of precisely $p$ vertices of $G$, and each edge of $H$ is the image of precisely $q$ edges of $G$.

**Theorem 6.2.1.** *Let $G, H$ be graphs with $n, m$ vertices respectively. If there exists a uniform homomorphism from $G$ to $H$, with parameters $p, q$, then*

$$\lambda_{n-m+i}(G) \le \frac{q}{p}\lambda_i(H) \le \lambda_i(G) \quad (i = 1, \ldots, m). \tag{6.5}$$

**Proof.** We have $S^\top S = pI$ and $S^\top A(G)S = qA(H)$. Thus if $Q = \frac{1}{\sqrt{p}}S$ then $Q^\top Q = I$ and $Q^\top A(G)Q = \frac{q}{p}A(H)$. Now the result follows by applying Theorem 1.3.11 to $A(G)$ and $\frac{q}{p}A(H)$. $\qquad\square$

If $K_{10}$ is the edge-disjoint union of three copies of the Petersen graph $P$, then there is a natural uniform homomorphism $\sigma$ from $G = 2P$ to $H = \overline{P}$ with $p = 2$ and $q = 1$. (In each of $2P$ and $\overline{P}$, we may colour the edges of one Petersen graph red, and the edges of the other green; then $\sigma$ maps red edges to red edges and green edges to green edges.) Now $\overline{P} = L(K_5)$, with spectrum $6, 1^4, (-2)^5$, while $2P$ has spectrum $3^2, 1^{10}, (-2)^8$. Since $\lambda_{12}(G) = \lambda_2(H) = 1$, the eigenvalues of $\frac{1}{2}A(H)$ do not interlace those of $A(G)$ in accordance with (6.5). Thus there is no 3-decomposition of $K_{10}$ into Petersen graphs.

In the case of a uniform homomorphism from a regular graph $G$ to a regular graph $H$, the inequalities (6.6) may be recast in terms of the eigenvalues of the Laplacian $D - A$, or of the signless Laplacian $D + A$. If the eigenvalues of $D - A$ are denoted by $v_i^*$ in non-decreasing order, and the eigenvalues of $D + A$ are denoted by $\xi_i^*$ in non-decreasing order then the inequalities (6.5) yield

$$v_i^*(G) \le \frac{q}{p}v_i^*(H) \quad \text{and} \quad \xi_i^*(G) \le \frac{q}{p}\xi_i^*(H) \quad (i = 1, \ldots, m). \tag{6.6}$$

These inequalities may be generalized to the case of an arbitrary homomorphism $\sigma$ from a graph $G$ onto a graph $H$ without isolated vertices, as follows.

Let $p_\sigma$ be the smallest of the numbers $\sigma^{-1}(u)$ $(u \in V(H))$, and let $q^\sigma$ be the largest of the numbers $N(u, v)$. Then

$$v_i^*(G) \leq \frac{q^\sigma}{p_\sigma} v_i^*(H) \quad \text{and} \quad \xi_i^*(G) \leq \frac{q^\sigma}{p_\sigma} \xi_i^*(H) \quad (i = 1, \dots, m),$$

where again $m = |V(H)|$. These are among the spectral inequalities established in [DanHa1] and [DanHa2].

We conclude this section with a second application of Theorem 6.2.1, this time to designs. Recall that a $2\text{-}(v, k, \lambda)$ design consists of a family $\mathcal{B}$ of $k$-subsets (or *blocks*) of a $v$-subset $V$ such that any two elements (or *points*) of $V$ lie in precisely $\lambda$ blocks. If $|\mathcal{B}| = b$ then each point lies in $r$ blocks, where $bk = vr$ and $r(k - 1) = (v - 1)\lambda$. Note that $r \geq \lambda$. Let $G$ be the graph whose components are the complete graphs on $\{B\} \times B$ $(B \in \mathcal{B})$, let $H$ be the complete graph on $V$, and define $\sigma(B, u) = u$ $(u \in V)$. Then $\sigma$ is a uniform homomorphism from $G$ to $H$ with parameters $r, \lambda$.

**Example 6.2.2.** Using the homomorphism $\sigma$ defined above, we prove *Fisher's inequality:* if $k < v$ then $b \geq v$. We have $G = bK_k$ and $H = K_v$. Thus if $b < v$ then $\lambda_{v-b+1}(H) = -1$ and $\lambda_{kb-b+1}(G) = -1$. Applying Theorem 6.2.1 (with $n = bk$, $m = v$, $i = v - b + 1$)), we have $\frac{\lambda}{r}(-1) \leq (-1)$, that is, $r \leq \lambda$. Hence $r = \lambda$, and so $k = v$, contrary to assumption.                    $\square$

## 6.3 The Friendship Theorem

For our third application, consider an *acquaintance graph* with $n$ vertices representing $n$ persons $(n > 1)$. An edge between two vertices indicates that the two persons are acquainted. The Friendship Theorem is often formulated as follows: if any two persons have exactly one common acquaintance then one person is acquainted with everybody else. It is easy to see that then the acquaintance graph must be a *windmill*, that is a graph of the form $K_1 \triangledown r K_2$ $(r \in \mathbb{N})$. The complete result may be stated as follows.

**Theorem 6.3.1.** *Let $G$ be a non-trivial graph in which any two vertices have a unique common neighbour. Then $G$ is a windmill.*

**Proof.** Let $A(G) = A = (a_{ij})$, and let $A^2 = (a_{ij}^{(2)})$. Thus our hypothesis is that $a_{ij}^{(2)} = 1$ whenever $i \neq j$. Since $a_{ii}^{(2)}$ is the degree $d_i$ of vertex $i$, we have

$$A^2 - D = J - I, \tag{6.7}$$

where, as usual, $D$ is the diagonal matrix of vertex degrees and $J$ is an all-1 matrix. It follows that $A$ commutes with $J + D - I$, and hence that

$$AJ + AD = JA + DA. \tag{6.8}$$

Equating $(i, j)$-entries in Equation (6.8), we have $d_i + a_{ij}d_j = d_j + d_i a_{ij}$, or

$$(d_i - d_j)(a_{ij} - 1) = 0.$$

Since $a_{ij} = 1$ whenever $d_i \neq d_j$, we label vertices with the same degree consecutively to obtain $A$ in the form

$$A = \begin{pmatrix} * & J_{k_1,k_2} & \cdots & J_{k_1,k_r} \\ J_{k_2,k_1} & * & \cdots & J_{k_1,k_r} \\ \cdots & \cdots & \cdots & \cdots \\ J_{k_r,k_1} & J_{k_r,k_2} & \cdots & * \end{pmatrix},$$

where $k_1, \ldots, k_r$ are the frequencies of the distinct degrees.

We may suppose that $k_1 > 1$ because any non-trivial connected graph has (at least) two vertices of the same degree. Then $a_{12}^{(2)} \geq n - k_1$, and since $a_{12}^{(2)} = 1$, we conclude that $k_1 \geq n - 1$. Moreover, either (a) $k_1 = n - 1$, $r = 2$ and $k_2 = 1$, or (b) $k_1 = n$, $r = 1$ and all vertices have the same degree. We consider these cases in turn.

In case (a), $A$ has the form

$$A = \begin{pmatrix} A^* & \mathbf{j} \\ \mathbf{j}^\top & 0 \end{pmatrix}.$$

Our hypothesis ensures that each row of $A^*$ has exactly one entry equal to 1, and so without loss of generality,

$$A = \begin{pmatrix} 0 & 1 & 0 & 0 & \cdots & 0 & 1 \\ 1 & 0 & 0 & 0 & \cdots & 0 & 1 \\ 0 & 0 & 0 & 1 & \cdots & 0 & 1 \\ 0 & 0 & 1 & 0 & \cdots & 0 & 1 \\ \cdots & \cdots & \cdots & \cdots & \cdots & \cdots & \cdots \\ 1 & 1 & 1 & 1 & \cdots & 1 & 0 \end{pmatrix}.$$

This is the adjacency matrix of a windmill.

In case (b), $G$ is regular, say of degree $d$, and we have $D = dI$, $AJ = dJ$. It follows from Equation (6.7) that

$$(A - dI)(A^2 - (d-1)I) = O.$$

We may suppose that $G$ is not complete (and hence that $d > 2$), because the only complete graph that satisfies our hypothesis is $K_3$ (a windmill).

From Theorem 3.6.5, we know that $G$ is strongly regular, with eigenvalues $d, \sqrt{d-1}, -\sqrt{d-1}$ of multiplicities $1, k, l$, where

$$k, l = \frac{1}{2}\left(n - 1 \mp \frac{d}{\sqrt{d-1}}\right).$$

Thus

$$n - 2k - 1 = \frac{d\sqrt{d-1}}{d-1}. \tag{6.9}$$

It is easy to see that there is no value of $d$ $(d > 2)$ for which the right-hand side of (6.9) is an integer. Thus the windmill $K_3$ is the only graph that arises in case (b), and the proof is complete. $\square$

## 6.4 Moore graphs

A *Moore graph* is a graph with diameter $d$ and girth $2d + 1$, for some $d > 1$. The 5-cycle and the Petersen graph are two of the three known examples with $d = 2$; we describe the third example later in this section.

**Lemma 6.4.1.** *A Moore graph is regular.*

**Proof.** Let $G$ be a Moore graph with diameter $d$. We show first that any two vertices $u, v$ of $G$ at distance $d$ have the same degree. Let $P(u, v)$ be the unique path of length $d$ from $u$ to $v$, and let $w$ be any neighbour of $v$ not on $P(u, v)$. Then $d(u, w) = d$ and the path $P(u, w)$ includes a neighbour $w'$ of $u$ not on $P(u, v)$. Different $w$ determine different $w'$, and so $\deg(v) \leq \deg(u)$. Similarly, $\deg(u) \leq \deg(v)$.

Next, let $Z$ be a cycle of length $2d + 1$ in $G$. If $x, y$ are adjacent vertices of $Z$ then there exists a vertex $z$ of $Z$ such that $d(x, z) = d(y, z) = d$, and so $\deg(x) = \deg(y)$. It follows that all vertices of $Z$ have the same degree.

Finally, consider a vertex $t$ not on $Z$, and a shortest path, of length $j$ say, from $t$ to $Z$. We may add $d - j$ consecutive edges of $Z$ to this path to reach a vertex $t'$ of $Z$ at distance $d$ from $t$. Then $\deg(t) = \deg(t')$, and it follows that all vertices of $G$ have the same degree. $\square$

It can be shown that a Moore graph $G$ is even distance-regular (Exercise 6.3), and this is the first step in a proof that $d = 2$ unless $G$ is an odd cycle of length $> 5$. We omit this proof, but show instead that there are at most four possibilities for the degree of a Moore graph $G$ with $d = 2$. Note that in this case, if $G$ is $r$-regular with $n$ vertices then $n = r^2 + 1$, because the number of vertices at distance 2 from a given vertex is $r(r - 1)$.

**Theorem 6.4.2.** *If G is a Moore graph of diameter* 2 *then G is r-regular, with* $r \in \{2, 3, 7, 57\}$.

**Proof.** For any two non-adjacent vertices $u$, $v$, there exists a unique walk of length 2 between $u$ and $v$. It follows that the adjacency matrix $A$ of $G$ satisfies

$$A^2 + A - (r - 1)I = J.$$

Since also $(A - rI)J = O$, $G$ is strongly regular with eigenvalues $r$, $\mu_2$, $\mu_3$, where $\mu_2$, $\mu_3$ are the roots of $x^2 + x - (r - 1)$. Thus $\mu_2, \mu_3 = \frac{1}{2}(-1 \pm s)$, where $s = \sqrt{4r - 3}$. If $k_2$, $k_3$ are the multiplicities of $\mu_2$, $\mu_3$ then (considering spectral moments) we have

$$1 + k_2 + k_3 = r^2 + 1 \quad \text{and} \quad r + k_2\mu_2 + k_3\mu_3 = 0. \tag{6.10}$$

It follows that

$$k_2 + k_3 = r^2 \quad \text{and} \quad s(k_2 - k_3) = r^2 - 2r.$$

If $4r - 3$ is not a perfect square then $s$ is irrational and necessarily $k_2 = k_3$, $r^2 = 2r$. In this case, $r = 2$.

If $4r - 3$ is a perfect square, then we substitute $\frac{1}{4}(s^2 + 3)$ for $r$ in (6.10) and eliminate $k_3$ to obtain

$$s^5 - s^4 + 6s^3 - 2s^2 + (9 - 32k_2)s - 15 = 0.$$

It follows that the integer $s$ is a divisor of 15. Since $r > 1$, we have $s > 1$. Hence $s \in \{3, 5, 15\}$, and so $r \in \{3, 7, 57\}$ in this case. $\qquad\square$

If $n$ is the number of vertices in an $r$-regular Moore graph $G$ (of diameter 2) then the possibilities for $(r, n)$ are $(2, 5)$, $(3, 10)$, $(7, 50)$ and $(57, 3250)$. It is not known whether the last possibility arises. The 5-cycle and the Petersen graph are the unique Moore graphs with parameters $(r, n) = (2, 5)$, $(3, 10)$ respectively, and the unique Moore graph with $(r, n) = (7, 50)$ is the *Hoffman–Singleton graph HoS*, which we now describe. For this purpose, recall that the *Fano plane* is the unique 2-$(7, 3, 1)$ design illustrated in Fig. 6.1, where the blocks are represented by the circle and the straight lines. The graph $HoS$ may be constructed as follows, where a *heptad* is a set of 7 triples which may be taken as the blocks of a Fano plane whose points are $1, 2, 3, 4, 5, 6, 7$. The vertices of $HoS$ are the 15 heptads in an orbit of the alternating group $A_7$ together with the 35 triples in $\{1, 2, 3, 4, 5, 6, 7\}$. There are edges in $HoS$ between disjoint triples, and between a heptad and each of its triples.

We note that $HoS$ has an induced subgraph $H_0$ illustrated in Fig. 6.2, where the vertices of degree 1 and 7 are the 15 independent heptads. Now

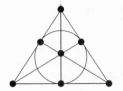

Figure 6.1 The Fano Plane.

Figure 6.2 The graph $H_0$.

the spectrum of $H_0$ is $3^1$, $\sqrt{2}\,^6$, $0^8$, $(-\sqrt{2})^6$, $(-3)^1$, while that of $HoS$ is $7^1$, $2^{28}$, $(-3)^{21}$. Thus $H_0$ is a star complement for 2 in $HoS$. It can be shown that $HoS$ is the unique regular graph with such a star complement (see [RowSc]).

An $r$-regular graph of diameter 2 has at most $r^2 + 1$ vertices, because the number of vertices at distance 2 from a given vertex is at most $r(r - 1)$; and when this bound is attained, the graph has girth 5. On the other hand, an $r$-regular graph of girth 5 has at least $r^2 + 1$ vertices, and when this bound is attained, the graph has diameter 2. Thus a Moore graph is extremal in both contexts. The technique used to prove Theorem 6.4.2 may be extended to prove the following.

**Theorem 6.4.3** [Brow]. *There is no $r$-regular graph of girth 5 on $r^2 + 2$ vertices.*

**Proof.** Suppose that $G$ is an $r$-regular graph of girth 5 with $n = r^2 + 2$ vertices. Then $r$ is even, and the number of vertices at distance $\leq 2$ from any given vertex $v$ is $r^2 + 1$. Accordingly, there is just one further vertex, $v^*$ say, in $G$, and $d(v, v^*) = 3$. Since $v^{**} = v$, we may label the vertices of $G$ so that

$$A^2 + A - rI = J - B - I, \qquad (6.11)$$

where $B$ is a direct sum of matrices $\begin{pmatrix} 0 & 1 \\ 1 & 0 \end{pmatrix}$. Now $J - B - I$ is the adjacency matrix of a cocktail-party graph, with spectrum $n - 2, 0^{\frac{n}{2}}, -2^{\frac{n}{2}-1}$. On diagonalizing $A$ and $J$ simultaneously, we now see from (6.11) that $G$ has $\frac{1}{2}n$ eigenvalues $\lambda$ satisfying

$$\lambda^2 + \lambda - r = 0, \quad \text{that is,} \quad \lambda = \frac{1}{2}(-1 \pm s) \quad \text{where} \quad s = \sqrt{4r + 1},$$

and $\frac{1}{2}n - 1$ eigenvalues $\lambda$ satisfying

$$\lambda^2 + \lambda - r + 2 = 0, \quad \text{that is,} \quad \lambda = \frac{1}{2}(-1 \pm t) \quad \text{where} \quad t = \sqrt{4r - 7}.$$

There are four cases to consider.

*Case 1: s and t both rational.* Here, $s$ and $t$ are odd positive integers such that $s^2 - t^2 = 8$, and so $s = 3, t = 1$. Then $r = 2$ and $G$ is a 6-cycle, a contradiction since $G$ has girth 5.

*Case 2: s and t both irrational.* Here $s/t$ is irrational, for otherwise $st$ is an integer such that $(st)^2 = (4r - 3)^2 - 16$ and again $s = 3, t = 1, r = 2$. Thus $s$ and $t$ are linearly independent over the rationals. Therefore the eigenvalues $\frac{1}{2}(-1 \pm s)$ appear in pairs, and the eigenvalues $\frac{1}{2}(-1 \pm t)$ appear in pairs. This is impossible since one of $\frac{1}{2}n, \frac{1}{2}n - 1$ is odd.

*Case 3: s is irrational and t is rational.* In this case, the eigenvalues $\frac{1}{2}(-1 \pm t)$ are integers and (since $\mathrm{tr}(A) = 0$) it follows that the eigenvalues $\frac{1}{2}(-1 \pm s)$ sum to an integer. This sum is $-\frac{1}{4}n$ since the eigenvalues $\frac{1}{2}(-1 \pm s)$ appear in pairs. Thus 4 divides $n$ and we have the contradiction $r^2 \equiv 2 \bmod 4$.

*Case 4: s is rational and t is irrational.* Here the eigenvalues $\frac{1}{2}(-1 \pm t)$ appear in pairs and so their sum is $-\frac{1}{4}n + \frac{1}{2}$. Now let $\frac{1}{2}(-1 + s)$ have multiplicity $m$. Since $\mathrm{tr}(A) = 0$ we have

$$r + m\frac{1}{2}(-1 + s) + \left(\frac{1}{2}n - m\right)\frac{1}{2}(-1 - s) - \frac{1}{4}n + \frac{1}{2} = 0. \qquad (6.12)$$

Since $n = r^2 + 2$ and $r = \frac{1}{4}(s^2 - 1)$, we obtain a quintic equation from (6.11):

$$s^5 + 2s^4 - 2s^3 - 20s^2 + (33 - 64m)s + 50 = 0. \qquad (6.13)$$

Thus $s$ divides 50. The only possibilities for $(s, m, r)$ ($s > 1$) arising from (6.12) are $(5, 12, 6)$ and $(25, 6565, 156)$. In both cases, $\mathrm{tr}(A^3) \neq 0$, a contradiction since $G$ has no triangles. $\qquad \square$

## 6.5 Generalized quadrangles

A *generalized polygon* is a bipartite graph with diameter $d$ and girth $2d$ for some integer $d > 1$. A refinement of the arguments used to prove Lemma 6.4.1 shows that if $G$ is a generalized polygon with minimal degree $\delta(G) > 2$ then $G$ is either regular or semi-regular. In this situation, the only possible values of $d$ are 3, 4, 6 and 8 (and all arise): the proof of this result, obtained by Feit and Higman [FeHi] in 1964, is outwith the scope of this book. It can also be shown that always $\delta(G) \geq 2$, and if $\delta(G) = 2$ then one of the following holds: (a) $G$ is an even cycle, (b) $G$ is the $k$-fold subdivision of multiple edges between two vertices, (c) $G$ is the $k$-fold subdivision of a generalized polygon $G'$ with $\delta(G') > 2$.

We say that a generalized polygon $G$ has *order* $(s, t)$ if the vertices in the two parts of $V(G)$ have degrees $s + 1$ and $t + 1$. The terminology reflects the fact that the incidence graph on the points and lines of a projective plane of order $s$ is a generalized polygon $G$ with $d = 3$ and order $(s, s)$. Similarly, the incidence graph on the vertices and edges of a $d$-cycle is a generalized polygon of order $(1, 1)$; this is just a $2d$-cycle, constructed as the total graph of a $d$-cycle.

A *generalized quadrangle* is a generalized polygon with $d = 4$. We show in Theorem 6.5.4 that if a generalized quadrangle has order $(2, t)$ then the only possible values of $t$ are 1, 2 and 4, and that a unique graph arises in each case. We begin by determining constraints on $s$ and $t$ for any generalized quadrangle $G$ of order $(s, t)$. Let $V(G) = S \cup T$, where vertices in $S$ have degree $s + 1$ and vertices in $T$ have degree $t + 1$. Let $G^*$ be the graph with $V(G^*) = T$, and with vertices $p, q$ adjacent if and only if $p, q$ are at distance 2 in $G$.

**Lemma 6.5.1.** *The graph $G^*$ is strongly regular with parameters*

$$((s + 1)(st + 1), s(t + 1), s - 1, t + 1).$$

**Proof.** Let $p \in T$, and let $\Gamma_i(p)$ be the set of vertices of $G$ at distance $i$ from $p$ in $G$ ($i = 1, 2, 3, 4$). Since $G$ is bipartite with diameter 4, these sets are independent and $|\Gamma_1(p)| = t+1, |\Gamma_2(p)| = s(t+1), |\Gamma_3(p)| = st(t+1)$. Counting in two ways the edges between $\Gamma_3(p)$ and $\Gamma_4(p)$, we have $|\Gamma_4(p)|(t + 1) = |\Gamma_3(p)|s$, and so $|\Gamma_4(p)| = s^2 t$. Hence $|V(G^*)| = 1 + |\Gamma_2(p)| + |\Gamma_4(p)| = (s + 1)(st + 1)$, and each vertex of $G^*$ has degree $|\Gamma_2(p)| = s(t + 1)$.

If $p, q$ are adjacent vertices of $G^*$ then $q \in \Gamma_2(p)$ and $p, q$ have a unique common neighbour, $x$ say, in $G$. Now the common neighbours of $p$ and $q$ in $G^*$ are precisely the $s - 1$ vertices other than $q$ in $\Gamma_2(p) \cap \Gamma_1(x)$.

Finally, suppose that $p, q$ are non-adjacent vertices of $G^*$, and let $\Gamma_1(p) = \{x_1, \ldots, x_{t+1}\}$. Then $q \in \Gamma_4(p)$ and for each $i \in \{1, \ldots, t + 1\}$, there exists a unique $q$-$x_i$ path in $G$ of length 3, say $q y_i q_i x_i$. Now the vertices $q_1, \ldots, q_{t+1}$ are distinct and they are all the common neighbours of $p$ and $q$ in $G^*$. $\square$

Note that the generalized quadrangle $G$ is determined by the graph $G^*$: the vertices of $S$ may be identified with the maximal cliques of $G^*$, since the neighbours of a vertex of $S$ induce a clique in $G^*$, and the vertices of any non-trivial clique in $G^*$ have unique common neighbour in $G$. Thus we construct $G$ as the incidence graph on the vertices and maximal cliques of $G^*$.

**Lemma 6.5.2.** *If $G$ is a generalized quadrangle of order $(s, t)$ then the eigenvalues of $G^*$ are $s(t+1)$, $s - 1$ and $-t - 1$, with respective multiplicities*

$$1, \quad \frac{st(s+1)(t+1)}{s+t}, \quad \frac{s^2(st+1)}{s+t}.$$

**Proof.** The result follows from Theorem 3.6.5 when we take $n = (s+1)$ $(st+1)$, $r = s(t+1)$, $e = s-1$ and $f = t+1$.                    □

Since the multiplicities here are integers, Lemma 6.5.2 imposes a constraint on $s$ and $t$. The Krein inequalities provide a further restriction:

**Lemma 6.5.3.** *If $G$ is a generalized quadrangle of order $(s, t)$ with $s > 1$ and $t > 1$, then $s \le t^2$ and $t \le s^2$.*

**Proof.** If we apply the Krein inequalities (Theorem 3.6.8) to the graph $G^*$, we obtain

$$(s^2 - t)(t+1)(s-1) \ge 0 \quad \text{and} \quad (t^2 - s)(s+1)(t-1) \ge 0.$$

The result follows since $s > 1$ and $t > 1$.                    □

**Theorem 6.5.4.** *If there exists a generalized quadrangle of order $(2, t)$ then $t \in \{1, 2, 4\}$. Conversely, if $(s, t) \in \{(2, 1), (2, 2), (2, 4)\}$ then there exists a unique generalized quadrangle of order $(s, t)$.*

**Proof.** By Lemma 6.5.3, we have $t \le 4$. Taking $s = 2$ in Lemma 6.5.2, we see that for the multipicities of the eigenvalues of $G^*$ to be integers, $t+2$ must be a divisor of 12. Hence $t \in \{1, 2, 4\}$. In these cases $G^*$ has parameters $(9, 4, 1, 3)$, $(15, 6, 1, 3)$, $(27, 10, 1, 5)$ respectively. We have seen in Chapter 3 that in each case there exists a unique strongly regular graph with the given parameters. Since $G^*$ determines $G$, we are done.                    □

From Chapter 3, the graphs $G^*$ that arise in Theorem 6.5.4 are $L(K_{3,3})$, $\overline{L(K_6)}$ and $Sch_{10}(= \overline{Sch_{16}})$. Setting $s = 2$ in the proof of Lemma 6.5.1, we see that $G$ has $(2t+1)(t+4)$ vertices. In each case, a maximal clique of $G^*$ has $s + 1 = 3$, vertices and so $G$ is the incidence graph on the vertices and triangles of $G^*$. In this way we obtain generalized quadrangles with 15, 30 and 72 vertices respectively as the only generalized quadrangles of order $(2, t)$.

# 6.6 Equiangular lines

Distinct concurrent lines in the Euclidean space $\mathbb{R}^t$ are said to be *equiangular* if the angle between any two of them is the same. For example, the six lines through antipodal pairs of vertices of an icosahedron are equiangular; the angle between any two of them is $\cos^{-1}(1/\sqrt{5})$.

Let $\mathcal{L}$ be a system of $n$ equiangular lines in $\mathbb{R}^t$ at angle $\alpha > 0$, and let $\mathbf{u}_1, \dots, \mathbf{u}_n$ be unit vectors along the lines of $\mathcal{L}$. The Gram matrix of these vectors has the form

$$(\mathbf{u}_i^\top \mathbf{u}_j) = (\cos \alpha) T + I,$$

where $T$ is the Seidel matrix of a graph $H$ with vertices $1, \dots, n$. For $i \neq j$, the $(i, j)$-entry of $T$ is $\pm 1$ according as the angle between $\mathbf{u}_i$ and $\mathbf{u}_j$ is acute or obtuse. If $\mathbf{u}_i$ is replaced with $-\mathbf{u}_i$ for each $i \in U$ then the system of lines is unchanged but $H$ is replaced with $H_U$, the graph obtained from $H$ by switching with respect to $U$. Thus $\mathcal{L}$ determines a switching class $\mathcal{S}(\mathcal{L})$ of graphs on $n$ vertices.

If the vectors $\mathbf{u}_1, \dots, \mathbf{u}_n$ are linearly dependent, we say that the lines are *dependent*. In this case, the Gram matrix $(\mathbf{u}_i^\top \mathbf{u}_j)$ is singular, and so $-1/\cos \alpha$ is an eigenvalue of $T$; it is the least eigenvalue of $T$ because the matrix $(\mathbf{u}_i^\top \mathbf{u}_j)$ is positive semi-definite. Note also that if $n > t$, then the multiplicity of $-1/\cos \alpha$ is at least $n - t$.

Conversely, if an $n \times n$ Seidel matrix $T$ has least eigenvalue $-\rho$ with multiplicity $k$ then $T + \rho I$ is a positive semi-definite matrix of rank $t = n - k$. Hence $T + \rho I = C^\top C$ for some matrix $C$ of size $t \times n$. Thus $C^\top C$ is the Gram matrix of $n$ vectors in $\mathbb{R}^t$; these vectors have length $\sqrt{\rho}$ and the angle between any two of them is $\cos^{-1}(\pm 1/\rho)$. The columns of $C$ are linearly dependent and determine $n$ equiangular lines in $\mathbb{R}^t$. Consequently we have:

**Proposition 6.6.1** [LinSe]. *There is a one-to-one correspondence between the switching classes of graphs on $n$ vertices and the dependent sets of $n$ equiangular lines.*

The following result gives a restriction on the angle $\alpha$ between sufficiently dense equiangular lines.

**Theorem 6.6.2.** *If $\mathbb{R}^t$ contains $n$ equiangular lines at angle $\alpha$, and if $n > 2t$ then $1/\cos \alpha$ is an odd integer.*

**Proof.** We have seen that $-1/\cos \alpha$ is an eigenvalue of $T$ with multiplicity at least $n - t$. Here, $n - t > \frac{1}{2}n$ and so $-1/\cos \alpha$ is an integer, $m$ say. Further, $-1/\cos \alpha$ is a multiple eigenvalue of $T$ and so $\mathbf{j}^\perp$ contains an eigenvector $\mathbf{x}$ with eigenvalue $m$. Now $\mathbf{x}$ is an eigenvector of the adjacency matrix $\frac{1}{2}(J - I - T)$ with eigenvalue $-\frac{1}{2}(m + 1)$. Since this rational eigenvalue is necessarily an integer, $m$ is odd and the theorem is proved. $\qquad \square$

Our example of six equiangular lines determined by an icosahedron in $\mathbb{R}^3$ shows that Theorem 6.6.2 cannot be improved in general. It is also the case

that $I\!R^3$ cannot contain more than six equiangular lines; more generally, we
have:

**Theorem 6.6.3.** *If $\mathcal{L}$ is a system of $n$ equiangular lines in $I\!R^t$ then*
$n \leq \frac{1}{2}t(t+1)$.

**Proof.** With $\alpha$ and $\mathbf{u}_1, \ldots, \mathbf{u}_n$ as above, define functions $f_1, \ldots, f_n$ on the
unit sphere in $I\!R^t$ by

$$f_i(\mathbf{x}) = (\mathbf{u_i}^\top \mathbf{x})^2 - \cos^2 \alpha \quad (i = 1, \ldots, n).$$

We have $f_i(\mathbf{u}_j) = \delta_{ij} \sin^2 \alpha$, and so $f_1, \ldots, f_n$ are linearly independent. On
the other hand, all $f_i$ lie in the space of functions of the form $\sum_{i=1}^t a_i x^i +$
$\sum_{i<j} b_{ij} x_i x_j$ $(a_i, b_{ij} \in I\!R)$ because $1 = x_1^2 + \cdots + x_t^2$. This space of
homogeneous quadratic functions $I\!R^t \rightarrow I\!R$ has dimension $t + \binom{t}{2}$, and so
$n \leq \frac{1}{2}t(t+1)$.                                                                         $\square$

Let $v(t)$ be the maximal number of equiangular lines in $I\!R^t$; clearly, $t \leq$
$v(t) \leq \frac{1}{2}t(t+1)$. We show next how strongly regular graphs can be used
to construct systems of equiangular lines, thereby obtaining improved lower
bounds for $v(t)$. To exclude trivial cases, we suppose that $n > t > 1$; in
particular, $\alpha < \pi/2$ and our lines are dependent.

Suppose that the eigenvalues of the Seidel matrix $T$ above are $\eta_1 \geq \eta_2 \geq$
$\cdots \geq \eta_n$. Since $(\mathbf{u}_i^\top \mathbf{u}_j)$ has rank $\leq n - t$, we have $\eta_{t+1} = \cdots = \eta_n = -\rho$,
where $\rho = 1/\cos\alpha$. Since $\operatorname{tr} T = 0$ and $\operatorname{tr} T^2 = n(n-1)$, we have

$$\eta_1 + \cdots + \eta_t - (n-t)\rho = 0 \quad \text{and} \quad \eta_1^2 + \cdots + \eta_t^2 + (n-t)\rho^2 = n(n-1). \quad (6.14)$$

Let $\eta = \frac{1}{t}(\eta_1 + \cdots + \eta_t)$. It follows from (6.14) that

$$t \sum (\eta_i - \eta)^2 = n(n-1)t - n(n-t)\rho^2.$$

Hence $\rho^2(n-t) \leq t(n-1)$, with equality if and only if $\eta_1 = \cdots = \eta_t$.
The case of equality is of particular interest, and in this situation we say that $\mathcal{L}$
is *extremal*. Thus $\mathcal{L}$ is extremal if and only if the graphs in $\mathcal{S}(\mathcal{L})$ have exactly
two Seidel eigenvalues. (Recall that swtiching-equivalent graphs have the same
Seidel spectrum.) If $\mathcal{L}$ is extremal then the distinct eigenvalues of $T$ are $\eta$ and
$-\rho$, and we have:

$$t\eta - (n-t)\rho = 0 \quad \text{and} \quad t\eta^2 + (n-t)\rho^2 = n(n-1).$$

On eliminating $t$ from these equations, we find that

$$n = 1 + \eta\rho.$$

The next result provides the link with strongly regular graphs; here 'strongly regular with $r = 2f$' means 'strongly regular with parameters $(n - 1, 2f, e, f)$'.

**Theorem 6.6.4.** *The line system $\mathcal{L}$ is extremal if and only if the switching class $\mathcal{S}(\mathcal{L})$ contains a graph $K_1 \mathbin{\dot{\cup}} G$, where $G$ is strongly regular with $r = 2f$.*

**Proof.** [$\Rightarrow$] We show a little more, namely that for any vertex $v$ of $H$, the graph $H'$ obtained from $H$ by switching with respect to the neighbourhood of $v$ has the required form. Clearly $H' = K_1 \mathbin{\dot{\cup}} G$ for some graph $G$, and we may take $T$ to be the Seidel matrix of $H'$. If $v$ is taken as the first vertex of $H'$ then

$$T = \begin{pmatrix} 0 & \mathbf{j}^\top \\ \mathbf{j} & S \end{pmatrix}, \tag{6.15}$$

where $S$ is the Seidel matrix of $G$. We show that $G$ is strongly regular with $r = 2f$. Note that $T^2 = \begin{pmatrix} n - 1 & \mathbf{j}^\top S \\ S\mathbf{j} & J + S^2 \end{pmatrix}$, and since $T$ has minimal polynomial $x^2 - (\eta - \rho)x - \eta\rho$, we have

$$S\mathbf{j} = (\eta - \rho)\mathbf{j} \quad \text{and} \quad J + S^2 - (\eta - \rho)S - \eta\rho I = O.$$

Writing $S = J - I - 2A$, we see from the first of these equations that $G$ is regular of degree $r = \frac{1}{2}(n - 2 + \rho - \eta) = \frac{1}{2}(\rho - 1)(\eta + 1)$. Since $AJ = JA = rJ$, the second equation yields

$$4A^2 + 2(\eta - \rho + 2)A - (\rho - 1)(\eta + 1)I = (\eta\rho + \eta + \rho)J.$$

Hence $G$ is strongly regular with $r - f = \frac{1}{4}(\rho - 1)(\eta + 1)$; in particular, $r = 2f$. Note that the eigenvalues of $G$ other than $r$ are the roots of $(2x + \eta + 1)(2x - \rho + 1)$, namely $-\frac{1}{2}(\eta + 1)$ and $\frac{1}{2}(\rho - 1)$.

[$\Leftarrow$] For the converse, suppose that $G$ is strongly regular with parameters $(n - 1, r, e, f)$, where $r = 2f$. Let $r, \lambda, \mu$ be the distinct eigenvalues of $G$. If $S$ is the Seidel matrix of $G$ then we may take the Seidel matrix of $K_1 \mathbin{\dot{\cup}} G$ to be the matrix $T$ of (6.15). Now $S$ has $n - 2$ linearly independent eigenvectors $\mathbf{x}$ in $\mathbf{j}^\perp$ with corresponding eigenvalues $-1 - 2\lambda$ or $-1 - 2\mu$. Hence $T$ has $n - 2$ linearly independent eigenvectors $\begin{pmatrix} 0 \\ \mathbf{x} \end{pmatrix}$, also with corresponding eigenvalues $-1 - 2\lambda$ or $-1 - 2\mu$. To see that there are two further eigenvectors of the form $\begin{pmatrix} 1 \\ a\mathbf{j} \end{pmatrix}$, note that

$$\begin{pmatrix} 0 & \mathbf{j}^\top \\ \mathbf{j} & S \end{pmatrix} \begin{pmatrix} 1 \\ a\mathbf{j} \end{pmatrix} = \begin{pmatrix} a(n - 1) \\ \{1 + a(n - 2 - 2r)\}\mathbf{j} \end{pmatrix}.$$

Thus $\begin{pmatrix} 1 \\ a\mathbf{j} \end{pmatrix}$ is an eigenvector of $T$, with corresponding eigenvalue $\theta = a(n-1)$, if and only if $\theta$ is a solution of the equation

$$x^2 - x(n - 2 - 2r) - (n - 1) = 0. \tag{6.16}$$

We express the coefficients here in terms of $\lambda$ and $\mu$. First, recall that these eigenvalues are the roots of $x^2 - (e - f)x - (r - f)$, and since $r = 2f$ we have $\lambda + \mu = e - f$, $f = -\lambda\mu$. Secondly, the parameters of $G$ satisfy (see Section 3.6):

$$r(r - e - 1) = f(n - 2 - r).$$

Using the relation $r = 2f$ once more, we deduce that $n - 2 - r = r - 2e - 2 = -2\lambda - 2\mu - 2$. Now $n - 1 = 3r - 2e - 1 = 4f + (2f - 2e) - 1 = -4\lambda\mu - 2\lambda - 2\mu - 1$, and (6.16) becomes:

$$x^2 + (2\lambda + 2\mu + 2)x + (2\lambda + 1)(2\mu + 1) = 0.$$

Therefore the remaining two eigenvalues are also $-1 - 2\lambda$ or $-1 - 2\mu$. Since $T$ has just two eigenvalues, $\mathcal{L}$ is extremal.                                   □

For an extremal system $\mathcal{L}$ of equiangular lines, we have $t\eta - (n - t)\rho = 0$, equivalently,

$$(n - 2t)\rho = t(\lambda - \rho).$$

It follows that either (a) $n = 2t$ and $\eta = \rho$, or (b) $n \neq 2t$ and $\lambda, \rho$ are integers (each being an eigenvalue of $T$ of unique multiplicity). In case (a), $T^2 = (n - 1)I$, that is, $T$ is an $n \times n$ symmetric *conference matrix* . Such matrices can exist only when $n \equiv 2 \pmod 4$ and $n - 1$ is the sum of two squares (see [Bele] or [LinSe]); they exist when $n - 1$ is a prime power congruent to $1 \bmod 4$ and for some other values of $n$. In case (b), the positive integers $\eta, \rho$ are odd integers because (as we saw in the proof of Theorem 6.6.4) $-\frac{1}{2}(\eta + 1)$ and $\frac{1}{2}(\rho - 1)$ are eigenvalues of an adjacency matrix. Note also that $\rho > 1$ and $\eta > 1$ because $t > 1$: this follows from the equations $t\eta = (n - t)\rho$ and $n = 1 + \eta\rho$. If we eliminate $n$ from these equations, we obtain:

$$t = \rho^2 + \frac{\rho - \rho^3}{\rho + \eta}.$$

Thus for given $\rho$ in case (b), there are only finitely many possibilities for $\eta$, hence for $t$ and $n$. We list the feasible parameters when $\rho < 7$. Note that there are no symmetric conference matrices with $n = 22$ or $34$; and that the cases $(\rho, \eta) \in \{(3, 15), (5, 115)\}$ are excluded by Theorem 6.6.3.

| $n$ | 6 | 10 | 14 | 16 | 18 | 26 | 28 | 30 | 36 | 38 | 42 | 46 | 76 | 96 | 126 | 176 | 276 |
|---|---|---|---|---|---|---|---|---|---|---|---|---|---|---|---|---|---|
| $t$ | 3 | 5 | 7 | 6 | 9 | 13 | 7 | 15 | 15 | 19 | 21 | 23 | 19 | 20 | 21 | 22 | 23 |
| $\rho$ | $\sqrt{5}$ | 3 | $\sqrt{13}$ | 3 | $\sqrt{17}$ | 5 | 3 | $\sqrt{29}$ | 5 | $\sqrt{37}$ | $\sqrt{41}$ | $\sqrt{45}$ | 5 | 5 | 5 | 5 | 5 |
| $\eta$ | $\sqrt{5}$ | 3 | $\sqrt{13}$ | 5 | $\sqrt{17}$ | 5 | 9 | $\sqrt{29}$ | 7 | $\sqrt{37}$ | $\sqrt{41}$ | $\sqrt{45}$ | 15 | 19 | 25 | 35 | 55 |

In all of these cases except $n = 76$ and $n = 96$, a strongly regular graph with the requisite parameters is known to exist, and so the corresponding system of equiangular lines exists. Since any such system in $\mathbb{R}^t$ may be embedded isometrically in $\mathbb{R}^{t+1}$, we may extract lower bounds $\ell_t$ for $v(t)$ as follows:

| $t$ | 2 | 3 | 4 | 5 | 6 | 7 | ... | 14 | 15 | 16 | 17 | 18 | 19 | 20 | 21 | 22 | 23 |
|---|---|---|---|---|---|---|---|---|---|---|---|---|---|---|---|---|---|
| $\ell_t$ | 3 | 6 | 6 | 10 | 16 | 28 | ... | 28 | 36 | 40 | 48 | 48 | 72 | 90 | 126 | 176 | 276 |
| $\sec \alpha$ | 2 | $\sqrt{5}$ | 3 | 3 | 3 | 3 | ... | 3 | 5 | 5 | 5 | 5 | 5 | 5 | 5 | 5 | 5 |

Here $\alpha$ is the angle corresponding to a known example of $\ell_t$ equiangular lines in $\mathbb{R}^t$. In the cases $t = 7, 8, \ldots, 13$, it is known that $v(t) = \ell_t$; moreover any set of 28 equiangular lines in $\mathbb{R}^{13}$ at angle $\cos^{-1}(1/3)$ span a 7-dimensional subspace [LemSe, Theorems 4.5 and 4.6].

We conclude by mentioning one general existence result: for any odd prime power $q$ there exists an equiangular system of lines with $n = q^3 + 1$, $t = q^2 - q + 1$, $\rho = q$, $\eta = q^2$ and $\alpha = \cos^{-1}(1/q)$. Thus $\ell_t > t\sqrt{t}$ in this case. The result is a consequence of the following example, described without proof of the details.

**Example 6.6.5.** Let $V$ be the vector space of triples over $GF(q^2)$, where $q$ is an odd prime power. For $\mathbf{x} = (x_1, x_2, x_3)^\top$ and $\mathbf{y} = (y_1, y_2, y_3)^\top$ in $V$, define

$$h(\mathbf{x}, \mathbf{y}) = x_1 y_1^q + x_2 y_2^q + x_3 y_3^q.$$

Let $\Omega$ be the set of 1-dimensional subspaces $\langle \mathbf{x} \rangle$ of $V$ such that $h(\mathbf{x}, \mathbf{x}) = 0$. Then $|\Omega| = q^3 + 1$. Next, let $\Delta$ be the set of 3-subsets $\{\langle \mathbf{x} \rangle, \langle \mathbf{y} \rangle, \langle \mathbf{z} \rangle\}$ of $\Omega$ for which $h(\mathbf{x}, \mathbf{y})h(\mathbf{y}, \mathbf{z})h(\mathbf{z}, \mathbf{x})$ is a square in $GF(q^2)$. Now fix $\langle \mathbf{x} \rangle \in \Omega$, and let $G$ be the graph with $V(G) = \Omega \setminus \{\langle \mathbf{x} \rangle\}$ and vertices $\langle \mathbf{y} \rangle, \langle \mathbf{z} \rangle$ adjacent if and only if $\{\langle \mathbf{x} \rangle, \langle \mathbf{x} \rangle, \langle \mathbf{z} \rangle\} \in \Delta$. Then $G$ is strongly regular with parameters $(n - 1, r, e, f)$, where $n - 1 = q^3$, $r = \frac{1}{2}(q - 1)(q^2 + 1)$, $e = \frac{1}{4}(q^3 - 3q^2 + 3q - 5)$ and $f = \frac{1}{4}(q - 1)(q^2 + 1)$. The eigenvalues of $G$ are $r$, $\frac{1}{2}(q - 1)$, $-\frac{1}{2}(q^2 + 1)$. $\square$

## 6.7 Counting walks

In this section we show how to calculate the number of walks of prescribed length in a graph, and we illustrate the technique by finding a formula for the number of walks that can be traversed by a king in $k$ moves on a chessboard.

Recall from Equation (1.8) that the number of walks of length $k$ in a graph $G$ is given by

$$N_k = \sum_{i=1}^{m} \mu_i^k \|P_i \mathbf{j}\|^2.$$

In practice it is convenient to reformulate this expression as follows. Let $\{\mathbf{u}_1, \dots, \mathbf{u}_n\}$ be an orthonormal basis of $I\!R^n$, with $A\mathbf{u}_h = \lambda_h \mathbf{u}_h$ ($h = 1, \dots, n$), and let

$$\mathbf{j} = \gamma_1 \mathbf{u}_1 + \cdots + \gamma_n \mathbf{u}_n.$$

Then $P_i \mathbf{j}$ is the sum of those $\gamma_h \mathbf{u}_h$ for which $\lambda_h = \mu_i$. Thus $\| P_i \mathbf{j} \|^2 = \sum_h \{ \gamma_h^2 : \lambda_h = \mu_i \}$ and we have

$$N_k = \sum_{h=1}^{n} \gamma_h^2 \lambda_h^k \quad \text{where} \quad \gamma_h = \mathbf{j}^\top \mathbf{u}_h \ (h = 1, \dots, n). \tag{6.17}$$

**Example 6.7.1.** For a path $P_n$ with adjacency matrix

$$\begin{pmatrix} 0 & 1 & 0 & \cdots & 0 \\ 1 & 0 & 1 & \cdots & 0 \\ \vdots & \ddots & \ddots & \ddots & \vdots \\ 0 & \cdots & 1 & 0 & 1 \\ 0 & \cdots & 0 & 1 & 0 \end{pmatrix},$$

we have $\lambda_h = 2\cos\frac{h\pi}{n+1}$ ($h = 1, \dots, n$). It is easy to verify that the numbers $\sqrt{\frac{2}{n+1}} \sin\frac{hi\pi}{n+1}$ ($i = 1, \dots, n$) are the entries $u_{ih}$ of the normalized eigenvector $\mathbf{u}_h$ corresponding to $\lambda_h$. Thus

$$\gamma_h = \sqrt{\frac{2}{n+1}} \sum_{i=1}^{n} \sin\frac{hi\pi}{n+1},$$

which is 0 for even $h$, and $\cot\dfrac{h\pi}{2(n+1)}$ for odd $h$. It follows from (6.17) that

$$N_k = \frac{2^{k+1}}{n+1} \sum_{l=1}^{\lceil \frac{n+1}{2} \rceil} \cot^2\frac{2l-1}{n+1}\frac{\pi}{2} \cos^k\frac{2l-1}{n+1}\pi. \tag{6.18}$$

$\square$

The number $N_k$ in (6.18) is the number of all zig-zag lines in the plane which (i) consist of segments of length $\sqrt{2}$ with direction $\binom{\pm 1}{1}$, (ii) start from one of the points $(0, 0), (1, 0), \dots, (n-1, 0)$ and, without leaving the rectangle $\{(x, y) \in I\!R^2 : 0 \le x \le n-1, 0 \le y \le k\}$, terminate in one of the points

$(0, k), (1, k), \ldots, (n-1, k)$. (The calculation of this number arises in certain problems in the theory of the function spaces.) If instead we wish to know the number of walks of length $k$ in the integer lattice on $\{(x, y) \in \mathbb{Z}^2 : 0 \le x \le n_1 - 1, 0 \le y \le n_2 - 1\}$ then we need to calculate $N_k$ for the graph $P_{n_1} + P_{n_2}$.

Another interpretation of (6.18) is as the number of possible walks in $k$ moves by a king on a one-dimensional chessboard. For a two-dimensional chessboard of size $n_1 \times n_2$, we need to calculate $N_k$ for the graph $P_{n_1} * P_{n_2}$.

Both the sum $P_{n_1} + P_{n_2}$ and the strong product $P_{n_1} * P_{n_2}$ are examples of the NEPS considered in Section 2.5, and so we extend our remarks to an arbitrary non-complete extended $p$-sum $G$ of graphs $G_1, \ldots, G_s$, say with basis $\mathcal{B}$. For an orthonormal basis of eigenvectors of $G$ we may take the vectors $\mathbf{u}_{1i_1} \otimes \cdots \otimes \mathbf{u}_{si_s}$, where the vectors $\mathbf{u}_{ji_j}$ form an orthonormal basis of $\mathbb{R}^{n_j}$ consisting of eigenvectors of $G_j$. We have

$$\mathbf{j}^\top (\mathbf{u}_{1i_1} \otimes \cdots \otimes \mathbf{u}_{si_s}) = \gamma_{1i_1} \cdots \gamma_{si_s},$$

where $\gamma_{ji_j}$ is the sum of entries of $\mathbf{u}_{ji_j}$. Hence the number of walks of length $k$ in $G$ is given by:

$$N_k = \sum_{i_1, \ldots, i_s} \gamma_{1i_1}^2 \cdots \gamma_{si_s}^2 \left( \sum_{\mathcal{B}} \lambda_{1i_1}^{\beta_1} \cdots \lambda_{si_s}^{\beta_s} \right)^k,$$

where $\sum_{\mathcal{B}}$ denotes the sum over all $(\beta_1, \ldots, \beta_s) \in \mathcal{B}$.

For an $s$-dimensional chessboard of size $n_1 \times \cdots \times n_s$, we have $G_j = P_{n_j}$ and $\mathcal{B} = \{0, 1\}^s \setminus \{(0, \ldots, 0)\}$. Then the number of possible walks traversed by a king in $k$ moves is given by:

$$N_k = \sum_{i_1, \ldots, i_s} \gamma_{1i_1}^2 \cdots \gamma_{si_s}^2 \left( -1 + \prod_{j=1}^{s} (\lambda_{ji_j} + 1) \right)^k,$$

where

$$\gamma_{ji_j}^2 = \frac{2}{n_j + 1} \cot^2 \frac{2i_j - 1}{n_j + 1} \frac{\pi}{2} \quad \text{and} \quad \lambda_{ji_j} = 2 \cos \frac{2i_j - 1}{n_j + 1} \pi.$$

We make one remark on the number $a_{jj}^{(k)}$ of walks of length $k$ starting and terminating at a given vertex $j$ in an arbitrary graph $G$. From Equation (2.21) we have

$$a_{jj}^{(k)} = \sum_{i=1}^{m} \mu_i^k \|P_i \mathbf{e}_j\|^2.$$

Proceeding as before, and with the same notation, we may reformulate this equation as

$$a_{jj}^{(k)} = \sum_{h=1}^{n} u_{jh}^2 \lambda_h^k.$$

The calculation of $a_{jj}^{(k)}$ when $G = P_n$ is left as an exercise.

We conclude with two results of a different nature, concerning $\sum_{h=1}^{n} a_{jh}^{(k)}$ and $a_{ij}^{(k)}$.

**Theorem 6.7.2** [Wei]. *Let $N_k(j)$ be the number of walks of length $k$ starting at the vertex $j$ of a non-bipartite connected graph $G$ with vertices $1, 2, \ldots, n$. Let $s_k(j) = N_k(j) \cdot \left(\sum_{j=1}^{n} N_k(j)\right)^{-1}$. Then, as $k \to \infty$, the vector $(s_k(1), s_k(2), \ldots, s_k(n))^\top$ approaches an eigenvector corresponding to the index of $G$.*

**Proof.** As before, let $\{\mathbf{u}_1, \ldots, \mathbf{u}_n\}$ be an orthonormal basis of $I\!R^n$ such that $A\mathbf{u}_h = \lambda_h \mathbf{u}_h$, and let $\gamma_h = \mathbf{j}^\top \mathbf{u}_h$ $(h = 1, \ldots, n)$. Here we take $\lambda_1 \geq \lambda_2 \geq \cdots \geq \lambda_n$, with $\mathbf{u}_1$ the principal eigenvector of $G$. The vector under consideration is $(\mathbf{j}^\top A^k \mathbf{j})^{-1} A^k \mathbf{j}$, or

$$\frac{\gamma_1 \lambda_1^k \mathbf{x}_1 + \gamma_2 \lambda_2^k \mathbf{x}_2 + \cdots + \gamma_n \lambda_n^k \mathbf{x}_n}{\gamma_1^2 \lambda_1^k + \gamma_2^2 \lambda_2^k + \cdots + \gamma_n^2 \lambda_n^k}.$$

By Theorems 1.3.6 and 3.2.4 we have $\gamma_1 > 0$ and $\lambda_1 > |\lambda_h|$ for all $h > 1$. Consequently the vector $(\gamma_1^2 \lambda_1^k + \cdots + \theta_n^2 \lambda_n^k)^{-1} \gamma_h \lambda_h^k \mathbf{u}_h$ approaches $\gamma_1^{-1} \mathbf{u}_h$ if $h = 1$, and approaches $\mathbf{0}$ if $h > 1$. The result follows.  $\square$

Note that Theorem 6.7.2 holds also for connected *regular* bipartite graphs because then $\gamma_n = 0$ (by Proposition 1.1.2) while $\lambda_1 > |\lambda_i|$ for all $i \in \{2, \ldots, n-1\}$.

The following result has a similar proof.

**Theorem 6.7.3** [LiFe]. *Let $G$ be a connected non-bipartite graph with index $\lambda_1$ and principal eigenvector $(x_1, x_2, \ldots, x_n)^\top$. For fixed vertices $i$ and $j$, the number of $i$-$j$ walks of length $k$ is asymptotic to $\lambda_1^k x_i x_j$ as $k \to \infty$.*

# Exercises

**6.1** Show that $K_{55}$ is not the edge-disjoint union of three copies of $L(K_{11})$.

**6.2** Show that, for a uniform homomorphism from a regular graph $G$ to a regular graph $H$, equations (6.5) and (6.6) are equivalent.

**6.3** Show that a Moore graph is distance-regular.

**6.4** Verify that $HoS$ is a Moore graph.

**6.5** Verify that the graph $H_0$ illustrated in Fig. 6.2 is an induced subgraph of $HoS$.

**6.6** Verify that the spectrum of the graph $H_0$ (Fig. 6.2) is $3^1, \sqrt{2}^{\ 6}$, $0^8, (-\sqrt{2})^6, -3^1$, and that the spectrum of $HoS$ is $7^1, 2^{28}, -3^{21}$.

**6.7** Find the parameters and eigenvalues of the possible strongly regular graphs on 76 and 96 vertices that arise in Section 6.6.

**6.8** Find a formula for the number of walks of length $k$ in an $n$-cycle.

**6.9** Find a formula for the number of $j$-$j$ walks of length $k$ in the path $P_n$.

**6.10** Prove Theorem 6.7.3.

**6.11** Show that the Petersen graph is non-Hamiltonian by applying the Interlacing Theorem to its line graph [GoRo].

# Notes

The first part of Section 6.1 is taken from [CvRS2, Chapter 9]. Generalizations of Theorem 6.1.1 may be found in [Dam3]. Example 6.1.2 appears in [Row2], while the examples in Section 6.2 appear in [DanHa2] in the context of Laplacian eigenvalues. The proof of Lemma 6.1.4 is taken from course notes of Brouwer and Haemers. The proof of Theorem 6.3.1 is derived from a discussion of the 'ordered love problem' in [Ham, Section 7]. One of the first proofs of the Friendship Theorem can be found in [ErRS].

A proof that a Moore graph other than an odd cycle has diameter 2 may be found in [Big2, Chapter 23], along with references to the original papers and an alternative construction of the Hoffman–Singleton graph $HoS$. The uniqueness of $HoS$ as a Moore graph of diameter 2 and degree 7 is established in [HofSi, Section 5]. Aschbacher [Asch] proved that a Moore graph of diameter 2 and degree 57 cannot be a rank three graph, and subsequently G. Higman showed that such a graph cannot be vertex-transitive (see [Cam1, Section 3.7]).

The basic properties of generalized polygons are established in [GoRo, Section 5.6]. Generalized quadrangles may be defined in terms of partial linear spaces: see [GoRo, Section 5.4]. More details of those constructed from $L(K_{3,3})$, $\overline{L(K_6)}$ and $Sch_{10}$ can be found in [BroCN, Section 1.15]. Further examples are constructed in [GoRo, Section 5.5]. A discussion of equiangular lines in the context of two-graphs may be found in [GoRo, Chapter 11]. Our treatment of extremal sets of equiangular lines is based on notes of lectures by Seidel on geometrical configurations. The results of Section 6.7 appear in [CvRS2, Section 2.2].

# 7

# Laplacians

Let $A$ be the adjacency matrix of a graph, and $D$ the diagonal matrix of vertex degrees. In this chapter we discuss the Laplacian $L = D - A$, the signless Laplacian $Q = D + A$, and the normalized Laplacian $\hat{L} = D^{-\frac{1}{2}}LD^{-\frac{1}{2}}$ (defined initially for graphs without isolated vertices). In the literature, $L$ is also referred to as the *Kirchhoff matrix* or *admittance matrix*, $Q$ is sometimes called the *co-Laplacian*, and $\hat{L}$ the *correlation matrix* or *transition matrix*. The Laplacian arises naturally in the study of electrical circuits, and the normalized Laplacian is closely related to random walks on a graph (Section 7.7). Both $L$ and $\hat{L}$ have a strong pedigree as discrete analogues of certain operators in differential geometry, and they are well suited to the spectral investigation of expansion and separation properties of a graph. We have already noted in Section 4.2 some evidence that the spectra of $L$ and $Q$ can be more effective than the spectrum of $A$ in distinguishing non-isomorphic graphs.

## 7.1 The Laplacian spectrum

Let $L$ ($= L_G$) be the Laplacian matrix of a graph $G$ with $n$ vertices and $m$ edges. We write $\nu_i$ ($= \nu_i(G)$) for the $i$-th largest eigenvalue of $L$, so that

$$\nu_1(G) \geq \nu_2(G) \geq \cdots \geq \nu_n(G).$$

We show first that $L$ is a positive semi-definite matrix by assigning an arbitrary orientation to the edges of $G$. The vertex–arc incidence matrix of the corresponding digraph $\vec{G}$ is the $n \times m$ matrix $R = (r_{ie})$ where

$$r_{ie} = \begin{cases} -1 & \text{if } i \text{ is an initial vertex of the arc } e, \\ 0 & \text{if } i \text{ and } e \text{ are not incident}, \\ +1 & \text{if } i \text{ is a terminal vertex of the arc } e. \end{cases}$$

We refer to $R$ as the *gradient matrix* of $\vec{G}$ (and a gradient matrix of $G$).

It is straightforward to verify that $L = RR^{\top}$, whatever the orientation of $G$. Hence $L$ is a positive semi-definite matrix, and so all its eigenvalues are non-negative. Note that $\nu_n = 0$ since $L\mathbf{j} = \mathbf{0}$, where $\mathbf{j}$ is the all-1 vector in $\mathbb{R}^n$.

Since $\mathbf{j}$ is an eigenvector, $L$ has $n - 1$ linearly independent eigenvectors in $\mathbf{j}^{\perp}$. This is an attractive feature of $L$ which means that we need not dwell on main eigenvalues, and that we can deal easily with complements:

**Proposition 7.1.1.** *We have* $\nu_n(\overline{G}) = 0$ *and* $\nu_i(\overline{G}) = n - \nu_{n-i}(G)$ *($i = 1, 2, \ldots, n - 1$).*

**Proof.** Let $\{\mathbf{x}_1, \mathbf{x}_2, \ldots, \mathbf{x}_n\}$ be an orthogonal basis of $\mathbb{R}^n$ such that $L_G\mathbf{x}_i = \nu_i\mathbf{x}_i$ ($i = 1, 2, \ldots, n$) and $\mathbf{x}_n = \mathbf{j}$. Since $L_{\overline{G}} = nI - L_G - J$ we have $L_{\overline{G}}\mathbf{x}_n = \mathbf{0}$ and $L_{\overline{G}}\mathbf{x}_i = (n - \nu_i)\mathbf{x}_i$ ($i = 1, 2, \ldots, n - 1$). The result follows. $\square$

When $n > 1$, Rayleigh's Principle yields the following expression for $\nu_{n-1}(G)$:

$$\nu_{n-1}(G) = \inf_{\mathbf{x}\in\mathbb{R}^n\setminus\{\mathbf{0}\},\ \mathbf{x}\perp\mathbf{j}} \frac{\mathbf{x}^{\top}L\mathbf{x}}{\mathbf{x}^{\top}\mathbf{x}}. \tag{7.1}$$

In addition we have

$$\mathbf{x}^{\top}L\mathbf{x} = \mathbf{x}^{\top}RR^{\top}\mathbf{x} = \|R^{T}\mathbf{x}\|^2 = \sum_{uv\in E}(x_u - x_v)^2, \tag{7.2}$$

and consequently

$$\nu_{n-1}(G) = \inf_{\mathbf{x}\in\mathbb{R}^n\setminus\{\mathbf{0}\},\ \mathbf{x}\perp\mathbf{j}} \frac{\sum_{uv\in E(G)}(x_u - x_v)^2}{\sum_{v\in V(G)}x_u^2}. \tag{7.3}$$

Now $\sum_{uv\in E(G)}(x_u - x_v)^2 = 0$ if and only if, for each component $H$ of $G$, the entries $x_u$ ($u \in V(H)$) are the same. Such a non-zero vector exists in $\mathbf{j}^{\perp}$ if and only if $G$ has more than one component. Hence $\nu_{n-1}(G) \neq 0$ if and only if $G$ is connected, and by considering components in the general case, we have:

**Theorem 7.1.2.** *The multiplicity of* 0 *as an eigenvalue of* $L_G$ *is equal to the number of components in* $G$.

Thus the spectrum of $L$, unlike the spectrum of $A$, determines the number of components in a graph. In what follows, we shall explore some parallels between the Laplacian spectrum and the adjacency spectrum. By considering the trace of $L$, we obtain:

$$\nu_1 + \nu_2 + \cdots + \nu_n = d_1 + d_2 + \cdots + d_n,$$

where $d_1, d_2, \ldots, d_n$ are the vertex degrees. Thus the number $m$ of edges is determined by the Laplacian spectrum:

$$m = \frac{1}{2}(\nu_1 + \nu_2 + \cdots + \nu_n). \tag{7.4}$$

Since $L$ is positive semi-definite, Theorem 1.3.2 yields:

**Theorem 7.1.3.** *Let $G$ be a graph with Laplacian eigenvalues $\nu_1 \geq \nu_2 \geq \cdots \geq \nu_n$. If the vertex degrees of $G$ are $d_1 \geq d_2 \geq \cdots \geq d_n$ then*

$$\sum_{i=1}^{k} \nu_i \geq \sum_{i=1}^{k} d_i \quad (k = 1, 2, \ldots, n), \tag{7.5}$$

*with equality when $k = n$.*

**Remarks 7.1.4.** (i) From (7.4) we see that

$$\frac{n-1}{n}\nu_{n-1}(G) \leq \bar{d} \leq \frac{n-1}{n}\nu_1(G),$$

where $\bar{d}$ denotes the mean degree. These two inequalities were improved by Fiedler [Fie1] as follows (see Exercise 7.10):

$$\nu_{n-1}(G) \leq \frac{n}{n-1}\delta \quad \text{and} \quad \nu_1(G) \geq \frac{n}{n-1}\Delta, \tag{7.6}$$

where $\delta$ and $\Delta$ are minimum and maximum degree, respectively. It is shown in [GroMe2] that if $G$ is not a null graph then $\nu_1 \geq \Delta + 1$; equality holds if and only if $\Delta = n - 1$. Further bounds for $\nu_1$ and $\nu_{n-1}$ are discussed in Sections 7.3 and 7.4.

(ii) The inequalities (7.5) are strengthened in [Gro] as follows. Let $G$ be a nontrivial connected graph with $d_1 \geq d_2 \geq \cdots \geq d_n$, and let $t_k$ be the number of components of the subgraph $G$ induced by the vertices $1, 2, \ldots, k$. Then

$$\sum_{i=1}^{k} \nu_i \geq t_k + \sum_{i=1}^{k} d_i \quad (k = 1, 2, \ldots, n-1). \qquad \square$$

We say that two graphs are $L$-cospectral if they have the same Laplacian spectrum. From what we have seen so far, we know that $L$-cospectral graphs have the same numbers of vertices, edges and components; the smallest pair of $L$-cospectral graphs is shown in Fig. 7.1 (see the Appendix, Table A1). Further examples of $L$-cospectral graphs may be constructed using the results on characteristic polynomials that appear later in this section.

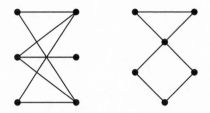

Figure 7.1 The smallest pair of $L$–cospectral graphs.

Next observe that we cannot invoke an analogue of the Interlacing Theorem when we delete vertices because a principal submatrix of $L$ is not the Laplacian matrix of the corresponding induced subgraph. However we do have an analogue in respect of edges:

**Theorem 7.1.5.** *If $e$ is an edge of the graph $G$ and $G' = G - e$ then*

$$0 = \nu_n(G') = \nu_n(G) \le \nu_{n-1}(G') \le \nu_{n-1}(G) \le \cdots \le \nu_2(G) \le \nu_1(G') \le \nu_1(G).$$

**Proof.** If $L$ is the Laplacian of $G - e$ then the Laplacian of $G$ has the form $L + M$, where $M$ is positive semi-definite of rank 1; the largest eigenvalue of $M$ is 2, and all other eigenvalues are 0. The result now follows by applying the Courant–Weyl inequalities (Theorem 1.3.15) to the matrix $L + M$. $\qquad\square$

If the graph $H$ is obtained from $G$ by deleting $k$ edges ($k < n$), then $k$ applications of Theorem 7.1.5 yield the interlacing property

$$\nu_{i+k}(G) \le \nu_i(H) \le \nu_i(G) \quad (i = 1, 2, \ldots, n - k).$$

A special case of Theorem 7.1.5 was noted in [So]: if $G' = G - uv$, where $u$ and $v$ are co-duplicate vertices then $\nu_j(G - uv) = \nu_j(G) - 2$ for some $j$, while $\nu_i(G - uv) = \nu_i(G)$ for all $i \ne j$. In the general case, we have $\sum_{i=1}^n \nu_i(G) - \sum_{i=1}^n \nu_i(G-e) = 2$ by (7.4), and so always $\nu_j(G) > \nu_j(G-e)$ for at least one value of $j$.

Next we point out that the divisor technique (Section 3.9) can be used in the Laplacian context. Recall that if $G$ is a graph with an equitable partition $V(G) = V_1 \dot{\cup} V_2 \dot{\cup} \cdots \dot{\cup} V_k$ then, for every $i, j \in \{1, 2, \ldots, k\}$, there exists a number $d_{ij}$ such that each vertex in $V_i$ is adjacent to exactly $d_{ij}$ vertices in $V_j$.

**Theorem 7.1.6.** *Let $V_1 \dot{\cup} V_2 \dot{\cup} \cdots \dot{\cup} V_k$ be an equitable partition of the graph $G$, with parameters $d_{ij}$ ($i, j \in \{1, 2, \ldots, k\}$), and let $B$ be the $k \times k$ matrix $(b_{ij})$ defined by*

$$b_{ij} = \begin{cases} -d_{ij} & \text{if } i \neq j, \\ \sum_{s=1}^{k} d_{is} - d_{ij} & \text{if } i = j. \end{cases}$$

*If $v$ is an eigenvalue of $B$ then $v$ is also an eigenvalue of $L_G$.*

**Proof.** Let $By = vy$, where $y = (y_1, y_2, \ldots, y_k)^\top$. Suppose that $|V(G)| = n$ and define $x = (x_1, x_2, \ldots, x_n)^\top$ by the relation: if $v \in V_i$ then $x_v = y_i$. Now $L_G x = v x$, for if $v \in V_i$ then

$$(L_G x)_v = \deg(v) x_v - \sum_{u \sim v} x_u = \sum_{j=1}^{k} d_{ij} y_i - \sum_{j=1}^{k} d_{ij} y_j = (By)_i = v y_i = v x_v.$$

This completes the proof.                                                                $\square$

Following [CvDSa], we write $C_G(x)$ for $\det(xI - L_G)$, called the Laplacian characteristic polynomial or *L-polynomial* of the graph $G$. We conclude this section by expressing the $L$-polynomials of certain compound graphs in terms of the $L$-polynomials of the constituent graphs. The first observation reflects a simple property of determinants.

**Theorem 7.1.7.** *If $G$ is the disjoint union of graphs $G_1, G_2, \ldots, G_k$ then*

$$C_G(x) = \prod_{i=1}^{k} C_{G_i}(x).$$

For the complement of a graph, we have immediately from Proposition 7.1.1:

**Theorem 7.1.8** [Kel1, Kel2]. *If $G$ is a graph with $n$ vertices then*

$$C_{\overline{G}}(x) = (-1)^{n-1} \frac{x}{n - x} C_G(n - x).$$

Since the join $G_1 \triangledown G_2$ is the complement of $\overline{G_1} \,\dot\cup\, \overline{G_2}$, three applications of Theorem 7.1.7 yield:

**Theorem 7.1.9.** *If $G_1, G_2$ are graphs with $n_1, n_2$ vertices respectively, then*

$$C_{G_1 \triangledown G_2}(x) = \frac{x - n_1 - n_2}{(x - n_1)(x - n_2)} C_{G_1}(x - n_2) C_{G_2}(x - n_1).$$

For the line graph $L(G)$, subdivision graph $S(G)$ and total graph $T(G)$ of a regular graph $G$, the following formulae (see [Fie1]) are straightforward analogues of corresponding results in Chapter 2. (The first formula is found also in [Vah].)

**Theorem 7.1.10.** *Let* $G$ *be an* $r$*-regular graph with* $n$ *vertices and* $m$ *edges. Then*

(i) $C_{L(G)}(x) = (x - 2r)^{m-n} C_G(x)$;

(ii) $C_{S(G)}(x) = (-1)^m (2 - x)^{m-n} C_G(x(r+2-x))$;

(iii) $C_{T(G)}(x) = (-1)^m (r+1-x)^n (2r+2-x)^{m-n} C_G(\frac{x(2r+2-x)}{r+1-x})$.

**Theorem 7.1.11.** *Let* $G$ *be a semi-regular bipartite graph with* $n$ *vertices,* $m$ *edges and parameters* $(n_1, n_2, r_1, r_2)$. *Then* $n = n_1 + n_2$, $m = n_1 r_1 = n_2 r_2$ *and*

$$C_{L(G)}(x) = (-1)^m (r_1 + r_2 - x)^{m-n} C_G(r_1 + r_2 - x).$$

For graphs obtained as NEPS, fewer results carry over from $A$ to $L$, but we can deal with sums by observing that $L_{G+H}$ has the form $L_G \otimes I + I \otimes L_H$. Accordingly, we have:

**Theorem 7.1.12** [Fie1]. *If* $G$ *has* $m$ *vertices and* $H$ *has* $n$ *vertices then the Laplacian eigenvalues of* $G + H$ *are the* $mn$ *numbers*

$$v_i(G) + v_j(H) \quad (i = 1, 2, \ldots, m; \quad j = 1, 2, \ldots, n).$$

## 7.2 The Matrix-Tree Theorem

We shall see that, for any graph $G$, the eigenvalues of $L_G$ determine the number of spanning trees in $G$. This number is called the *complexity* of $G$, denoted by $\tau(G)$. The result follows from a classical theorem of algebraic graph theory known as the 'Matrix-Tree Theorem'. This theorem says that for any connected graph $G$, all cofactors of $L_G$ are equal, and their common value is $\tau(G)$.

We write $L = L_G$ and assume first that $G$ is connected. It is easy to see that the cofactors of $L$ are all the same: we have $L \operatorname{adj}(L) = \det(L)I = O$, where the adjoint $\operatorname{adj}(L)$ is the matrix of cofactors. Since $G$ is connected, we know from the proof of Theorem 7.1.1 that the nullspace of $L$ is spanned by the all-1 vector $\mathbf{j}$. Thus each column of $\operatorname{adj}(L)$ is a scalar multiple of $\mathbf{j}$. Since $\operatorname{adj}(L)$ is symmetric, $\operatorname{adj}(L)$ has the required form $\alpha J$, where $J$ is the all-1 matrix. It remains to be shown that $\alpha = \tau(G)$.

**Lemma 7.2.1.** *Let* $R$ *be the gradient matrix of a non-trivial oriented tree. If* $R'$ *is obtained from* $R$ *by deleting any row then* $\det(R') = \pm 1$.

**Proof.** The proof is by induction on the number $n$ of vertices in a non-trivial oriented tree $T$. The result is immediate if $n = 2$ and so we assume that $n > 2$.

Suppose that $R'$ is obtained from $R$ by deleting row $v$, and let $u$ be a neighbour of $v$. We take $V(T) = \{1, 2, \ldots, n\}$, with $v = n$, $u = n-1$. Without loss of generality, we assume that the last column of $R$ is indexed by the edge $uv$.

Let $R^*$ be the $(n-1) \times (n-2)$ matrix obtained from $R$ by adding the $n$-th row to the $(n-1)$-th row and then deleting the last row and column. Then $R^*$ is the gradient matrix of the oriented tree $T^*$ obtained from $T$ by contracting the edge $uv$ to a vertex $v^*$. Now $\det(R') = \pm \det(R'')$ where $R''$ is obtained from $R^*$ by deleting row $v^*$. By our induction hypothesis, $\det(R'') = \pm 1$, and so $\det(R') = \pm 1$. The result follows.                                    □

**Theorem 7.2.2** (The Matrix-Tree Theorem). *If $L$ is the Laplacian matrix of a graph $G$ then each cofactor of $L$ is equal to $\tau(G)$, the number of spanning trees of $G$.*

**Proof.** If $G$ is not connected then $\tau(G) = 0$, while each cofactor of $L$ is $0$ because $L$ has rank at most $n-2$. Accordingly, we suppose that $G$ is connected.

Let $R$ be a gradient matrix of $G$, and for any set $F$ of $n-1$ edges of $G$, let $R(F)$ be the $n \times (n-1)$ matrix consisting of the columns of $R$ indexed by $F$. For any vertex $i$, let $R_i(F)$ be the matrix obtained from $R(F)$ by deleting row $i$, and let $R_i$ be the matrix obtained from $R$ by deleting row $i$. The $i$-th diagonal entry of $\mathrm{adj}(L)$ is $\det(R_i R_i^\top)$, and by the Binet–Cauchy formula (Theorem 1.3.18) we have

$$\det(R_i R_i^\top) = \sum_F \det(R_i(F)) \det(R_i(F)^\top). \qquad (7.7)$$

We show that for a fixed set $F$ of $n-1$ edges, we have $\det(R_i(F)) = \pm 1$ if the edges of $F$ determine a spanning tree in $G$, and $\det(R_i(F)) = 0$ otherwise.

Suppose first that $F$ does not determine a spanning tree of $G$. Then some subset of $F$, say $C$, forms a cycle in $G$. Without loss of generality we may assume that all edges of $C$ are oriented to create a directed cycle. Then the sum of the corresponding columns of $R(F)$ is zero, and so $\det(R_i(F)) = 0$ as required. On the other hand, if $F$ determines a spanning tree $T$, then it has $R(F)$ as a gradient matrix, and so $\det(R_i(F)) = \pm 1$ by Lemma 7.2.1. It follows that the number of non-zero summands in (7.7) is $\tau(G)$, and each such summand is equal to 1. Hence the diagonal entries of $\mathrm{adj}(L)$ are all equal to $\tau(G)$. We have already seen that all entries of $\mathrm{adj}(L)$ are the same, and so the result follows.                                    □

**Corollary 7.2.3.** *Let $C_G(x)$ be the characteristic polynomial of the Laplacian matrix of $G$. Then*

$$\tau(G) = \frac{(-1)^{n-1}}{n} C'_G(0) = \frac{1}{n} \Pi_{i=1}^{n-1} \nu_i(G).$$

**Proof.** If $G$ is not connected then $\tau(G) = 0$, $\nu_2 = 0$ and $C'_G(0) = 0$. In the case that $G$ is connected, we use the fact that $\det(xI - R_i R_i^\top)$ is the $(i, i)$-cofactor of $xI - L$. Then the result follows from Theorem 7.2.2 because $C'_G(x) = \sum_{i=1}^{n} \det(xI - R_i R_i^\top)$. $\square$

For many classes of graphs the number of spanning trees can be calculated directly, but almost all existing results can be derived using spectral techniques. For regular graphs, we can reformulate Corollary 7.2.3 in terms of the characteristic polynomial or eigenvalues of the adjacency matrix (cf. [Hut]):

**Proposition 7.2.4.** *For an $r$-regular graph $G$ we have*

$$\tau(G) = \frac{1}{n} P'_G(r) = \frac{1}{n} \prod_{i=2}^{n} (r - \lambda_i).$$

In the following examples, we use Proposition 7.2.4 in conjunction with characteristic polynomials given in Section 2.6.

**Examples 7.2.5.** (i) For complete graphs, we have Cayley's formula [Cay]:

$$\tau(K_n) = n^{n-2}.$$

(ii) For cocktail party graphs, we have:

$$\tau((CP(k)) = 2^{2k-2}(k-1)^k k^{k-2}.$$

(iii) If $G$ is the $k$-dimensional lattice of size $n$ (that is, the $k$-fold sum $K_n + \cdots + K_n$) then [Cve2]:

$$\tau(G) = n^{n^k - k - 1} \prod_{i=1}^{k} i^{\binom{k}{i}(n-1)^i}.$$

(iv) For Möbius ladders, we have (cf. p. 49):

$$\tau(M_n) = \frac{1}{2n} \prod_{j=1}^{2n-1} \left( 3 - 2\cos\frac{\pi j}{n} - (-1)^j \right).$$

(v) Let $G$ be a graph with eigenvalues $\lambda_1, \ldots, \lambda_n$, and let $G^{(2)} = G + K_2$. By Theorem 2.5.4 the eigenvalues of $G^{(2)}$ are $\lambda_1 + 1, \ldots, \lambda_n + 1, \lambda_1 - 1, \ldots, \lambda_n - 1$, and so

$$P_{G^{(2)}}(x) = P_G(x-1) P_G(x+1).$$

If $G$ is regular of degree $r$ with $n$ vertices then $G^{(2)}$ is regular of degree $r+1$ with $2n$ vertices, and Proposition 7.2.4 yields

$$\tau(G^{(2)}) = \frac{1}{2n}P'_{G^{(2)}}(r+1) = \frac{1}{2n}P'_G(r)P_G(r+2) = \frac{1}{2}\tau(G)P_G(r+2).$$

Now let $G = C_n$, so that $G^{(2)}$ is the graph of an $n$-faced prism. Clearly, $\tau(C_n) = n$. Since $P_{C_n}(x) = 2T_n(\frac{x}{2}) - 2$, where $T_n(x)$ is the Chebyshev polynomial of the first kind, we obtain

$$\tau(C_n^{(2)}) = nT_n(2) - n.$$

(vi) To find the number of spanning trees in a complete bipartite graph, recall that

$$K_{m,n} = \overline{K_m \dot\cup K_n}.$$

The Laplacian spectrum of $K_m \dot\cup K_n$ is $0^2, m^{m-1}, n^{n-1}$, and by Proposition 7.1.1 its complement has Laplacian spectrum $0, m+n, n^{m-1}, m^{n-1}$. Finally, using Corollary 7.2.3 we obtain

$$\tau(K_{m,n}) = m^{n-1}n^{m-1}.$$

$\square$

We can also deal with graphs in which all vertices but one have a fixed degree $r$; such graphs are called *nearly $r$-regular*, and the vertex not of degree $r$ is called the *exceptional* vertex.

**Proposition 7.2.6** [CvGu2]. *Let $G$ be a nearly regular graph of degree $r$ and let $H$ be the subgraph obtained by removing the exceptional vertex. Then $\tau(G) = P_H(r)$.*

**Proof.** If $L$ is the Laplacian of $G$, and the $i$-th vertex of $G$ is exceptional then the $i$-th diagonal entry of adj$(L)$ is $C_H(0)$, and this is equal to $P_H(r)$.   $\square$

For instance, the wheel $W_{n+1}$ is a nearly regular graph, obtained from $C_n$ by adding a vertex adjacent to all others. Applying Proposition 7.1.14 in conjunction with Example 7.1.13(v) we obtain (cf. [Nos]):

$$\tau(W_{n+1}) = 2T_n(\tfrac{3}{2}) - 2.$$

Next consider a plane graph $G$, with dual $G^*$. The *inner dual $G^{**}$* is obtained from $G^*$ by deleting the vertex corresponding to the infinite region of $G$. It is well known that $G$ and its dual $G^*$ have the same number of spanning trees [Big2, p. 43]. In the case that each finite region of $G$ is bounded by an $r$-cycle, $G^*$ is nearly $r$-regular, and so we may apply Proposition 7.2.6 to $G^*$ to obtain:

**Proposition 7.2.7** [CvGu2]. *Let $G$ be a plane graph, and let $G^{**}$ be its inner dual. If every finite region of $G$ is bounded by an $r$-cycle then $\tau(G) = P_{G^{**}}(r)$.*

More recently, Hammer and Kel'mans [HaKe] have investigated the Laplacian eigenvalues of *threshold graphs*; such graphs are constructed from a trivial graph by the successive addition of vertices adjacent to no other vertex or all other vertices. In this case, the Laplacian spectrum is close to the degree sequence (cf. Equation (7.5)), and the number of spanning trees can be expressed in terms of vertex degrees.

The Matrix-Tree Theorem was generalized by Kel'mans and Chelnokov, who gave an interpretation of the coefficients of $C_G(x)$ in terms of spanning subforests of $G$:

**Theorem 7.2.8** [KelCh]. *If $C_G(x) = x^n + c_1 x^{n-1} + \cdots + c_{n-1}x$ then*

$$c_i = (-1)^i \sum_{|E(F)|=i} p(F) \quad (i = 1, 2, \ldots, n-1),$$

*where the sum is taken over all spanning forests $F$, and $p(F)$ is the product of the numbers of vertices in the components of $F$.*

We state, also without proof, a version for multigraphs due to Kel'mans. (Here, the amalgamation of vertices with a common neighbour results in multiple edges.)

**Theorem 7.2.9** [Kel3]. *If $C_G(x) = x^n + c_1 x^{n-1} + \cdots + c_{n-1}x$ then*

$$c_i = (-1)^i \sum_{U \subseteq V,\ |U|=n-i} \tau(G_{[U]}) \quad (i = 1, 2, \ldots, n-1),$$

*where $G_{[U]}$ is obtained from $G$ by amalgamating all vertices of $U$.*

## 7.3 The largest eigenvalue

Since the supremum of $x^\top L x / x^\top x$ ($x \neq 0$) is attained when $x$ is orthogonal to $j$, we have:

$$\nu_1(G) = \sup_{x \in \mathbb{R}^n \setminus \{0\},\ x \perp j} \frac{\sum_{uv \in E}(x_u - x_v)^2}{\sum_{v \in V} x_u^2}.$$

For an alternative formula due to Fiedler, observe first that if $x = (x_1, \ldots, x_n)^\top \in j^\perp$ then

$$\sum_{u \in V} \sum_{v \in V} (x_u - x_v)^2 = 2n \sum_{i=1}^n x_i^2,$$

and so

$$v_1(G) = 2n \sup_{\mathbf{x} \in \mathbb{R}^n \setminus \{\mathbf{0}\}, \, \mathbf{x} \perp \mathbf{j}} \frac{\sum_{uv \in E} (x_u - x_v)^2}{\sum_{u \in V} \sum_{v \in V} (x_u - x_v)^2}. \qquad (7.8)$$

Secondly, if $\mathbf{x} \in \mathbb{R}^n \setminus \langle \mathbf{j} \rangle$ then $\mathbf{x} = \mathbf{x}' + \alpha \mathbf{j}$ for some $\alpha$, where $\mathbf{x}'$ is a non-zero vector orthogonal to $\mathbf{j}$. But now the quotient in (7.8) has the same value when $\mathbf{x}$ is replaced with $\mathbf{x}'$, and so

$$v_1(G) = 2n \sup_{\mathbf{x} \in \mathbb{R}^n \setminus \langle \mathbf{j} \rangle} \frac{\sum_{uv \in E} (x_u - x_v)^2}{\sum_{u \in V} \sum_{v \in V} (x_u - x_v)^2}. \qquad (7.9)$$

The following inequalities for $v_1$ follow directly from Rayleigh's Principle, using the relation

$$v_1(G) = \sup_{\mathbf{x} \in \mathbb{R}^n \setminus \{\mathbf{0}\}} \frac{\mathbf{x}^\top L \mathbf{x}}{\mathbf{x}^\top \mathbf{x}}.$$

**Theorem 7.3.1** [Moh2]. *If $G_1$ and $G_2$ are graphs with the same vertex set (but not necessarily with disjoint edge sets) then*

$$\max\{v_1(G_1), v_1(G_2)\} \leq v_1(G_1 \cup G_2) \leq v_1(G_1) + v_1(G_2).$$

**Corollary 7.3.2.** *If $H$ is a spanning subgraph of $G$, then*

$$v_1(H) \leq v_1(G).$$

Since $v_1(G) + v_{n-1}(\overline{G}) = n$, Theorem 7.1.2 yields the following upper bound for $v_1$:

**Proposition 7.3.3.** *If $G$ is a graph on $n$ vertices, then*

$$v_1(G) \leq n,$$

*with equality for a connected graph $G$ if and only if $\overline{G}$ is not connected.*

Sometimes we obtain a better bound as follows:

**Theorem 7.3.4** [AnMo]. *For any graph $G$,*

$$v_1(G) \leq \max\{d_u + d_v : u \sim v\}.$$

**Proof.** Since $R R^\top (= L)$ and $R^\top R$ have the same non-zero eigenvalues, there exists a non-zero vector $\mathbf{y}$ such that $v_1 \mathbf{y} = R^\top R \mathbf{y}$. Let $y_e$ be the entry of $\mathbf{y}$ with maximum modulus, and let $e$ be the arc $hk$. We have

$$v_1 y_e = \sum_f \left( \sum_i r_{ie} r_{if} \right) y_f,$$

while $r_{ie}r_{if} \neq 0$ if and only if the vertex $i$ is common to the the arcs $e$ and $f$. Therefore,

$$v_1 y_e = \sum_f r_{hf} y_f + \sum_f r_{kf} y_f,$$

whence $v_1|y_e| \leq d_h|y_e| + d_k|y_e|$. The result follows. □

The last bound can be expressed in the form $v_1(G) \leq 2 + \max\{\deg(e) : e \in E(G)\}$, where $\deg(e)$ denotes the degree of the edge $e$ in $G$. This was improved in [LiZh1] to:

$$v_1(G) \leq 2 + \max\{\sqrt{\deg(e)\deg(f)} : e, f \in E(G), e \neq f\}.$$

Next, let $m_v$ be the average degree of the neighbours of the vertex $v$. The following bound (see [Mer4]) is in many situations better than that of Theorem 7.3.4:

$$v_1(G) \leq \max\{d_u + m_v : u \sim v\}.$$

This in turn can be improved as follows:

**Theorem 7.3.5** [LiZh2]. *For any graph $G$ on $n$ vertices,*

$$v_1(G) \leq \max_{i \sim j} \frac{d_i(d_i + m_i) + d_j(d_j + m_j)}{d_i + d_j}.$$

**Proof.** Let $\mathbf{y}$ be a unit eigenvector of $R^\top R$ corresponding to $v_1$, and let $\mathbf{z}$ be obtained from $\mathbf{y}$ by taking absolute values of entries. If $B$ is the incidence matrix of $G$ then $B^\top B = A(L(G)) + 2I$ and we have

$$v_1 = \mathbf{y}^\top R^\top R \mathbf{y} \leq \mathbf{z}^\top B^\top B \mathbf{z} \leq \lambda_1(A(L(G)) + 2I).$$

Now let $\mathbf{w}$ be the vector with entries $d_u + d_v$ ($uv \in E(G)$), and write $\Gamma(u)$ for the neighbourhood of $u$ ($u \in V(G)$). The entry of $(A(L(G) + 2I)\mathbf{w}$ indexed by the edge $ij$ is

$$\sum_{u \in \Gamma(j)\setminus\{i\}} (d_u + d_j) + \sum_{v \in \Gamma(i)\setminus\{j\}} (d_i + d_v) + 2(d_i + d_j) = d_i(d_i + m_i) + d_j(d_j + m_j).$$

If $G$ is connected then we may apply Corollary 1.3.7 to $(A(L(G)) + 2I)$ to obtain

$$\lambda_1(A(L(G)) + 2I) \leq \max_{i \sim j} \frac{d_i(d_i + m_i) + d_j(d_j + m_j)}{d_i + d_j}.$$

and the result follows. If $G$ is not connected then it suffices to invoke the result for each component of $G$. □

We mention without proof several results in similar vein. The first two bounds were obtained by Zhang [Zha], and the third by Das [Das2]:

$$\nu_1(G) \leq \max\left\{\sqrt{d_i(d_i + m_i) + d_j(d_j + m_j)} : i \sim j\right\},$$

$$\nu_1(G) \leq \max\left\{2 + \sqrt{d_i(d_i + m_i - 4) + d_j(d_j + m_j - 4) + 4} : i \sim j\right\},$$

$$\nu_1(G) \leq \max\left\{\frac{1}{2}\left(d_i + d_j + \sqrt{(d_i - d_j)^2 + 4m_i m_j}\right) : i \sim j\right\}.$$

For upper bounds in which the maximum is taken over vertices rather than edges we have the following, which are due to Zhang [Zha], Li and Pan [LiPa], and Guo [Guo1] respectively:

$$\nu_1(G) \leq \max_{1 \leq i \leq n}\left\{d_i + \sqrt{d_i m_i}\right\},$$

$$\nu_1(G) \leq \max_{1 \leq i \leq n}\left\{\sqrt{2d_i(d_i + m_i)}\right\},$$

$$\nu_1(G) \leq \max_{1 \leq i \leq n}\left\{\frac{1}{2}\left(d_i + \sqrt{d_i^2 + 8d_i m_i}\right)\right\}.$$

We have already noted in Remark 7.1.4 that $\nu_1(G) \geq 1 + \Delta(G)$ for a non-null graph $G$. Here we establish a lower bound which follows from a general result for Hermitian matrices [Mir, Theorem 2]:

**Theorem 7.3.6.** *If $G$ is a graph with vertex degrees $d_1, \ldots, d_n$ and adjacency matrix $(a_{ij})$, then*

$$\nu_1(G) \geq \max\left\{\sqrt{(d_i - d_j)^2 + 4a_{ij}} : i, j \in V(G), \ i \neq j\right\}.$$

**Proof.** Consider the principal submatrix $M = \begin{pmatrix} d_i & -a_{ij} \\ -a_{ji} & d_j \end{pmatrix}$ of the Laplacian $D - A$. Let $\nu_1'$, $\nu_2'$ be the eigenvalues of $M$, with $\nu_1' \geq \nu_2'$. By interlacing, we have $\nu_1(G) \geq \nu_1'$ and $\nu_2' \geq \nu_n(G)$. Hence $\nu_2' \geq 0$ and

$$\nu_1(G) \geq \nu_1' - \nu_2' = \sqrt{(d_i - d_j)^2 + 4a_{ij}^2}.$$

The result follows since $a_{ij}^2 = a_{ij}$. □

## 7.4 Algebraic connectivity

For reasons explained further below, the second smallest eigenvalue $v_{n-1}(G)$ is usually called the *algebraic connectivity* of $G$, denoted by $a(G)$. From Proposition 7.1.1 we know already that $0 \leq a(G) \leq n$, that $a(G) = 0$ if and only if $G$ is not connected, and that $a(G) = n$ if and only if $G$ is complete. Moreover, if $G$ is $r$-regular then $a(G) = r - \lambda_2(G)$, and in this case we have already noted a connection between $\lambda_2(G)$ and the connectivity of $G$ (see Section 3.5).

Always $a(G) = n - v_1(\overline{G})$, and so the results of the previous section provide bounds on the algebraic connectivity of $G$ in terms of $\overline{G}$. In this section we investigate bounds for $a(G)$ in terms of $G$. From Equation (7.1), we have:

$$a(G) = \inf_{\mathbf{x} \in \mathbb{R}^n \setminus \{\mathbf{0}\}, \; \mathbf{x} \perp \mathbf{j}} \frac{\mathbf{x}^\top L \mathbf{x}}{\mathbf{x}^\top \mathbf{x}}.$$

Now the following expression for $a(G)$ is derived in exactly the same way as Equation (7.9):

$$a(G) = 2n \inf_{\mathbf{x} \in \mathbb{R}^n \setminus \langle \mathbf{j} \rangle} \frac{\sum_{uv \in E(G)} (x_u - x_v)^2}{\sum_{u \in V(G)} \sum_{v \in V(G)} (x_u - x_v)^2}. \tag{7.10}$$

This last equation may be rewritten as

$$a(G) = \inf_{\mathbf{x} \in \mathbb{R}^n \setminus \langle \mathbf{j} \rangle} \sup_{t \in \mathbb{R}} \frac{\sum_{uv \in E(G)} (x_u - x_v)^2}{\sum_{u \in V(G)} (x_u - t)^2}. \tag{7.11}$$

since $\sum_{u \in V(G)} (x_u - t)^2$ is least when $t$ is the mean of the $x_u$.

**Examples 7.4.1.** $a(P_n) = 2(1 - \cos \frac{\pi}{n})$, $a(C_n) = 2(1 - \cos \frac{2\pi}{n})$, $a(Q_m) = 2$, $a(K_{m,n}) = \min\{m, n\}$ and $a(K_n) = n$ $(n > 1)$. $\qquad\square$

**Theorem 7.4.2** [Fie1]. *If $G_1$ and $G_2$ are edge-disjoint graphs with the same vertex set then*

$$a(G_1) + a(G_2) \leq a(G_1 \cup G_2).$$

**Proof.** Let $G = G_1 \cup G_2$, with Laplacian $L_G$, and write $U = \{\mathbf{x} \in \mathbb{R}^n : \|\mathbf{x}\| = 1, \; \mathbf{x} \perp \mathbf{j}\}$. By (7.1) we have

$$a(G) = \min_{\mathbf{x} \in U} \mathbf{x}^\top L_G \mathbf{x} = \min_{\mathbf{x} \in U} (\mathbf{x}^\top L_{G_1} \mathbf{x} + \mathbf{x}^\top L_{G_2} \mathbf{x}) \geq$$
$$\min_{\mathbf{y} \in U} \mathbf{y}^\top L_{G_1} \mathbf{y} + \min_{\mathbf{z} \in U} \mathbf{z}^\top L_{G_2} \mathbf{z} = a(G_1) + a(G_2).$$

$\qquad\square$

We deduce the following useful property of algebraic connectivity:

**Corollary 7.4.3.** *If $H$ is a spanning subgraph of $G$, then $a(H) \leq a(G)$.*

**Theorem 7.4.4.** *If u and v are two non–adjacent vertices of a graph G on n vertices, then*

$$a(G) \le \tfrac{1}{2}(d_u + d_v).$$

*In particular, if G is not complete then $a(G) \le n - 2$.*

**Proof.** Let $\mathbf{y} = (y_1, y_2, \ldots, y_n)^\top$ be defined as follows:

$$y_i = \begin{cases} 1 & \text{if } i = u, \\ -1 & \text{if } i = v, \\ 0 & \text{otherwise.} \end{cases}$$

Now $\mathbf{y}^\top \mathbf{j} = 0$ and the result follows by substituting the vector $\mathbf{y}$ in (7.3).  □

The interest in the parameter $a(G)$ stems from the following inequalities which relate it to the vertex connectivity and edge connectivity of $G$. Recall that the vertex connectivity $\kappa(G)$ is the smallest number of vertices whose removal results in a disconnected or trivial graph, and the edge connectivity $\kappa'(G)$ is the smallest number of edges whose removal results in a disconnected graph. Always, $\kappa(G) \le \kappa'(G) \le \delta(G)$ [Har2, Theorem 5.1].

**Proposition 7.4.5.** *For any graph G and any $U \subseteq V(G)$, we have*

$$a(G) \le a(G - U) + |U|.$$

**Proof.** Let $G' = G - U$, $V(G) = \{1, \ldots, n\}$ and $V(G') = \{1, \ldots, k\}$. From (7.3) we know that $a(G') = \sum_{uv \in E(G')} (y_u - y_v)^2$ for some unit vector $(y_1, \ldots, y_k)^\top$ orthogonal to the all-1 vector in $\mathbb{R}^k$. If $\mathbf{x} = (y_1, \ldots, y_k, 0, \ldots, 0)^\top$ then $\mathbf{x} \perp \mathbf{j}$ and so

$$a(G) \le \sum_{uv \in E(G)} (x_u - x_v)^2.$$

Hence

$$a(G) \le \sum_{uv \in E(G')} (y_u - y_v)^2 + \sum_{u \in U} \sum_{v \sim u} y_v^2 \le a(G') + |U|.$$

□

In particular, if $G$ is connected then $a(G - v) \ge a(G) - 1$ for all $v \in V(G)$.

**Corollary 7.4.6.** *For any non-complete graph G we have $a(G) \le \kappa(G)$.*

**Proof.** In Proposition 7.4.5, take $U$ to be a set of $\kappa(G)$ vertices whose removal results in a disconnected graph.  □

**Proposition 7.4.7.** *If $T$ is a tree with diameter $d$ then*

$$a(T) \leq 2\left(1 - \cos\frac{\pi}{d+1}\right).$$

**Proof.** The tree $T$ can be constructed from the path $P_{d+1}$ by adding pendant edges. By Theorem 7.1.5, the addition of a pendant edge does not increase the algebraic connectivity, and so $a(T) \leq a(P_{d+1})$. Since $a(P_{d+1}) = 2(1 - \cos\frac{\pi}{d+1})$, the result follows. $\qquad\square$

**Remarks 7.4.8.** Fiedler [Fie1] established the followiing lower bound, where $n = |V(G)|$:

$$a(G) \geq 2\kappa'(G)\left(1 - \cos\frac{\pi}{n}\right).$$

Note that $1 - \cos(\pi/n) > \pi^2/2n^2$. Oshikiri [Osh] obtained the sharp lower bound

$$a(G) \geq 2\frac{\kappa'(G)}{n}.$$

$\qquad\square$

The following result provides another upper bound for $a(G)$; it can also be viewed as an upper bound on the diameter of a graph (cf. Theorem 7.5.11).

**Theorem 7.4.9** [Nil]. *If $G$ is a connected graph with maximum degree $\Delta$ and diameter $d$, then*

$$a(G) \leq \Delta - 2\sqrt{\Delta - 1} + \frac{2\sqrt{\Delta - 1} - 1}{\lfloor \frac{d}{2} \rfloor}.$$

## 7.5 Laplacian eigenvalues and graph structure

In this section we will examine how Laplacian eigenvalues are related to certain graph invariants or properties which, in most situations, are NP-hard to determine. If $S$ and $T$ are disjoint subsets of $V(G)$, then we define

$$E(S, T) = \{st \in E(G) : s \in S,\ t \in T\}.$$

If $S \cup T$ is a bipartition of $V(G)$ then $E(S, T)$ is called an *edge cut* of $G$. We write $\overline{S} = V(G) \setminus S$ and define the *(edge) boundary* $\partial S$ of $S$ as the edge-set $E(S, \overline{S})$. Note that if $S \neq \emptyset$ and $\mathbf{x}$ is the characteristic vector of $S$, then $\mathbf{x}^\top L\mathbf{x} = \sum_{uv \in E}(x_u - x_v)^2$ and (7.2) yields

$$\frac{\mathbf{x}^\top L\mathbf{x}}{\mathbf{x}^\top \mathbf{x}} = \frac{|\partial S|}{|S|}. \tag{7.12}$$

This explains why Laplacian eigenvalues are closely related to edge cuts.

### 7.5.1 Separation problems

Our first result provides bounds on the number of edges we need to delete to separate a set of vertices from the rest of the graph.

**Theorem 7.5.1.** *If G is a graph wth n vertices ($n \geq 2$) and $\emptyset \neq S \subset V(G)$, then*

$$\nu_{n-1}(G)\frac{|S||\overline{S}|}{n} \leq |\partial S| \leq \nu_1(G)\frac{|S||\overline{S}|}{n}.$$

**Proof.** Again let $\mathbf{x}$ be the characteristic vector of $S$. Since $\sum_{uv \in E}(x_u - x_v)^2 = |\partial S|$ and $\sum_{u \in V}\sum_{v \in V}(x_u - x_v)^2 = 2|S||\overline{S}|$, the upper bound follows from (7.9) and the lower bound from (7.10). $\qquad\square$

It follows from this theorem that the number $\nu_1 - \nu_{n-1}$ restricts the range of the cardinality of the cut $\partial S$. If this number is small, then for sets $S$ of fixed size, all boundaries $\partial S$ have approximately the same cardinality. As noted by Mohar [Moh4], this is the case for random graphs, and it explains why some algorithms dealing with cuts perform well on randomly chosen graphs. An application of Theorem 7.5.1 will be given in Section 7.6 in the context of graph expansion.

The max-cut problem is to find

$$\mathrm{mc}(G) = \max\{|\partial S| \ : \ \emptyset \neq S \subset V\},$$

and the min-cut problem is defined analogously; both problems are NP-hard. The *bipartition width* is defined as

$$\mathrm{bw}(G) = \min\{|\partial S| \ : \ S \subset V, \ |S| = \left\lfloor\frac{n}{2}\right\rfloor\}.$$

Thus determination of bipartition width, itself an NP-hard problem, is a restricted version of the min-cut problem. From Theorem 7.5.1 it is straight-forward to derive the following bounds:

**Corollary 7.5.2** [MohPo1]. *Let G be a graph on n vertices. Then*

$$\mathrm{mc}(G) \leq \frac{n}{4}\nu_1(G).$$

**Corollary 7.5.3.** *Let G be a graph on n vertices. Then*

$$\mathrm{bw}(G) \geq \begin{cases} \frac{n}{4}\nu_{n-1}(G) & \text{if } n \text{ is even,} \\ \frac{n^2-1}{4n}\nu_{n-1}(G) & \text{if } n \text{ is odd.} \end{cases}$$

### 7.5.2 Metric problems

We first address the problem of estimating the diameter of a graph by spectral means. We find basic upper and lower bounds and note without proof just some of the improved bounds which appear in the literature. We also establish an upper bound for the distance between two sets of vertices. For our first bound we require the following lemma:

**Lemma 7.5.4** [Moh3]. *Let $G$ be a graph with $n$ vertices, and for each pair $u$, $v$ of distinct vertices of $G$, choose a shortest $u$-$v$ path $P_{uv}$. Then any edge of $G$ lies in at most $\frac{1}{4}n^2$ of the paths $P_{uv}$.*

**Proof.** For fixed $e \in E(G)$, let $H_e$ be the graph on $V(G)$ with $u \sim v$ if and only if $e \in E(P_{uv})$. A graph with $n$ vertices and no triangles has at most $\frac{1}{4}n^2$ edges (see [Har2, Theorem 2.3]), and so it suffices to show that $H_e$ has no triangles. Suppose by way of contradiction that $uvw$ is a triangle in $H_e$, and orient the paths $P_{uv}$, $P_{uw}$, $P_{vw}$ from $u$ to $v$, $u$ to $w$, $v$ to $w$ respectively. Two of these paths, say $P_{uw}$ and $P_{vw}$, include $e$ in the same direction, say from $x$ to $y$ where $e = xy$. Thus $d(u, y) > d(u, x)$ and $d(v, y) > d(v, x)$. Hence $d(u, v) \leq d(u, x) + d(x, v) < d(u, x) + d(y, v)$, and secondly $d(u, v) < d(u, y) + d(x, v)$. It follows that $e \notin E(P_{uv})$, a contradiction. □

**Theorem 7.5.5** [Moh3]. *If $G$ is a connected graph on $n$ vertices, then*

$$\text{diam}(G) \geq \left\lceil \frac{4}{n\,a(G)} \right\rceil.$$

**Proof.** Let $\mathbf{x} = (x_1, x_2, \ldots, x_n)^T$ be an eigenvector of $G$ corresponding to $a(G)$. Since $\mathbf{x} \perp \mathbf{j}$, we have from (7.10):

$$2n \sum_{uv \in E(G)} (x_u - x_v)^2 = a(G) \sum_{u \in V(G)} \sum_{v \in V(G)} (x_u - x_v)^2. \qquad (7.13)$$

As in Lemma 7.5.4, we choose a shortest $u$-$v$ path $P_{uv}$ for each pair of distinct vertices $u$, $v$. Now $x_u - x_v$ is expressible in the form $\sum_{ij \in E(P_{uv})} (x_i - x_j)$ and the Cauchy–Schwarz inequality yields

$$(x_u - x_v)^2 \leq d(u, v) \sum_{ij \in E(P_{uv})} (x_i - x_j)^2.$$

We write $d = \text{diam}(G)$ and if $e = ij$, we write $q(e) = (x_i - x_j)^2$. Thus

$$(x_u - x_v)^2 \leq d \sum_{e \in E(P_{uv})} q(e).$$

Now let

$$\chi_{uv}(e) = \begin{cases} 1 & \text{if } e \in E(P_{uv}), \\ 0 & \text{otherwise.} \end{cases}$$

We have

$$\sum_{u\in V(G)} \sum_{v\in V(G)} (x_u - x_v)^2 \le d \sum_{u\in V(G)} \sum_{v\in V(G)} \sum_{e\in E(G)} q(e)\chi_{uv}(e)$$

$$= d \sum_{e\in E(G)} q(e) \sum_{u\in V(G)} \sum_{v\in V(G)} \chi_{uv}(e). \quad (7.14)$$

By Lemma 7.5.4, $\sum_{u\in V(G)} \sum_{v\in V(G)} \chi_{uv}(e) \le 2(\frac{1}{4}n^2)$, and so the result follows from (7.13) and (7.14). $\qquad\square$

Turning to upper bounds for the diameter, we note first the Laplacian counterpart of Theorem 3.3.5:

**Proposition 7.5.6.** *If $G$ is a connected graph with $r$ distinct eigenvalues in its Laplacian spectrum, then* $\text{diam}(G) \le r - 1$.

This is proved in the same way as the analogous result for the adjacency matrix $A$, that is by considering entries in $m_G(A)$, where $m_G(x)$ is the minimal polynomial of $A$ (Exercise 7.16). As an extension of this method, we can consider a polynomial $p_t(x)$ of degree $t$: if each entry of $p_t(L)$ is positive then $\text{diam}(G) \le t$. The next result is proved in this way (cf. [Chu2], [DamHa1]).

**Theorem 7.5.7.** *If $G$ is a connected graph on $n$ vertices, and $G \ne K_n$, then*

$$\text{diam}(G) \le 1 + \left\lfloor \frac{\log(n-1)}{\log \frac{\nu_1 + \nu_{n-1}}{\nu_1 - \nu_{n-1}}} \right\rfloor.$$

**Proof.** Note that since $G \ne K_n$, we have $\nu_{n-1} < \nu_1$ by Corollary 7.4.6. Let $\{x_1, \ldots, x_n\}$ be an orthonormal basis of $\mathbb{R}^n$ such that $Lx_i = \nu_i x_i$ ($i = 1, \ldots, n$) and $\sqrt{n}x_n = \mathbf{j}$. Let $u, v$ be distinct vertices of $G$, and let

$$\mathbf{e}_u = \sum_{i=1}^{n} a_i x_i, \quad \mathbf{e}_v = \sum_{i=1}^{n} b_i x_i.$$

Note that $a_n = b_n = 1/\sqrt{n}$. Now let

$$p_t(x) = \left(1 - \frac{2x}{\nu_1 + \nu_{n-1}}\right)^t.$$

We have $|p_t(v_i)| \leq (1-v)^t$ $(i = 1, \ldots, n-1)$, where $v = \frac{2v_{n-1}}{v_1+v_{n-1}}$. Using the Cauchy–Schwarz inequality, we can now derive a lower bound for the $(u, v)$-entry of $p_t(L)$:

$$
\mathbf{e}_u^\top p_t(L)\mathbf{e}_v = p_t(0)a_n b_n + \sum_{i=1}^{n-1} p_t(v_i)a_i b_i \geq a_n b_n - \left| \sum_{i=1}^{n-1} p_t(v_i)a_i b_i \right|
$$

$$
\geq \frac{1}{n} - (1-v)^t \sqrt{\sum_{i=1}^{n-1} a_i^2 \sum_{i=1}^{n-1} b_i^2} = \frac{1}{n} - (1-v)^t \frac{n-1}{n}. \quad (7.15)
$$

To complete the proof, we observe that this last term is positive whenever $t$ exceeds $\log(n-1)/\log\frac{v_1+v_{n-1}}{v_1-v_{n-1}}$. $\qquad\qquad\square$

If we take

$$
p_t(x) = \frac{T_t\left(\frac{v_1+v_{n-1}-2x}{v_1-v_{n-1}}\right)}{T_t\left(\frac{v_1+v_{n-1}}{v_1-v_{n-1}}\right)},
$$

where $T_t$ is a Chebyshev polynomial of the first kind, we obtain:

**Theorem 7.5.8** [ChuFM]. *If $G$ is a connected graph on $n$ vertices, and $G \neq K_n$, then*

$$
\mathrm{diam}(G) \leq 1 + \left\lfloor \frac{\cosh^{-1}(n-1)}{\cosh^{-1}\frac{v_1+v_{n-1}}{v_1-v_{n-1}}} \right\rfloor.
$$

We establish one upper bound for the distance between two sets of vertices; here $d(X, Y) = \min\{d(u, v) : u \in X, v \in Y\}$.

**Proposition 7.5.9** [AloMi2]. *Let $G$ be a connected graph on $n$ vertices, with maximum degree $\Delta$. Let $X, Y$ be (disjoint) non-empty subsets of $V(G)$ with $d(X, Y) = \rho > 1$. If $|X| = an$ and $|Y| = bn$, then*

$$
\rho^2 \leq \frac{\Delta}{a(G)}\left(\frac{1}{a} + \frac{1}{b}\right)(1 - a - b).
$$

**Proof.** We define $y_v$ $(v \in V(G))$ by

$$
y_v = \frac{1}{a} - \frac{1}{\rho}\left(\frac{1}{a} + \frac{1}{b}\right)\min\{d(v, X), \rho\}.
$$

Then $y_v = \frac{1}{a}$ if $v \in X$ and $y_v = -\frac{1}{b}$ if $v \in Y$; moreover, $|y_u - y_v| \leq \frac{1}{\rho}(\frac{1}{a} + \frac{1}{b})$ whenever $u \sim v$. Now let $\mathbf{x} = (x_1, \ldots, x_n)^\top$, where $x_v = y_v - \alpha$ and $\alpha = \frac{1}{n}\sum_{v \in V(G)} y_v$. Then $\mathbf{x} \perp \mathbf{j}$ and, making use of (7.3), we have:

$$a(G)n\left(\frac{1}{a} + \frac{1}{b}\right) \leq a(G)\left(\left(\frac{1}{a} - \alpha\right)^2 na + \left(\frac{1}{b} + \alpha\right)^2 nb\right)$$

$$\leq a(G)\sum_{v \in V(G)} x_v^2 \leq \sum_{uv \in E(G)} (x_u - x_v)^2$$

$$= \sum_{uv \in E(G)} (y_u - y_v)^2. \qquad (7.16)$$

Now $y_u = y_v$ when $uv \in E(X) \cup E(Y)$, while the condition $\rho > 1$ ensures that any edge outwith $E(X) \cup E(Y)$ is incident with at least one of the $n(1 - a - b)$ vertices in $V(G) \setminus (X \,\dot\cup\, Y)$. It follows from (7.16) that

$$a(G) \leq \frac{1}{\rho^2}\left(\frac{1}{a} + \frac{1}{b}\right)(1 - a - b)\Delta$$

as required. $\qquad\square$

**Corollary 7.5.10.** *With the notation of Theorem 7.5.9,*

$$b \leq \frac{1 - a}{1 + \frac{a(G)}{\Delta}a\rho^2}. \qquad (7.17)$$

**Proof.** Since $a > 0$, $b > 0$ and $a + b \leq 1$, we have $\frac{1}{a} + \frac{1}{b} \leq \frac{1}{ab}$. Hence

$$a(G) \leq \frac{\Delta}{\rho^2 ab}(1 - a - b),$$

which is equivalent to (7.17). $\qquad\square$

Corollary 7.5.10 will be used in the next section, in the context of graph expansion. Here we note without proof that Alon and Milman [AloMi2] make repeated use of Corollary 7.5.10 to obtain the following bound on the diameter of a graph.

**Theorem 7.5.11.** *If $G$ is a connected graph on $n$ vertices, with maximum degree $\Delta$, then*

$$\mathrm{diam}(G) \leq 2\left\lfloor \sqrt{\frac{2\Delta}{a(G)}}\log_2(n)\right\rfloor.$$

### 7.5.3 Isoperimetric problems

The classical isoperimetric problem (in Euclidean space) is to find the maximum area with given perimeter, or the maximum volume with given surface area. In a graph, an analogue is to find the maximum number of vertices in a set with a boundary of prescribed cardinality. The *isoperimetric number* (or *conductance*) of the non-trivial graph $G$ is the minimum of $|E(S, T)|/\min\{|S|, |T|\}$ taken over all bipartitions $S \cup T$ of $V(G)$. In other words,

$$i(G) = \min_{0 < |S| \le \frac{n}{2}} \frac{|\partial S|}{|S|}, \tag{7.18}$$

where $n = |V(G)| \ge 2$. Since $i(G) = 0$ if and only if $G$ is disconnected, we usually consider only connected graphs in the context of isoperimetric problems. If $i(G)$ is small then a relatively large set can be separated by relatively few edges, and so the isoperimetric number can be seen as a measure of connectivity. It is also a measure of graph expansion, the topic of the next section.

**Examples 7.5.12.** The isoperimetric number of some common graphs are:
$i(P_n) = \frac{1}{\lfloor \frac{n}{2} \rfloor}$, $i(C_n) = \frac{2}{\lfloor \frac{n}{2} \rfloor}$, $i(Q_n) = 1$, $i(K_n) = \lceil \frac{n}{2} \rceil$, $i(K_{m,n}) = \frac{\lceil \frac{mn}{2} \rceil}{\lfloor \frac{m+n}{2} \rfloor}$,
$i(S_n) = 1$.
□

**Remarks 7.5.13.** In view of (7.12) we have

$$i(G) = \min \left\{ \frac{\mathbf{x}^\top L \mathbf{x}}{\mathbf{x}^\top \mathbf{x}} : \mathbf{x} \in \{0, 1\}^n, \ 1 \le \mathbf{x}^\top \mathbf{j} \le \frac{n}{2} \right\}.$$

Thus $i(G)$ and $\nu_2(G)$ are obtained by optimizing the same function on different subsets of $\mathbb{R}^n$. Also noteworthy is the fact that:

$$i(G) = \inf_{\mathbf{x} \in \mathbb{R}^n \setminus \langle \mathbf{j} \rangle} \sup_{t \in \mathbb{R}} \frac{\sum_{uv \in E(G)} |x_u - x_v|}{\sum_{u \in V(G)} |x_u - t|},$$

which is very similar to (7.11), but with a different norm. Here the supremum is attained not when $t$ is the mean of the $x_v$, but when $t$ is a median value (which is not necessarily uniquely determined).
□

Before discussing spectral bounds for $i(G)$ we note from the definition (7.18) that when $G$ is connected, $\lfloor \frac{n}{2} \rfloor^{-1}$ is a lower bound, while $\delta(G)$, $\delta(L(G))$, $\lceil \frac{n}{2} \rceil$ are upper bounds. (The latter are obtained from (7.18) by taking $S$ to have cardinality 1, 2 and $\lceil \frac{n}{2} \rceil$, respectively.) We also have an upper bound which is approximately half the mean vertex degree:

**Theorem 7.5.14** [Moh1]. *For any graph with n vertices and m edges* $(n \geq 2)$,

$$i(G) \leq \frac{2m \left\lceil \frac{n}{2} \right\rceil}{n(n-1)}.$$

**Proof.** Fix $r \in \{1, \ldots, \lfloor \frac{n}{2} \rfloor\}$. For each edge $e$ of $G$ there exist $2\binom{n-2}{r-1}$ subsets $S$ of $V(G)$ of size $r$ such that $e \in E(S, \overline{S})$. Hence the mean of the corresponding $|\partial S|$ is

$$m \frac{2\binom{n-2}{r-1}}{\binom{n}{r}} = \frac{2mr(n-r)}{n(n-1)}.$$

The result follows by taking $r = \lfloor \frac{n}{2} \rfloor$.                                    $\square$

Turning now to spectral spectral bounds, we establish upper and lower bounds for the isoperimetric number in terms of algebraic connectivity.

**Theorem 7.5.15** [Moh1]. *For any graph G on n vertices* $(n \geq 2)$,

$$i(G) \geq \frac{a(G)}{2}.$$

**Proof.** Suppose that $i(G) = \frac{|\partial S|}{|S|}$ $(|S| \leq \lfloor \frac{n}{2} \rfloor)$. The lower bound of Theorem 7.5.1 shows that $i(G) \geq a(G) \frac{|\overline{S}|}{n}$. Since $|\overline{S}| \geq \lceil \frac{n}{2} \rceil$, the result follows.                                    $\square$

**Theorem 7.5.16** [Moh1]. *Let G be a graph with maximum degree* $\Delta$. *If* $G \neq K_1, K_2, K_3$ *then*

$$i(G) \leq \sqrt{a(G)(2\Delta - a(G))}.$$

**Proof.** It is straightforward to verify the inequality when $G = K_n$ $(n \geq 4)$, and so we suppose that $G$ is not complete. Then $a(G) \leq \delta$ by Corollary 7.4.6. If $a(G) = \delta$ then

$$\sqrt{a(G)(2\Delta - a(G))} \geq \sqrt{\delta\Delta} \geq \delta \geq i(G).$$

Accordingly, we suppose that $a(G) < \delta$.

Let $\mathbf{x} = (x_1, \ldots, x_n)^\top$ be an eigenvector of $L_G$ corresponding to $a(G)$, and let $U = \{v \in V(G) : x_v > 0\}$. Replacing $\mathbf{x}$ with $-\mathbf{x}$ if necessary, we may assume that $|U| \leq \frac{1}{2}n$. We define $y_1, \ldots, y_n$ by

$$y_v = \begin{cases} x_v & \text{if } v \in U, \\ 0 & \text{otherwise.} \end{cases}$$

Writing $E(U)$ for the set of edges joining vertices of $U$, we have

$$a(G) \sum_{v \in U} x_v^2 = \sum_{v \in U} \left( d_v x_v - \sum_{u \sim v} x_u \right) x_v = \sum_{v \in U} \sum_{u \sim v} (x_v - x_u) x_v$$

$$= \sum_{uv \in E(U)} \{(x_v - x_u)x_v + (x_u - x_v)x_u\} + \sum_{uv \in \partial U} (x_v - x_u)x_v$$

$$= \sum_{uv \in E(G)} (y_u - y_v)^2 - \sum_{uv \in \partial U} x_u x_v. \tag{7.19}$$

Similarly,

$$(2\Delta - a(G))) \sum_{v \in U} x_v^2 \geq \sum_{v \in U} \left( d_v x_v + \sum_{u \sim v} x_u \right) x_v$$

$$= \sum_{uv \in E(G)} (y_u + y_v)^2 + \sum_{uv \in \partial U} x_u x_v. \tag{7.20}$$

Let $\alpha = \sum_{uv \in \partial U} x_u x_v$. It follows from (7.19) and (7.20) that

$$a(G)(2\Delta - a(G)) \left( \sum_{v \in U} y_v^2 \right)^2 \geq \sum_{uv \in E(G)} (y_u - y_v)^2 \sum_{uv \in E(G)} (y_u + y_v)^2$$

$$- \alpha \left( 4 \sum_{uv \in E(U)} x_u x_v + \alpha \right).$$

Now $\alpha \leq 0$ and

$$4 \sum_{uv \in E(U)} x_u x_v + \alpha = 2 \sum_{uv \in E(U)} x_u x_v + \sum_{v \in U} x_v \sum_{u \sim v} x_u$$

$$= 2 \sum_{uv \in E(U)} x_u x_v + \sum_{v \in U} (d_v - a(G)) x_v^2 \geq 0.$$

Hence

$$a(G)(2\Delta - a(G)) \left( \sum_{v \in U} y_v^2 \right)^2 \geq \sum_{uv \in E(G)} (y_u - y_v)^2 \sum_{uv \in E(G)} (y_u + y_v)^2.$$

Now let $\beta = \sum_{uv \in E(G)} |y_u^2 - y_v^2|$. By the Cauchy–Schwarz inequality, we have

$$a(G)(2\Delta - a(G)) \left( \sum_{v \in U} y_v^2 \right)^2 \geq \beta^2$$

and so it suffices to show that $\beta \geq i(G) \sum_{v \in U} y_v^2$.

Let $0 = t_0 < t_1 < \cdots < t_m$ be the different values taken by $y_v$ ($v \in V(G)$), and define $V_k = \{v \in V(G) : y_v \geq t_k\}$ ($k = 0, \ldots, m$), $V_{m+1} = \emptyset$. For $k = 1, \ldots, m$ we have $|V_k| \leq \frac{1}{2}|U|$ and so $|\partial V_k| \geq i(G)|V_k|$ ($k = 1, \ldots, m$). Therefore,

$$
\begin{aligned}
\beta &= \sum_{k=1}^{m} \left( \sum \{y_v^2 - y_u^2 : uv \in E(G),\ y_u < y_v = t_k\} \right) \geq \sum_{k=1}^{m} \sum_{uv \in \partial V_k} (t_k^2 - t_{k-1}^2) \\
&= \sum_{k=1}^{m} |\partial V_k|(t_k^2 - t_{k-1}^2) \geq i(G) \sum_{k=1}^{m} |V_k|(t_k^2 - t_{k-1}^2) \\
&= i(G) \sum_{k=0}^{m} t_k^2 (|V_k| - |V_{k+1}|) = i(G) \sum_{v \in V(G)} y_v^2 = i(G) \sum_{v \in U} y_v^2
\end{aligned}
$$

This completes the proof.                                                        □

## 7.6 Expansion

There are several measures of expansion of graph which appear in the literature. In Section 3.5, we have already encountered such a measure in terms of $|N(S)|/|S|$ ($S \subset V(G)$). The isoperimetric number, defined in terms of $|\partial S|/|S|$, is a measure of edge expansion. Here, instead of the edge boundary $\partial S$, we shall use the *vertex boundary* $\delta S$, defined as the set of vertices outside $S$ which are adjacent to some vertex inside $S$. Note that $|\delta S| \leq |\partial S| \leq |\Delta(G)||\delta S|$. The *vertex expansion* of $G$ is defined by

$$
j(G) = \min_{1 \leq |S| \leq \frac{n}{2}} \frac{|\delta S|}{|S|}.
$$

The differences between the various measures of expansion which are used is largely superficial, in that all conform to the general principle that expansion in graphs of bounded degree is controlled by algebraic connectivity. In the case of the isoperimetric number $i(G)$, this property is made explicit in Theorems 7.5.15 and 7.5.16. In Theorems 7.6.1 and 7.6.2 below we establish an analogous property for $j(G)$.

**Theorem 7.6.1** [Alo1]. *Let $G$ be a non-trivial graph $G$ with maximal degree $\Delta$. If $a(G) \geq \epsilon \geq 0$ then*

$$
j(G) \geq \frac{2\epsilon}{\Delta + 2\epsilon}. \tag{7.21}
$$

**Proof.** We may take $\epsilon > 0$, so that $G$ is connected. Let $n = |V(G)|$, $X \subseteq V(G)$, $1 \leq |X| \leq \frac{1}{2}n$. If $V(G) = X \dot{\cup} \delta X$ then clearly $|\delta X|/|X| \geq 1 > 2\epsilon/\Delta + 2\epsilon$. Otherwise, we apply Corollary 7.5.10 to the non-empty sets $A = X$ and $B = V(G) \setminus (X \dot{\cup} \delta X)$. Since $d(A, B) = 2$ and $a(G) \geq \epsilon$, we have

$$\frac{n - |X| - |\delta X|}{n} \leq \frac{1 - \frac{|X|}{n}}{1 + \frac{4\epsilon}{\Delta}\frac{|X|}{n}}, \quad \text{or} \quad \frac{|\delta X|}{n} \geq \left(1 - \frac{|X|}{n}\right)\left(1 - \frac{1}{1 + \frac{4\epsilon}{\Delta}\frac{|X|}{n}}\right).$$

Since $|X| \leq \frac{1}{2}n$, we deduce that

$$\frac{|\delta X|}{|X|} \geq \frac{\frac{2\epsilon}{\Delta}}{1 + \frac{4\epsilon}{\Delta}\frac{|X|}{n}} \geq \frac{2\epsilon}{\Delta + 2\epsilon}.$$

The result follows. $\qquad\square$

**Theorem 7.6.2** [Alo1]. *If $G$ is a non-trivial graph with $j(G) \geq c > 0$ then*

$$a(G) \geq \frac{c^2}{4 + 2c^2}. \tag{7.22}$$

**Proof.** Let $\mathbf{x} = (x_1, \ldots, x_n)^\top$ be an eigenvector of $L_G$ corresponding to $a(G)$, and let $U = \{v \in V(G) : x_v > 0\}$. Replacing $\mathbf{x}$ with $-\mathbf{x}$ if necessary, we may assume that $|U| \leq \frac{1}{2}n$. We define $y_1, \ldots, y_n$ by

$$y_v = \begin{cases} x_v & \text{if } v \in U, \\ 0 & \text{otherwise.} \end{cases}$$

As in the proof of Theorem 7.5.16, we have

$$a(G) \geq \frac{\sum_{uv \in E(G)} (y_u - y_v)^2}{\sum_{v \in V(G)} y_v^2} \tag{7.23}$$

(cf. Equation (7.19)). To exploit this inequality, we apply the Max-flow Min-cut Theorem (see, for example, [Gib, Chapter 4]) to the digraph (or flow network) $\mathcal{N}$ defined as follows. The vertex set of $\mathcal{N}$ is $\{s\} \dot{\cup} U' \dot{\cup} V \dot{\cup} \{t\}$, where $s$ is a source, $t$ is a sink, $V = V(G)$ and $U'$ consists of vertices $u'$ in one-one correspondence with vertices $u$ of $U$. The arcs of $\mathcal{N}$ are $su'$ ($u' \in U'$), $u'u$ ($u \in U$), $u'v$ ($v \sim u \in U$) and $vt$ ($v \in V$). The capacity of each arc $su'$ is $1 + c$, and all other arcs have capacity 1. The edges $su'$ ($u' \in U'$) form a cut of capacity $(1 + c)|U|$ separating $s$ and $t$, and we show that no other edge cut $C$ separating $s$ and $t$ has lower capacity.

Let $X = \{u \in U : su' \notin C\}$. For each $w \in X$, the cut $C$ contains $w'w$ or $wt$, and for each $w \in \delta X$, $C$ contains $wt$ or the edges $v'w$ ($w \sim v \in X$). Together, these number at least $|X| + |\delta X|$, and so the capacity of $C$ is at least

$(1+c)(|U|-|X|)+|X|+|\delta X|$. Since $|\delta X| \geq j(G)|X| \geq c|X|$, this is at least $(1+c)|U|$, as required.

In a maximum flow of $(1+c)|U|$ from $s$ to $t$, let $f(v_1, v_2)$ be the flow in arc $v_1 v_2$. Then $f(s, u') = 1+c$ for all $u \in U$ and $0 \leq f(v_1, v_2) \leq 1$ for all other arcs $v_1 v_2$ of $\mathcal{N}$. Note that

$$f(u', u)+\sum_{v \sim u} f(u', v) = 1+c \ \forall u \in U \text{ and } f(v', v)+\sum_{u \sim v} f(u', v) \leq 1 \ \forall v \in U.$$

Now we define the function $h : V^2 \to [0, 1]$ by:

$$h(u, v) = \begin{cases} f(u', v) & \text{if } u \in U, \ v \in V \text{ and } u \sim v, \\ 0 & \text{otherwise.} \end{cases}$$

Note that

$$\sum_{u \sim v} h(v, u)+f(v', v) = 1+c \ \forall v \in U \ \text{ and } \ \sum_{u \sim v} h(u, v)+f(v', v) \leq 1 \ \forall v \in U,$$

while any sum of the form $\sum_{v \in V} \alpha_v y_v^2$ may be written as $\sum_{v \in U} \alpha_v y_v^2$.

Now we have

$$\sum_{uv \in E(G)} h(u, v)^2 (y_u + y_v)^2 \leq 2 \sum_{uv \in E(G)} h(u, v)^2 (y_u^2 + y_v^2)$$

$$= 2 \sum_{u \in V} \left( \sum_{v \sim u} h(u, v)^2 \right) y_u^2 + 2 \sum_{v \in V} \left( \sum_{u \sim v} h(u, v)^2 \right) y_v^2$$

$$\leq 2(2 + c^2) \sum_{v \in V} y_v^2. \tag{7.24}$$

(Note that $\sum_{v \sim u} h(u, v)^2$ is maximized when the number of summands equal to 1 is maximized.) Secondly,

$$\sum_{uv \in E(G)} h(u, v)(y_u^2 - y_v^2) = \sum_{v \in V} \left( \sum_{u \sim v} h(v, u) - \sum_{u \sim v} h(u, v) \right) y_v^2 \geq c \sum_{v \in V} y_v^2. \tag{7.25}$$

Using (7.23) and the Cauchy–Schwarz inequality in conjunction with (7.24) and (7.25), we have

$$a(G) \geq \frac{\sum_{uv \in E(G)} (y_u - y_v)^2}{\sum_{v \in V} y_v^2}$$

$$= \frac{\sum_{uv \in E(G)} (y_u - y_v)^2 \sum_{uv \in E(G)} h(u, v)^2 (y_u + y_v)^2}{\sum_{v \in V} y_v^2 \sum_{uv \in E(G)} h(u, v)^2 (y_u + y_v)^2}$$

$$\geq \frac{\left(\sum_{uv \in E(G)} h(u,v)(y_u^2 - y_v^2)\right)^2}{2(2+c^2)\left(\sum_{v \in V} y_v^2\right)^2}$$

$$\geq \frac{1}{4+2c^2}\left(\frac{\sum_{uv \in E(G)} h(u,v)(y_u^2 - y_v^2)}{\sum_{v \in V} y_v^2}\right)^2 \geq \frac{c^2}{4+2c^2}.$$

This completes the proof. □

The inequalities of Theorems 7.6.1 and 7.6.2 are often couched in terms of magnifiers and enlargers, defined as follows. An $(n, \Delta, c)$-*magnifier* is a non-trivial graph $G$ with $|V(G)| = n$, $\Delta(G) = \Delta$ and $j(G) \geq c$. An $(n, \Delta, \epsilon)$-*enlarger* is a non-trivial graph $G$ with $|V(G)| = n$, $\Delta(G) = \Delta$ and $a(G) \geq \epsilon$. Thus Theorem 7.6.1 says that every $(n, \Delta, \epsilon)$-enlarger is an $(n, \Delta, c)$-magnifier, where $c = \epsilon/(\Delta + 2\epsilon)$; and Theorem 7.6.2 says that every $(n, \Delta, c)$-magnifier is an $(n, \Delta, \epsilon)$-enlarger, where $\epsilon = c^2/(4+2c^2)$.

In general terms, a graph is a 'good expander' if some measure of expansion is 'large'. It is known that a random regular graph is, with high probability, a good expander (see [MohPo3, Appendix B]) but the explicit construction of graphs with a prescribed measure of expansion is a non-trivial problem outwith the scope of this book; for constructional details, the reader is referred to [DavSV], [GabGa], [Mar], [LuPS], [Mor1] and [Mor2].

The technical definition of an expander applies only to bipartite graphs and employs a measure of expansion slightly different from those encountered so far:

**Definition 7.6.3.** An $(n, \Delta, c)$-*expander* is a bipartite graph $G$ on two sets of vertices, $I$ (inputs) and $O$ (outputs), with $|I| = |O| = n$ and maximum degree $\Delta$, such that

$$(\forall U \subseteq I)\left(|U| \leq \frac{n}{2} \Rightarrow |\delta U| \geq \left(1 + c\left(1 - \frac{|U|}{n}\right)|U|\right)\right).$$

If $\overline{U}$ denotes the complement of $U$ in $I$ then the condition becomes:

$$(\forall U \subseteq I)\left(|U| \leq \frac{n}{2} \Rightarrow |\delta U| \geq |U| + c\frac{|U||\overline{U}|}{n}\right).$$

**Example 7.6.4.** Let $G$ be a graph with $V(G) = \{v_1, \dots, v_n\}$. The *extended double cover* of $G$ is the bipartite graph $D(G)$ with vertex set $\{x_1, \dots, x_n\} \cup \{y_1, \dots, y_n\}$ and edge set $\{x_i y_j : i = j \text{ or } i \sim j\}$. In other words, $D(G)$ is the NEPS of $G$ and $K_2$ with basis $\{(0,1),(1,1)\}$.

In $D(G)$ the boundary of a set $\{x_i : i \in X\}$ is $\{y_i : i \in X\} \cup \{y_j : j \in \delta X\}$, of size $|X| + |\delta X|$. Thus if $G$ is an $(n, \Delta, c)$-magnifier then $D(G)$ is a

$(2n, 1 + \Delta, 1 + c)$-magnifier; moreover, $D(G)$ is an $(n, 1 + \Delta, c)$-expander. By Theorem 7.5.1,

$$|\delta X| \geq \frac{1}{\Delta}|\partial X| \geq \frac{a(G)}{\Delta}\frac{|X||\overline{X}|}{n},$$

and so $D(G)$ is also an $(n, 1 + \Delta, c^*)$-expander, where $c^* = a(G)/\Delta$.    $\square$

Expanders are used as models for robust networks in computer science, where one objective is to construct a sequence of graphs $(G_i)$ such that

  (i)  $G_i$ is an $(n_i, \Delta, c)$-expander,
  (ii)  $|V(G_i)| = n_i \to \infty$ and $\frac{n_{i+1}}{n_i} \to 1$ as $i \to \infty$.

Note that as $i \to \infty$, the edges of $G_i$ become more sparse, while the connectivity properties (quantified by $c$) are retained. It can be shown (see [Alo1, Lemma 3.3]) that if $G$ is an $r$-regular subgraph of $K_{n,n}$ with $a(G) = a$ then $G$ is an $(n, r, c)$-expander with $c = (2ra - a^2)/(r^2 - ra + \frac{1}{2}a^2)$. Thus $c > a/r$ and for good connectivity we want $a(G)$ to be large. On the other hand, in the case that the graphs $G_i$ are all $r$-regular, we know from (3.12) that $\limsup_{i \to \infty} a(G_i) \leq r - 2\sqrt{r - 1}$. Accordingly, Ramanujan graphs (the $r$-regular graphs with $a(G) \geq r - 2\sqrt{r - 1}$) are best possible candidates for the graphs $G_i$, and those described in Section 3.5 can indeed be used in this context.

## 7.7 The normalized Laplacian matrix

Recall that if $G$ is a graph without isolated vertices then its normalized Laplacian is $\hat{L} = D^{-\frac{1}{2}}LD^{-\frac{1}{2}}$ ($= \hat{L}_G$), where $L$ is the Laplacian of $G$ and $D = \text{diag}(d_1, \ldots, d_n)$. By way of motivation, we point out the connection between $\hat{L}$ and random walks on $G$.

**Remark 7.7.1.** The transition matrix of a random walk on $G$ is $P = (p_{ij})$, where $p_{ij}$ is the probability of moving to vertex $j$ from vertex $i$. Thus

$$p_{ij} = \begin{cases} \frac{1}{d_i} & \text{if } i \sim j, \text{ and} \\ 0 & \text{otherwise.} \end{cases}$$

Hence $P = D^{-1}A$ where $A$ is the adjacency matrix of $G$. It follows that $\hat{L} = I - M$ where $M = D^{\frac{1}{2}}PD^{-\frac{1}{2}}$. Therefore the eigenvalues of $P$ are

$$1 = \rho_1 \geq \rho_2 \geq \cdots \geq \rho_n$$

where $\rho_i = 1 - \hat{v}_{n-i+1}$ $(i = 1, 2, \ldots, n)$ and $\hat{v}_i (= \hat{v}_i(G))$ is the $i$-th largest eigenvalue of $\hat{L}$. The eigenvalue $\rho_2$ is used to establish properties of random walks on $G$. This remark is made precise in a more general context in Subsection 9.4.2. □

The definition of $\hat{L}$ may be extended to arbitrary graphs by taking $\hat{L} = (\hat{l}_{ij})$, where

$$\hat{l}_{ij} = \begin{cases} 1 & \text{if } i = j \quad \text{and} \quad d_i \neq 0, \\ -\dfrac{1}{\sqrt{d_i d_j}} & \text{if } i \text{ and } j \text{ are adjacent and,} \\ 0 & \text{otherwise.} \end{cases}$$

Let $T$ be the diagonal matrix whose $i$-th diagonal entry is $1/d_i$ if $d_i \neq 0$, and 0 otherwise. Then $\hat{L} = T^{\frac{1}{2}} L T^{\frac{1}{2}}$ and for any gradient matrix $R$ we have $\hat{L} = \hat{R}\hat{R}^T$, where $\hat{R} = T^{\frac{1}{2}} R$. Hence all the eigenvalues of $\hat{L}$ are non-negative. Moreover the least eigenvalue $v_n$ of $\hat{L}$ is 0 since $(\sqrt{d_1}, \sqrt{d_2}, \ldots, \sqrt{d_n})^T$ is a corresponding eigenvector.

If $\mathbf{x} = D^{\frac{1}{2}}\mathbf{y}$, the Rayleigh quotient $R(\mathbf{x}) = \mathbf{x}^T \hat{L}\mathbf{x}/\mathbf{x}^T\mathbf{x}$ may be written as $R^*(\mathbf{y}) = \mathbf{y}^T L\mathbf{y}/\mathbf{y}^T D\mathbf{y}$. Using (7.2) we obtain

$$R^*(\mathbf{y}) = \frac{\sum_{uv\in E(G)}(y_u - y_v)^2}{\sum_{v\in V(G)} d_v y_v^2}. \tag{7.26}$$

This form of the Rayleigh quotient enables us to give an alternative description of the eigenvalues of $\hat{L}$. For the largest and second smallest eigenvalues we have the following expressions, where $\mathbf{d} = D\mathbf{j} = (d_1, d_2, \ldots, d_n)^T$:

$$\hat{v}_1 = \sup_{\mathbf{y}\in \mathbb{R}^n\setminus\{\mathbf{o}\},\mathbf{y}\perp\mathbf{d}} \frac{\sum_{uv\in E(G)}(y_u - y_v)^2}{\sum_{v\in V(G)} d_v y_v^2}, \tag{7.27}$$

$$\hat{v}_{n-1} = \inf_{\mathbf{y}\in \mathbb{R}^n\setminus\{\mathbf{o}\},\mathbf{y}\perp\mathbf{d}} \frac{\sum_{uv\in E}(y_u - y_v)^2}{\sum_{v\in V} d_v y_v^2}. \tag{7.28}$$

Note that $\mathbf{y} \perp \mathbf{d}$ if and only if $\mathbf{x}$ is orthogonal to $(\sqrt{d_1}, \sqrt{d_2}, \ldots, \sqrt{d_n})^T$. Also, when $G$ has isolated vertices, $\hat{v}_{n-1} = 0$ while the supremum of $R(\mathbf{x})$ is attained when $\mathbf{x}$ has the form $D^{\frac{1}{2}}\mathbf{y}$.

The basic properties of the spectrum of $\hat{L}$ are summarized in the following theorems.

**Theorem 7.7.2** [Chu2]. *Let $G$ be a graph on $n$ vertices $(n \geq 2)$. Then*

(i) $\sum_{i=1}^{n} \hat{v}_i \leq n$ *with equality if and only if $G$ has no isolated vertices;*
(ii) *if $G \neq K_n$ then $\hat{v}_{n-1} \leq 1$;*

(iii) *if $G$ has no isolated vertices, then $\hat{v}_{n-1} \leq \frac{n}{n-1}$ with equality if and only if $G = K_n$;*

(iv) *if $G$ has no isolated vertices, then $\hat{v}_1 \geq \frac{n}{n-1}$ with equality if and only if $G = K_n$;*

(v) *$\hat{v}_1 \leq 2$, with equality if and only if $G$ has a non-trivial component which is bipartite.*

**Proof.** First, (i) follows from the relation $\operatorname{tr}(\hat{L}) = \operatorname{tr}(T^{\frac{1}{2}}LT^{\frac{1}{2}})$, since $\operatorname{tr}(T^{\frac{1}{2}}LT^{\frac{1}{2}})$ is the number of non-isolated vertices. To prove (ii), let $s$ and $t$ be non-adjacent vertices in $G$, and define $\mathbf{z} = (z_1, z_2, \ldots, z_n)^\top$ by:

$$z_i = \begin{cases} d_t & \text{if } i = s, \\ -d_s & \text{if } i = t, \text{ and} \\ 0 & \text{otherwise.} \end{cases}$$

Then $\mathbf{z}^\top \mathbf{d} = 0$ and $R^*(\mathbf{z}) = 1$. Now $\hat{v}_{n-1} \leq 1$ by (7.28).

The inequalities in (iii) and (iv) follow directly from (i) since $\hat{v}_n = 0$. In view of (ii), equality can hold only if $G = K_n$. But the normalized Laplacian spectrum of $K_n$ is $0^1$, $(\frac{n}{n-1})^{n-1}$, and so (iii) and (iv) are proved.

The inequality in (v) follows from (7.27) because we have, for $\mathbf{y} \perp \mathbf{d}$:

$$R^*(\mathbf{y}) = \frac{\sum_{uv \in E(G)} (y_u - y_v)^2}{\sum_{v \in V(G)} d_v y_v^2} \leq \frac{\sum_{uv \in E(G)} 2(y_u^2 + y_v^2)}{\sum_{v \in V(G)} d_v y_v^2} = 2.$$

If $R^*(\mathbf{y}) = 2$ then $y_u = -y_v$ whenever $u \sim v$; then in some component $H$ of $G$, $y_u = -y_v \neq 0$ whenever $u \sim v$, and the signs of $y_u$ ($u \in V(H)$) determine a 2-colouring of $H$. Conversely if $G$ has a non-trivial bipartite component with parts $U$, $V$, we define $\mathbf{y}$ by:

$$y_u = \begin{cases} 1 & \text{if } u \in U, \\ -1 & \text{if } u \in V, \\ 0 & \text{otherwise.} \end{cases}$$

Then $\mathbf{y} \perp \mathbf{d}$ and $R^*(\mathbf{y}) = 2$, whence $\hat{v}_1 = 2$.

This completes the proof. $\square$

Just as in Theorem 7.1.2, we have:

**Theorem 7.7.3.** *The multiplicity of $0$ as an eigenvalue of $\hat{L}_G$ is equal to the number of components in $G$.*

Now we can show that the property of being bipartite is recognizable from the spectrum of $\hat{L}$:

**Corollary 7.7.4.** *A graph $G$ is bipartite if and only if the eigenvalue $\hat{v}_1(G)$ is equal to $2$, with the same multiplicity as $\hat{v}_n(G)$.*

**Proof.** From the proof of Theorem 7.7.2(v) we see that, for a bipartite connected graph, 2 is a simple eigenvalue because $\mathcal{E}_{\hat{L}}(2)$ is spanned by a $(1, -1)$-vector. The result therefore follows from Theorem 7.7.3. $\qquad\square$

Most results concerning the spectrum of $L$ have analogues in the context of $\hat{L}$, and we mention three without proof. For this purpose, we define the *volume* of a subset $S$ of $V(G)$ by:

$$\text{vol}(S) = \sum_{v \in S} d_v.$$

When $\emptyset \subset S \subset V(G)$, we define:

$$h_G(S) = \frac{|E(S, \overline{S})|}{\min\{\text{vol}(S), \text{vol}(\overline{S})\}},$$

and this can be used to provide alternative definitions of expansion in a graph. The analogue of the isoperimetric number $i(G)$ is the *Cheeger constant* $h(G)$, defined by

$$h(G) = \min_{\emptyset \subset S \subset V(G)} h_G(S).$$

The terminology is borrowed from spectral Riemannian geometry. It follows from (7.28) that $\hat{v}_{n-1}(G) \leq 2h(G)$ (Exercise 7.16). As an analogue of Theorems 7.6.1, 7.6.2 (and of Theorems 7.5.15, 7.5.16) we have the *Cheeger inequality*:

**Theorem 7.7.5** [Che]. *For any non-trivial connected graph $G$,*

$$2h(G) \geq \hat{v}_{n-1}(G) \geq \frac{1}{2}h(G)^2.$$

The arguments for Theorems 7.5.7 and 7.5.8 may be extended to obtain upper bounds for $d(X, Y)$ by considering $(D^{\frac{1}{2}}\mathbf{x})^{\top} p_t(\hat{L})(D^{\frac{1}{2}}\mathbf{y})$, where $\mathbf{x}, \mathbf{y}$ are the characteristic vectors of $X, Y$ respectively (cf. Equation (7.15)). As an analogue of Theorem 7.5.7, we obtain (see [Chu2], [Kir2]):

**Theorem 7.7.6.** *Let $G$ be a connected graph on $n$ vertices ($G \neq K_n$), and let $X, Y$ be subsets of $V(G)$. Then*

$$d(X, Y) \leq 1 + \lfloor \alpha(X, Y) \rfloor, \quad \text{where} \quad \alpha(X, Y) = \frac{\log\sqrt{\frac{\text{vol}(\overline{X})\text{vol}(\overline{Y})}{\text{vol}(X)\text{vol}(Y)}}}{\log\frac{\hat{v}_1 + \hat{v}_{n-1}}{\hat{v}_1 - \hat{v}_{n-1}}}.$$

**Remark 7.7.7.** Kirkland [Kir2] improved the bound in Theorem 7.7.6 as follows: if $Y \neq X, \overline{X}$ (and $G \neq K_n$) then

$$d(X, Y) \leq \max\{\lceil \alpha(X, Y) \rceil, 2\}. \qquad \square$$

As an analogue of Theorem 7.5.8 we obtain (see [Chu2]):

**Theorem 7.7.8.** *Let G be a connected graph on n vertices ($G \neq K_n$), and let X, Y be subsets of $V(G)$. Then*

$$d(X, Y) \leq 1 + \lfloor \beta(X, Y) \rfloor, \quad \text{where} \quad \beta(X, Y) = \frac{\cosh^{-1} \sqrt{\frac{\mathrm{vol}(\overline{X})\mathrm{vol}(\overline{Y})}{\mathrm{vol}(X)\mathrm{vol}(Y)}}}{\cosh^{-1} \frac{\hat{v}_1 + \hat{v}_{n-1}}{\hat{v}_1 - \hat{v}_{n-1}}}.$$

**Example 7.7.9.** [Kir2] Let $G = \overline{K_p} \triangledown K_q$, so that the eigenvalues of $\hat{L}$ are 0, 1 (with multiplicity $p - 1$), $\frac{p+q}{p+q-1}$ (with multiplicity $q - 1$) and $1 + \frac{p}{p+q-1}$. Now suppose that $p$ is even and let $X, Y$ be disjoint sets of size $\frac{1}{2}p$ such that $X \mathbin{\dot{\cup}} Y$ is the independent set of vertices of degree $q$. Then $d(X, Y) = 2$ while $\alpha(X, Y) = \beta(X, Y) = 1$. $\qquad \square$

# 7.8 The signless Laplacian

In contrast to the Laplacian $L = D - A$, the signless Laplacian $Q = D + A$ of a graph $G$ has so far featured very rarely in published papers. In this section we survey the known properties of spectra of signless Laplacians and point to the possibilities for developing a spectral theory of graphs based on this matrix. The characteristic polynomial of $Q$ is called the *Q-polynomial* of $G$, denoted by $Q_G(x)$. The spectrum and the eigenvalues of $Q$ are called the *Q-spectrum* and *Q-eigenvalues* respectively. Table A1 of the Appendix contains the $Q$-spectra of the connected graphs with up to five vertices.

## 7.8.1 Basic properties of $Q$-spectra

Recall from Section 2.4 that if $B$ is the incidence matrix of a graph $G$ with $n$ vertices and $m$ edges then

$$BB^\top = Q, \quad B^\top B = A(L(G)) + 2I \qquad (7.29)$$

and so

$$P_{L(G)}(x) = (x + 2)^{m-n} Q_G(x + 2). \qquad (7.30)$$

In Theorem 2.4.4 we saw also that $P_{S(G)}(x) = x^{m-n} Q_G(x^2)$.

We denote the $i$-th largest eigenvalue of $Q$ by $\xi_i = \xi_i(G)$. Since $Q$ is a positive semi-definite matrix we have:

$$\xi_1 \geq \xi_2 \geq \cdots \geq \xi_n \geq 0.$$

Observe that $m = \frac{1}{2} \operatorname{tr}(Q) = \frac{1}{2} \sum_{i=1}^{n} \xi_i$.

We call $\xi_1$ the *Q-index* of $G$. If $G$ is connected then $Q$ is irreducible and so $Q$ has a unique positive unit eigenvector corresponding to $\xi_1$; we call this vector the *principal Q-eigenvector* of $G$.

Our first theorem concerns the least eigenvalue:

**Theorem 7.8.1.** *Let $G$ be a non-trivial connected graph with $n$ vertices. Then $G$ is bipartite if and only if $\xi_n(G) = 0$. In this situation, $0$ is a simple Q-eigenvalue.*

**Proof.** For a vector $\mathbf{x}^\top = (x_1, x_2, \ldots, x_n)^\top$ we have $Q\mathbf{x} = \mathbf{0}$ if and only if $B^\top \mathbf{x} = \mathbf{0}$. The latter holds if and only if $x_i = -x_j$ whenever $i \sim j$. Since $G$ is connected, it follows that if $0$ is a eigenvalue of $Q$ then $\mathcal{E}_Q(0)$ is spanned by a $(1, -1)$-vector $\mathbf{x}$; then the signs of the $x_i$ determine a 2-colouring of $G$. Conversely if $G$ has a 2-colouring, and we define $x_i = \pm 1$ accordingly, then $Q\mathbf{x} = \mathbf{0}$. $\square$

**Corollary 7.8.2.** *For any graph, the multiplicity of the Q-eigenvalue $0$ is equal to the number of components that are bipartite or trivial.*

**Remark 7.8.3.** From the spectrum of the adjacency matrix, we know whether a graph is bipartite (see Theorem 3.2.4), but not whether a graph is connected (see Fig. 1.3(a)). The spectrum of the Laplacian tells us whether a graph is connected (see Theorem 7.1.2), but not whether it is bipartite (see Fig. 7.1). Given the $Q$-spectrum of a graph $G$, we see from Corollary 7.8.2, that if $G$ is connected, we can say whether $G$ is bipartite; and if $G$ is bipartite, we can say whether $G$ is connected. On the other hand, the spectrum of the normalized Laplacian tells us whether $G$ is connected (see Theorem 7.7.3) and whether $G$ is bipartite (see Corollary 7.7.4). $\square$

In view of Remark 7.8.3, it is usual when discussing the relation between a graph $G$ and its $Q$-polynomial to specify the number of components in $G$.

**Proposition 7.8.4.** *For any bipartite graph, the Q-polynomial coincides with the L-polynomial.*

**Proof.** With a suitable labelling of vertices, $A$ has the form $\begin{pmatrix} O & B^\top \\ B & O \end{pmatrix}$. Then $D + A = T^{-1}(D - A)T$, where $T$ has the form $\begin{pmatrix} I & O \\ O & -I \end{pmatrix}$. $\square$

Two graphs are said to be *Q-cospectral* if they have the same *Q*-polynomial. By analogy with the definitions of PING and cospectral mate (see Chapter 4) we introduce the notions of *Q*-PING and *Q*-cospectral mate with the obvious meanings.

The graphs $K_{1,3}$ and $K_3 \dot\cup K_1$ represent the smallest *Q*-PING; no other *Q*-PINGs on four vertices exist. These graphs have the same line graph, namely $K_3$, with characteristic polynomial $(x - 2)(x + 1)^2$. By (7.30) they have the same *Q*-polynomial, namely $x(x - 4)(x - 1)^2$. By Corollary 7.8.2, a graph $G$ with this *Q*-polynomial has exactly one bipartite or trivial component but (as the examples show) $G$ may or may not be connected, and may or may not be bipartite.

There are two *Q*-PINGs on five vertices: one is provided by the graphs $K_{1,3} \dot\cup K_1$ and $K_3 \dot\cup 2K_1$ and the other by the graphs numbered 14 and 15 in Table A1 of the Appendix. Note that the smallest PING (Fig. 1.3(a)) and the smallest PING consisting of connected graphs (Fig. 1.3(b)) are not *Q*-PINGs. The paper [HaeSp] provides an example of two non-isomorphic (non-regular, non-bipartite) graphs on 10 vertices which are cospectral, *Q*-cospectral and *L*-cospectral, and which have cospectral complements.

Two graphs are called *line-cospectral* if their line graphs are cospectral.

**Proposition 7.8.5.** *If two graphs are Q-cospectral, then they are line-cospectral.*

**Proof.** Since *Q*-cospectral graphs have the same number of vertices and the same number of edges, their line-cospectrality follows from (7.30).                    □

However, line-cospectral graphs are not necessarily *Q*-cospectral, since the root graphs of cospectral line graphs need not have the same number of vertices. Such an example of cospectral line graphs is given in Fig. 7.2. Each of these graphs is a line graph with characteristic polynomial $x(x^2 - x - 4)(x - 1)^2(x + 1)^2$. The root graph of the first graph has *Q*-polynomial $x(x - 1)^2(x - 2)(x - 3)(x^2 - 5x + 2)$ while the root graph of the second has *Q*-polynomial $x^2(x - 1)(x - 2)(x - 3)(x^2 - 5x + 2)$.

This example suggests that the polynomial $Q_G(x)$ may be more useful than $P_{L(G)}(x)$. On the other hand, very few relations between $Q_G(x)$ and the

Figure 7.2 Cospectral line graphs.

structure of $G$ are known. Since we have just the opposite situation with eigen-values of the adjacency matrix of a line graph, we may prefer to use $P_{L(G)}(x)$ in spite of the fact that $L(G)$ usually has more vertices than $G$.

However, we have seen that $P_{L(G)}(x)$ contains less information on the struc-ture of $G$ than $Q_G(x)$. This disadvantage can be eliminated if, in addition to $P_{L(G)}(x)$, we know the number of vertices of $G$. Then our information about $G$ is the same as that provided by $Q_G(x)$, since $Q_G(x)$ can be calculated by formula (7.30), and either of the two polynomials can be considered.

In view of our remarks in this section, it is desirable when using the theory of $Q$-eigenvalues in the study of a graph $G$ to prescribe either

(a) $Q_G(x)$ and the number of components of $G$ or, equivalently,

(b) $P_{L(G)}(x)$, the number of vertices of $G$ and the number of components of $G$.

For regular graphs, there is no need to specify of the number of components, as the following result demonstrates.

**Theorem 7.8.6.** *Let $G$ be a graph with $n$ vertices and $m$ edges, and let $\xi_1$ be its largest $Q$-eigenvalue. Then $\xi_1 \geq 4m/n$, with equality if and only if $G$ is regular. If $G$ is regular then its degree is equal to $\frac{1}{2}\xi_1$, and the number of components equals the multiplicity of $\xi_1$.*

**Proof.** We have $\mathbf{j}^\top Q \mathbf{j} / \mathbf{j}^\top \mathbf{j} = 4m/n$. Hence $\xi_1 \geq 4m/n$, with equality if and only if $\mathbf{j}$ is an eigenvector of $Q$ corresponding to $\xi_1$. The first assertion follows because $Q\mathbf{j} = \xi_1 \mathbf{j}$ if and only if $G$ is regular. The second assertion follows from the analogous property of the adjacency matrix (see Corollary 1.3.8). $\square$

The largest $Q$-eigenvalue is discussed further in Subsection 7.8.3.

### 7.8.2 $Q$-eigenvalues and graph structure

Our first result is an analogue of Proposition 1.3.4, which says that the $(i, j)$-entry of $A(G)^k$ is the number of $i$-$j$ walks of length $k$ in $G$. We may regard such a walk as an alternating sequence $v_0, e_1, v_1, e_2, \ldots, v_{k-1}, e_k, v_k$ of ver-tices and edges such that for each $i = 1, \ldots, k$ the vertices $v_{i-1}$ and $v_i$ are distinct endvertices of the edge $e_i$.

In following this walk, a traveller traverses an edge from one endvertex to the other. Suppose instead that, on reaching the mid-point of an edge, the traveller is permitted to return to the initial endvertex. Then the basic constituent of a walk is no longer an edge but a *semi-edge*: a semi-edge is followed by either the other semi-edge in the same edge (in which case the traveller completes the edge) or the same semi-edge (in which case the traveller returns to the initial endvertex). We arrive at the following definition.

**Definition 7.8.7.** A semi-edge walk of length $k$ is an alternating sequence $v_1, e_1, v_2, e_2, \ldots, v_k, e_k, v_{k+1}$ of vertices and edges such that for each $i = 1, 2, \ldots, k$ the vertices $v_i$ and $v_{i+1}$ are endvertices (not necessarily distinct) of the edge $e_i$.

The following result has a straightforward proof by induction on $k$, or by consideration of the adjacency matrix of the multigraph obtained by adding $d_i$ loops to the vertex $i$ ($i = 1, 2, \ldots, n$).

**Theorem 7.8.8.** *Let $Q$ be the signless Laplacian of a graph $G$. The $(i, j)$-entry of the matrix $Q^k$ is equal to the number of semi-edge walks of length $k$ starting at vertex $i$ and terminating at vertex $j$.*

We write $\tau_k$ for the spectral moment $\sum_{i=1}^{n} \xi_i^k$ ($k = 0, 1, 2, \ldots$). Since $\tau_k = \mathrm{tr}(Q^k)$, it follows immediately from Theorem 7.8.8 that $\tau_k$ is equal to the number of closed semi-edge walks of length $k$.

**Corollary 7.8.9.** *Let $G$ be a graph with $n$ vertices, $m$ edges, $t$ triangles and vertex degrees $d_1, d_2, \ldots, d_n$. We have*

$$\tau_0 = n, \quad \tau_1 = \sum_{i=1}^{n} d_i = 2m, \quad \tau_2 = 2m + \sum_{i=1}^{n} d_i^2, \quad \tau_3 = 6t + 3\sum_{i=1}^{n} d_i^2 + \sum_{i=1}^{n} d_i^3.$$

**Proof.** The formulae for $\tau_0$ and $\tau_1$ are obvious. In the expression for $\tau_2$, the first term counts the closed semi-edge walks which traverse an edge while the second term counts those traversing two semi-edges. In the expression for $\tau_3$, the terms are related to walks around a triangle, walks along one edge and one semi-edge, and walks along three semi-edges. □

Alternatively, the formulae for $\tau_2$ and $\tau_3$ may be derived from the relations $\mathrm{tr}(A + D)^2 = \mathrm{tr}A^2 + 2\,\mathrm{tr}\,AD + \mathrm{tr}D^2$ and $\mathrm{tr}(A + D)^3 = \mathrm{tr}A^3 + 3\mathrm{tr}A^2D + 3\mathrm{tr}AD^2 + \mathrm{tr}D^3$.

Next we investigate the coefficients of the $Q$-polynomial. Let $G$ be a connected graph with $n$ vertices and $m$ edges where $m \geq n$, and let

$$Q_G(x) = \sum_{j=0}^{n} p_j x^{n-j} = p_0 x^n + p_1 x^{n-1} + \cdots + p_n.$$

A spanning subgraph of $G$ whose components are trees or odd-unicyclic graphs is called a *TU-subgraph* of $G$. Suppose that a $TU$-subgraph $Y$ of $G$ consists of $c$ unicyclic graphs and trees $T_1, T_2, \ldots, T_s$. Then the weight $w(Y)$ of $Y$ is defined by $w(Y) = 4^c \prod_{i=1}^{s}(1 + |E(T_i)|)$. Note that isolated vertices in $Y$ do not contribute to $w(Y)$ and may be ignored. To obtain expressions for the coefficients of $Q_G(x)$ in terms of weights of the $TU$-subgraphs of $G$, we require the following observation:

**Lemma 7.8.10.** *For a connected graph G with m edges,*

$$(-1)^m P_{L(G)}(-2) = \begin{cases} 4 & \text{if } G \text{ is odd unicyclic,} \\ m+1 & \text{if } G \text{ is a tree,} \\ 0 & \text{otherwise.} \end{cases}$$

**Proof.** By Corollary 3.4.10, $L(G)$ has $-2$ as an eigenvalue unless $G$ is a tree or an odd-unicyclic graph. In these remaining two cases, let $B$ be the incidence matrix of $G$, so that $(-1)^m P_{L(G)}(-2) = \det(B^\top B)$ by (7.29). If $G$ is odd-unicyclic then it is a straightforward exercise to show (by induction on $m$) that $\det(B) = \pm 2$ and hence that $(-1)^m P_{L(G)}(-2) = 4$. If $G$ a tree then (like any bipartite graph) it has a gradient matrix $R$ such that $R^\top R = B^\top B$. If $R_i$ is the matrix obtained from $R$ by deleting the $i$-th row then $\det(R_i) = \pm 1$ by Lemma 7.2.1. By the Binet–Cauchy formula (Theorem 1.3.18), $\det(B^\top B) = \sum_{i=1}^{m+1} \det(R_i^\top R_i) = m+1$, and this completes the proof. $\square$

**Theorem 7.8.11.** *With the above notation, we have $p_0 = 1$ and*

$$p_j = (-1)^j \sum_{Y_j} w(Y_j), \quad j = 1, 2, \ldots, n,$$

*where the summation runs over all $TU$-subgraphs of $G$ with $j$ edges.*

**Proof.** We first recall the formula of Exercise 2.11:

$$P_G^{(k)}(x) = k! \sum_{|S|=k} P_{G-S}(x), \tag{7.31}$$

where the summation runs over all $k$-subsets $S$ of $V(G)$. Using a Maclaurin expansion of $P_{L(G)}(x)$, we have from (7.30):

$$Q_G(x) = x^{n-m} P_{L(G)}(x - 2)$$

$$= x^{n-m} \sum_{k=0}^{m} P_{L(G)}^{(k)}(-2) \frac{x^k}{k!}$$

$$= x^{n-m} \sum_{k=m-n}^{m} x^k \frac{1}{k!} P_{L(G)}^{(k)}(-2).$$

Applying (7.31), we obtain

$$Q_G(x) = x^{n-m} \sum_{k=m-n}^{m} x^k \sum_{|S|=k} P_{L(G)-S}(-2). \tag{7.32}$$

A subgraph $L(G) - S$ is, of course, a line graph and it has $-2$ as an eigenvalue unless all components are line graphs of trees or of odd-unicyclic graphs.

Thus it follows from Lemma 7.8.10 that

$$\sum_{|S|=k} P_{L(G)-S}(-2) = \sum_{Y_{m-k}} (-1)^{m-k} w(Y_{m-k}),$$

where, in the second sum, the summation runs over all $TU$-subgraphs $Y_{m-k}$ of $G$ with $m - k$ edges. Now the formula (7.32) becomes

$$Q_G(x) = x^{n-m} \sum_{k=m-n}^{m} x^k (-1)^{m-k} \sum_{Y_{m-k}} w(Y_{m-k}),$$

By substituting $j$ for $m - k$ we obtain

$$Q_G(x) = \sum_{j=0}^{n} x^{n-j} (-1)^j \sum_{Y_j} w(Y_j).$$

This completes the proof.                                                    □

For $j = 1$ the only $TU$-subgraph $Y_1$ is equal to $K_2$, with $w(Y_1) = 2$, and we readily obtain $p_1 = -2m$, thereby recovering the formula $\tau_1 = 2m$. For $j = 2$, the possible $TU$-subgraphs $Y_2$ are $2K_2$ and $K_{1,2}$. Since $w(2K_2) = 4$ and $w(K_{1,2}) = 3$ we have $p_2 = 4a + 3b$ where $a$ is the number of pairs of non-adjacent edges and $b$ is the number of pairs of adjacent edges in $G$. Since $a + b = m(m - 1)/2$, we have the following result:

**Corollary 7.8.12.** *With the notation above,* $p_1 = -2m$ *and* $p_2 = a + \frac{3}{2}m$ $(m - 1)$, *where $a$ is the number of pairs of non-adjacent edges in $G$.*

An interlacing theorem holds for $Q$-eigenvalues in the same way as for Laplacian eigenvalues. Exactly as in Theorem 7.1.5, the $Q$-eigenvalues of an edge-deleted subgraph $G - e$ interlace those of $G$:

**Theorem 7.8.13.** *If $e$ is an edge of the graph $G$ and $G' = G - e$ then*

$$0 \leq \xi_n(G') \leq \xi_n(G) \leq \cdots \leq \xi_2(G') \leq \xi_2(G) \leq \xi_1(G') \leq \xi_1(G).$$

Theorem 7.8.13 may also be proved by applying Corollary 1.3.12 (the Interlacing Theorem) to $L(G)$. In fact, most of the results in this section are obtained either by considering line graphs or by replicating arguments for the adjacency matrix. We conclude this subsection by mentioning without proof two results which exhibit characteristics peculiar to the signless Laplacian.

For a subset $S$ of $V = V(G)$, let $e_{min}(S)$ be the minimum number of edges whose removal from the subgraph of $G$ induced by $S$ results in a bipartite graph. Let cut$(S)$ be the set of edges with one vertex in $S$ and the other in the complement $V \setminus S$. Thus $|$cut$(S)| + e_{min}(S)$ is the minimum number of edges whose removal from $E(G)$ disconnects $S$ from $V \setminus S$ and results in a bipartite subgraph induced by $S$. Let $\psi = \psi(G)$ be the minimum over all non-empty proper subsets $S$ of $V(G)$ of the quotient

$$\frac{|\mathrm{cut}(S)| + e_{\min}(S)}{|S|}.$$

The parameter $\psi$ was introduced in [DesRa] as a measure of non-bipartiteness. It is shown that the value of $\xi_n$ is controlled by $\psi$ (cf. Theorem 7.8.1). In particular, if $G$ is connected then

$$\frac{\psi^2}{4\Delta} \leq \xi_n \leq 4\psi,$$

where $\Delta$ is the maximal vertex degree.

Secondly, let $p$ be the number of endvertices in a graph, and let $q$ be the number of vertices adjacent to endvertices. It is proved in [Far] that the difference $p - q$ is equal to the multiplicity of the root 1 of the permanental polynomial $\mathrm{per}(xI - Q)$ of the signless Laplacian. Examples demonstrate that there is no analogous result for the Laplacian or adjacency matrix.

### 7.8.3 The largest $Q$-eigenvalue

In this final subsection we establish various bounds on the largest eigenvalue of the signless Laplacian. For the adjacency matrix of a graph $G$, we have

$$\delta(G) \leq \lambda_1(G) \leq \Delta(G). \tag{7.33}$$

For a connected graph $G$, equality holds in either place if and only if $G$ is regular. For $\xi_1(G)$, we have the following analogue, with a similar proof:

**Proposition 7.8.14.** *For any graph $G$, we have $2\delta(G) \leq \xi_1(G) \leq 2\Delta(G)$. For a connected graph $G$, equality holds in either place if and only if $G$ is regular.*

**Proof.** We may assume throughout that $G$ is connected. By Theorem 1.3.5, $G$ has a principal $Q$-eigenvector $(x_1, \ldots, x_n)^\top$ such that $x_1 \geq \cdots \geq x_n > 0$. The corresponding eigenvalue equations yield:

$$\xi_1 x_1 = d_1 x_1 + \sum_{i \sim 1} x_i \leq 2\Delta x_1 \quad \text{and} \quad \xi_1 x_n = d_n x_n + \sum_{j \sim n} x_j \geq 2\delta x_n,$$

where $\xi_1 = \xi_1(G)$, $\delta = \delta(G)$ and $\Delta = \Delta(G)$. The first assertion follows.

If $G$ is $r$-regular then $\xi_1(G) = 2r = 2\delta = 2\Delta$. If $\xi_1 = \delta$ or $\Delta$ then the $n$ eigenvalue equations force $x_1 = \cdots = x_n$ and $d_1 = \cdots = d_n$. This completes the proof. $\qquad \square$

Stronger inequalities can be obtained by applying (7.33) to the line graph of $G$:

**Theorem 7.8.15.** *Let $G$ be a graph on $n$ vertices, with vertex degrees $d_1, d_2, \ldots, d_n$ and largest $Q$-eigenvalue $\xi_1$. Then*

$$\min (d_i + d_j) \leq \xi_1 \leq \max (d_i + d_j),$$

*where $(i, j)$ runs over all pairs of adjacent vertices of $G$. For a connected graph $G$, equality holds in either place if and only if $G$ is regular or semi-regular bipartite.*

**Proof.** The graph $L(G)$ has index $\xi_1 - 2$, while the edge $ij$ has degree $d_i + d_j - 2$. By (7.33), we have

$$\min (d_i + d_j - 2) \leq \xi_1 - 2 \leq \max (d_i + d_j - 2),$$

and the result follows.                                                            $\square$

By applying Proposition 1.3.9 to the line graph of a connected graph $G$, we can also see that $\xi_1(H) < \xi_1(G)$ for any proper subgraph $H$ of $G$.

**Proposition 7.8.16.** *If $\xi_1$ is the largest $Q$-eigenvalue of a graph $G$, then:*

  (i) *$\xi_1 = 0$ if and only if $G$ has no edges;*
 (ii) *$\xi_1 < 4$ if and only if all components of $G$ are paths;*
(iii) *for a connected graph $G$ we have $\xi_1 = 4$ if and only if $G$ is a cycle or $K_{1,3}$.*

**Proof.** Statement (i) is immediate, since $G$ is a null graph if and only if all $Q$-eigenvalues of $G$ are zero.

The eigenvalues of $L(P_n) = P_{n-1}$ are $2 \cos \frac{\pi}{n} j$ ($j = 1, 2, \ldots, n - 1$) and so by (7.30) the $Q$-eigenvalues of $P_n$ are $2 + 2 \cos \frac{\pi}{n} j$ ($j = 1, 2, \ldots, n$). Hence for paths we have $\xi_1 < 4$. For cycles and for $K_{1,3}$ we have $\xi_1 = 4$. By interlacing, these graphs are forbidden subgraphs in graphs for which $\xi_1 < 4$, and this completes the proof of (ii).

To prove the sufficiency in (iii) we use the strict monotonicity of the largest $Q$-eigenvalue when adding edges to a connected graph. First, $G$ cannot contain a cycle $Z$ unless $G = Z$. If $G$ does not contain a cycle, it must contain $K_{1,3}$ since otherwise $G$ would be a path and we would have $\xi_1 < 4$. Finally $G$ must be $K_{1,3}$ since otherwise we would have $\xi_1 > 4$. This completes the proof.    $\square$

The proof of the next proposition can now be left to the reader.

**Proposition 7.8.17.** *The $Q$-index $\xi_1$ of a connected graph on $n$ vertices satisfies the inequalities*

$$2 + 2\cos\tfrac{\pi}{n} \le \xi_1 \le 2n - 2.$$

*The lower bound is attained for $P_n$, and the upper bound for $K_n$.*

## Exercises

**7.1** Determine the Laplacian eigenvalues of the graphs in Fig. 7.1.

**7.2** Prove that the Laplacian eigenvalues of $P_n$ are $4\sin^2\left(\frac{\pi(i-1)}{2n}\right)$ $(i = 1, 2, \ldots, n)$.

**7.3** Find the Laplacian spectrum of the lattice graph $P_m + P_n$.

**7.4** Show that the Laplacian spectrum of a graph determines $\sum_{i=1}^n d_i^2$, where $d_1, \ldots, d_n$ are the vertex degrees.

**7.5** Prove Theorem 7.2.3.

**7.6** Prove Theorem 7.2.4.

**7.7** Let $G$ be an $r$-regular graph with $n$ vertices and $m$ edges. Show that [Kel2]

$$\tau(L(G)) = 2^{m-n+1} r^{m-n-1} \tau(G).$$

**7.8** Determine the number of spanning trees in the graph obtained from $K_n$ by removing $m$ non-adjacent edges ($2m \le n$).

**7.9** Prove that the skeleta of the Platonic solids have the following numbers of spanning trees: (i) tetrahedron, $2^4$; (ii) cube and octahedron, $2^7 3$; (ii) icosahedron and dodecahedron, $2^9 3^4 5^3$.

**7.10** Use (7.3) and (7.9) to establish the inequalities (7.6).

**7.11** Verify the values of $a(G)$ for the graphs $G$ given in Examples 7.4.1.

**7.12** Prove that for any $r$-regular graph $G$ on $n$ vertices,
$$\nu_{n-1}(G) \ge \sqrt{r(1 - o_n(1))}.$$
(Hint: Consider the trace of the adjacency matrix of $G$.)

**7.13** Verify the values of $i(G)$ for the graphs $G$ given in Examples 7.5.12.

**7.14** Prove Proposition 7.5.6.

**7.15** Prove that the normalized Laplacian spectrum of (i) $K_n$, (ii) $K_{m,n}$, (iii) $P_n$, (iv) $C_n$ is given by:

(i) $0, \left(\frac{n}{n-1}\right)^{n-1}$;

(ii) $0^{m+n-2}, 1^2$;

(iii) $1 - \cos(\frac{\pi k}{n-1})$ $(k = 0, 1, \ldots, n-1)$;

(iv) $1 - \cos(\frac{2\pi k}{n-1})$ $(k = 0, 1, \ldots, n-1)$.

**7.16** Show that if $G$ is a graph on $n$ vertices ($n \ge 2$) then $\hat{\nu}_{n-1}(G) \le 2h(G)$.

**7.17** Determine the $Q$-eigenvalues of the following regular graphs: (i) the complete graph $K_n$, (ii) the cycle $C_n$, (iii) the cocktail-party graph $CP(k)$.

**7.18** Use the results of Chapter 2 to determine the $Q$-eigenvalues of (i) the path $P_n$, (ii) the complete bipartite graph $K_{m,n}$.

**7.19** Prove Theorem 7.8.8.

**7.20** Prove Proposition 7.8.17.

# Notes

Surveys of Laplacians include those by Merris [Mer1, Mer3] and Mohar [Moh2, Moh4]. These articles, together with Chung's monograph [Chu2] on normalized Laplacians, show clearly that we have merely scratched the surface of a topic which is both broad and deep. A geometric approach to Laplacians is described in [Fie5]. The role of Laplacian eigenvalues in combinatorial optimization is explored in [MohPo3] and [Moh5]. For more examples of non-isomorphic graphs with the same Laplacian spectrum, the reader is referred to three papers from 1977, namely [DinKZ], [Hat] and [GoHMK], where this phenomenon was first noted. Characterizations of certain trees by their Laplacian spectrum may be found in [OmTa] and [WaXu].

The Matrix-Tree Theorem is attributed to Kirchhoff [Kirc] and Trent [Tre]. Further results on the enumeration of spanning trees can be found in [CvDSa, Section 7.6] and the expository paper [Cve2]. Associated algorithms feature in [JoMa], [JoSa1] and [JoSa2], while an extension to weighted graphs was established by Fiedler and Sedláček [FieSe]. One problem which has received considerable attention is the determination of connected graphs, with a pre-scribed number of vertices and a prescribed number of edges, which have the smallest or largest number of spanning trees; relevant references include [BoLS], [Cheng], [Con1], [Kel4], [KelCh], [Shi] and [Wang]. Some bounds for the complexity of a graph are obtained in [Das3], [Gri] and [GroMe1]. Other results concern spanning trees in random regular graphs [McK] and the characterization of graphs in which each edge is contained in a constant num-ber of spanning trees [God]. A proof of Theorem 7.2.8 may be found in [Big2, Chapter 7]. The formula in Exercise 7.7 appears in [CvDSa, Theorem 7.24].

Further information on the largest Laplacian eigenvalue $\nu_1$ may be found in in [BrHS], [Das1], [LiPa] and [ShuHW]. For changes in $\xi_1$ resulting from certain graph modifications, see Exercises 8.1, 8.2, 8.4 and 8.5.

The pioneering work on algebraic connectivity was undertaken by Fiedler [Fie1, Fie2, Fie3, Fie4]. Related results concerning the diameter and other metric invariants of a graph may be found in [Chu1], [DamHa2], [DelSo] and [Moh3]. For a survey of results on algebraic connectivity, see [Abr].

The bounds for mc($G$) and bw($G$) mentioned in Section 7.5 can be improved by introducing certain correction functions; see [Bop] and [DedPo] for more details. For an extension to weighted graphs, see [MohPo2] and [MohPo3]. For a discussion of invariants that are NP-hard to determine, see, for example, [GarJo].

The papers [Alo1], [AloMi1], [AloMi2] contain more information on expansion properties; the article [Alo2] provides a useful overview. Theorems 7.7.6 and 7.7.8 appear in [Chu2] in a form which is not quite accurate; the corrected versions are due to Kirkland (see [Kir2]).

Few papers treating the signless Laplacian can be found in the literature; it appears that the only papers prior to 2003 which contain substantive results of this sort are [Ded], [DesRa] and [Far]. More recent observations may be found in [Cve14], [CvRS11], [CvRS10], [DanHa1], [DanHa2], [HaeSp] and [ZhWi], while several new papers are in preparation. The papers [CvSi5], [CvSi6] lay the foundations of a spectral theory of graphs based on the signless Laplacian.

The paper [CvRS11] discusses 30 computer-generated conjectures concerning the $Q$-eigenvalues of a graph, and several of the conjectures are confirmed there. A conjecture concerning a lower bound for the least $Q$-eigenvalue is confirmed in [CarCRS]. Further bounds for the largest $Q$-eigenvalue may be found in [OLAH]. The Coefficient Theorem for $Q$-polynomials (Theorem 7.8.11), for which we have given a recent proof from [CvRS10], features in [Ded] along with some results on the reconstructibility of the $Q$-polynomial from the deck of vertex-deleted subgraphs of $G$.

# 8

# Some additional results

This chapter is devoted to results which did not fit readily into earlier chapters. Section 8.1 is concerned with the behaviour of certain eigenvalues when a graph is modified, and with further bounds on the index of a graph. Section 8.2 deals with relations between the structure of a graph and the sign pattern of certain eigenvectors. Results from these first two sections enable us to give a general description of the connected graphs having maximal index or minimal least eigenvalue among those with a given number of vertices and edges. In Section 8.3 we discuss the reconstruction of the characteristic polynomial of a graph from the characteristic polynomials of its vertex-deleted subgraphs. In Section 8.4 we review what is known about graphs whose eigenvalues are integers.

## 8.1 More on graph eigenvalues

In this section we revisit two topics which have featured in previous chapters. The first topic concerns the relation between the spectrum of a graph $G$ and the spectrum of some modification $G'$ of $G$. When the modification arises as a small structural alteration (such as the deletion or addition of an edge or vertex), the eigenvalues of $G'$ are generally small perturbations of those of $G$, and we say that $G'$ is a *perturbation* of $G$. In Subsection 8.1.1, we use algebraic arguments to establish some general rules which determine whether certain eigenvalues increase or decrease under particular graph perturbations.

Many articles in the area of spectral graph theory are concerned with various bounds on the eigenvalues of graphs. We have already encountered some bounds on $\lambda_1$, $\nu_1$ and $\xi_1$ in Chapters 1, 3 and 7. In Subsection 8.1.2, we provide further bounds for $\lambda_1$ as a sample from the extensive literature on this topic.

### 8.1.1 Graph perturbations

We have already seen in Propositions 1.3.9 and 1.3.10 that the index of a graph decreases when a vertex or edge is deleted. Similar arguments show that the same is true of the largest eigenvalue of the signless Laplacian (Exercise 8.1):

**Theorem 8.1.1.** *If $G'$ is a graph obtained from a graph $G$ by deleting any vertex or any edge then $\lambda_1(G') \leq \lambda_1(G)$ and $\xi_1(G') \leq \xi_1(G)$; moreover, these inequalities are strict when $G$ is connected.*

Note that, in general, the corresponding assertion fails for the Laplacian because the arguments require matrix entries to be non-negative. However, for bipartite graphs, results concerning $\xi_1$ may be translated to analogous results for $\nu_1$ by means of Proposition 7.8.4.

In considering perturbations $G'$ of a graph $G$, we assume that $G$ is connected with $n$ vertices, and that $\mathbf{x} = (x_1, x_2, \ldots, x_n)^\top$ is its principal eigenvector (that is, the unique positive unit eigenvector corresponding to the index of $G$). To investigate the change in index we frequently invoke Rayleigh's Principle: if $A, A'$ are the adjacency matrices of $G, G'$ respectively, then

$$\lambda_1(G') - \lambda_1(G) = \max_{\|\mathbf{y}\|=1} \mathbf{y}^\top A' \mathbf{y} - \mathbf{x}^\top A \mathbf{x} \geq \mathbf{x}^\top A' \mathbf{x} - \mathbf{x}^\top A \mathbf{x} = \mathbf{x}^\top (A' - A)\mathbf{x}.$$

$$(8.1)$$

In some cases, we investigate the behaviour of the index $\lambda_1$ using characteristic polynomials. We consider the following perturbations: (i) the relocation of edges, (ii) local switching of two edges, (iii) the splitting of a vertex, (iv) the subdivision of an edge. Most arguments apply also to $\xi_1$, while the results for (i) and (ii) have analogues for $\lambda_n$; together they provide a useful tool for re-arranging edges so that the largest or least eigenvalue of the resulting graph is extremal in some family of graphs with prescribed numbers of vertices and edges.

**Theorem 8.1.2.** *Let $G'$ be the graph obtained from a connected graph $G$ by relocating the edge $rs$ to the position of a non-edge $tu$. If $x_t x_u \geq x_r x_s$ then $\lambda_1(G') > \lambda_1(G)$.*

**Proof.** By (8.1) we have $\lambda_1(G') - \lambda_1(G) \geq \Delta_1$, where $\Delta_1 = \mathbf{x}^\top (A' - A)\mathbf{x}$. Since $\Delta_1 = 2(x_t x_u - x_r x_s)$, we have $\lambda_1(G') \geq \lambda_1(G)$. Equality holds if and only if $\Delta_1 = 0$ and $\mathbf{x}$ is an eigenvector for $G'$. But then the eigenvalue equations do not hold for the vertices $r, s, t$ and $u$ of $G'$. This contradiction completes the proof. □

We record separately the important case of Theorem 8.1.2 in which $u = r$; in this situation, replacement of the edge $rs$ with $rt$ is called a *rotation* about $r$.

**Theorem 8.1.3.** *Let $G'$ be the graph obtained from a connected graph $G$ by rotating the edge $rs$ to the position of a non-edge $rt$. If $x_t \geq x_s$ then $\lambda_1(G') > \lambda_1(G)$.*

**Proof.** Since $x_r > 0$, the condition $x_t \geq x_s$ is equivalent to the condition $x_t x_r \geq x_r x_s$, and so the result follows from Theorem 8.1.2. $\qquad\square$

A useful consequence of Theorem 8.1.3 applies to the principal eigenvector $\mathbf{x}' = (x_1', x_2', \dots, x_n')^\top$ of $G'$:

**Corollary 8.1.4.** *In the situation of Theorem 8.1.3, if also $G'$ is connected, then $x_t' > x_s'$.*

**Proof.** Suppose by way of contradiction that $x_t' \leq x_s'$. We consider $G'$ and rotate the edge $rt$ to the non–edge position $rs$. Then we obtain $G$, and by Theorem 8.2.3 we have $\lambda_1(G) > \lambda_1(G')$, a contradiction. $\qquad\square$

We may generalize Theorem 8.1.3 as follows:

**Theorem 8.1.5.** *Let $s$ and $t$ be two vertices of a connected graph $G$, and let $R$ be a set of vertices adjacent to $s$ but not to $t$. Let $G'$ be a graph obtained from $G$ by replacing the edge $rs$ with $rt$ for each $r \in R$. If $x_t \geq x_s$ then $\lambda_1(G') > \lambda_1(G)$.*

The proof is left as an exercise: one can either extend the arguments for Theorem 8.1.3 or make repeated use of Corollary 8.1.4. (Note that $G'$ is obtained from $G$ by successive rotations about the vertices in $R$.) An exact analogue of Theorem 8.1.5 holds for the largest eigenvalue of the signless Laplacian (Exercise 8.2).

**Example 8.1.6.** Suppose that $G$ is a graph with a non-pendant edge $uv$ not belonging to a triangle of $G$. Let $G'$ be the graph obtained from $G$ by contracting the edge $uv$ to a vertex $w$ and adding a pendant edge at $w$. By rotating the edges incident to $u$, or to $v$ (according as $x_u \leq x_v$, or $x_u \geq x_v$), we deduce from Theorem 8.1.5 that $\lambda_1(G') > \lambda_1(G)$. $\qquad\square$

We can use Theorem 8.1.5 in exactly the same way to prove the following result, which is useful when we encounter a bridge in a graph whose index is assumed to be maximal among the connected graphs with a prescribed number of vertices and edges.

**Theorem 8.1.7.** *Suppose that the non-pendant edge $uv$ is a bridge in the connected graph $G$. Let $G'$ be the graph obtained from $G$ by contracting $uv$ to a vertex $w$ and adding a pendant edge at $w$. Then $\lambda_1(G') > \lambda_1(G)$ and $\xi_1(G') > \xi_1(G)$.*

A nice application of this result is encountered when $x_u \geq x_v$ and the component of $G - uv$ containing $v$ is a star with centre $v$. This situation arises typically in graphs with a pendant tree (see the treatment of unicyclic graphs in [Sim1]).

**Corollary 8.1.8.** *Let $G$ be a graph whose index is maximal among connected graphs with a prescribed number of vertices and edges. Then $G$ does not contain any of the graphs $2K_2$, $P_4$, $C_4$ as an induced subgraph.*

**Proof.** Suppose by way of contradiction that $G$ contains a graph $F \in \{P_4, 2K_2, C_4\}$ as an induced subgraph, say with vertices $r, s, t, w$. Without loss of generality, $x_s = \min_{v \in V(F)} x_v$. Additionally, the structure of $F$ allows us to assume that $r$ is a neighbour of $s$ but not of $t$. Now let $G'$ be the graph obtained from $G$ by rotating edge $rs$ to $rt$. By Theorem 8.1.3, we have $\lambda_1(G') > \lambda_1(G)$. If $G'$ is connected, this is a contradiction and we are done.

Accordingly, suppose that $G'$ is not connected. By Theorem 8.1.7, $rs$ is a pendant edge, and we may suppose that $\deg(s) = 1$.

We first observe that $x_u < x_r$ for any vertex $u \in V(G - r - s)$, for otherwise we may replace $rs$ with $us$ to obtain a connected graph with larger index. Now either $F$ is the path $srut$ or $F$ consists of the two independent edges $rs$, $tu$. In either case we may replace $tu$ with $tr$ to obtain a connected graph with larger index. This final contradiction completes the proof. □

**Remarks 8.1.9.** (i) The graphs without $2K_2$, $P_4$, $C_4$ as an induced subgraph are precisely the *threshold graphs* mentioned in Section 7.2 in the context of Laplacian eigenvalues. They also known as *nested split graphs* (see, for example, [ABCHRSS]) or *stepwise graphs* (see [CvRS2, Chapter 3]). They are best visualized as graphs with a stepwise adjacency matrix: in such a matrix $(a_{ij})$, the pattern of 0s and 1s has a stepped form determined by the condition:

$$\text{if } i < j \text{ and } a_{ij} = 1 \text{ then } a_{hk} = 1 \text{ whenever } h < k \leq j \text{ and } h \leq i.$$

(ii) If $G$ is a graph whose index is maximal among all graphs with just a prescribed number of edges then again $G$ is a threshold graph. This follows from Corollary 8.1.8 because $G$ has just one non-trivial component. To see this, let $H$ be a component with $\lambda_1(H) = \lambda_1(G)$. Then we may rotate an edge from a second component to construct a graph $G'$ with an induced subgraph $H'$ obtained from $H$ by adding a pendant edge.

(iii) An exact analogue of Corollary 8.1.8 holds for the signless Laplacian (Exercise 8.3). □

Next we turn to *local switching*, which concerns two non-adjacent edges $st$ and $uv$ such that $s \not\sim v$, $t \not\sim u$: we replace $st$, $uv$ with $sv$, $tu$. Note that local switching preserves degrees; moreover, if $G_1, G_2$ are graphs with the same degree sequence then $G_1$ can be transformed to $G_2$ by a succession of local switchings (see, for example, [Wes, p. 45]).

**Theorem 8.1.10.** *Let $G'$ be the graph obtained from a connected graph $G$ by the local switching of $st$, $uv$ to $sv$, $tu$. If $(x_s - x_u)(x_v - x_t) \geq 0$ then $\lambda_1(G') \geq \lambda_1(G)$, with $\lambda_1(G') = \lambda_1(G)$ if and only if $x_s = x_u$ and $x_v = x_t$.*

**Proof.** By (8.1), $\lambda_1(G') - \lambda_1(G) \geq \Delta_1$, where $\Delta_1 = \mathbf{x}^\top(A' - A)\mathbf{x}$. Since $\Delta_1 = 2(x_s - x_u)(x_v - x_t)$ we have $\lambda_1(G') \geq \lambda_1(G)$. Equality holds if and only if $\Delta_1 = 0$ and $\mathbf{x}$ is an eigenvector of $G'$. It remains to eliminate the possibility that exactly one of the factors $x_s - x_u$, $x_v - x_t$ is zero. Assume, without loss of generality, that $x_s = x_u$, while $x_t \neq x_v$. Then the eigenvalue equations hold (in respect of $G'$) for the vertices $t$ and $v$, but not for the vertices $s$ and $u$. This contradiction completes the proof.                                    □

An analogue of Theorem 8.1.10 holds for the signless Laplacian (Exercise 8.5).

In discussing two further perturbations of a connected graph, we use a comparison of vectors. Writing $\mathbf{x} \succ \mathbf{y}$ to mean that $\mathbf{x} - \mathbf{y}$ is a non-negative non-zero vector, we have:

$$\text{if } \mathbf{y} \succ \mathbf{0} \text{ and } A\mathbf{y} \succ \rho\mathbf{y} \text{ then } \lambda_1 > \rho, \tag{8.2}$$

and

$$\text{if } \mathbf{y} \succ \mathbf{0} \text{ and } A\mathbf{y} \prec \rho\mathbf{y} \text{ then } \lambda_1 < \rho. \tag{8.3}$$

In each case, the conclusion follows by taking the scalar product with the principal eigenvector $\mathbf{x}$.

Suppose first that $G'$ is obtained from $G$ by *splitting* the vertex $v$: thus if the edges incident with $v$ are $vw$ ($w \in W$) then $G'$ is obtained from $G - v$ by adding two new vertices $v_1, v_2$ and edges $v_1 w_1$ ($w_1 \in W_1$), $v_2 w_2$ ($w_2 \in W_2$), where $W_1 \dot\cup W_2$ is a non-trivial bipartition of $W$.

**Theorem 8.1.11 [Sim2].** *If $G'$ is obtained from the connected graph $G$ by splitting a vertex then $\lambda_1(G') < \lambda_1(G)$.*

**Proof.** We may assume that $G'$ is connected for otherwise $G'$ has two components each of which is a proper subgraph of $G$, with largest eigenvalue less than $\lambda_1(G)$. Accordingly we may apply (8.3) to the adjacency matrix $A'$ of $G'$.

Suppose that vertices are numbered so that vertex 1 of $G$ splits into vertices $0, 1$ of $G'$. Let $\lambda_1 = \lambda_1(G)$, and let $\mathbf{y} = (y_0, y_1, y_2, \ldots, y_n)^\top$ where $y_0 = x_1$

and $y_i = x_i$ $(i = 1, \ldots, n)$. Then $A'\mathbf{y} = (y_0', y_1', y_2', \ldots, y_n')^\top$ where $y_0' < \lambda_1 y_0$, $y_1' < \lambda_1 y_1$ and $y_i' = \lambda_1 y_i$ $(i = 2, \ldots, n)$. Thus $A'\mathbf{y} \prec \lambda_1 \mathbf{y}$ and by (8.3) we have $\lambda_1(G') < \lambda_1$. $\qquad\square$

The same result holds for the $Q$-index (Exercise 8.4).

Secondly we consider a graph obtained from the connected graph $G$ by *subdividing* an edge $uv$, that is by replacing $uv$ with edges $uw$ and $wv$, where $w$ is an additional vertex. We say that the walk $v_0 v_1 \ldots v_{k+1}$ is an *internal path* of $G$ if one of the following holds:

(a) $k \geq 2$, the vertices $v_0, v_1, \ldots, v_k$ are distinct, $v_{k+1} = v_0$, $v_i \sim v_{i+1}$ $(i = 0, \ldots, k)$, $\deg(v_0) \geq 3$ and $\deg(v_i) = 2$ $(i = 1, \ldots, k)$;

(b) $k \geq 0$, the vertices $v_0, v_1, \ldots, v_{k+1}$ are distinct, $v_i \sim v_{i+1}$ $(i = 0, \ldots, k)$, $\deg(v_0) \geq 3$, $\deg(v_{k+1}) \geq 3$ and $\deg(v_i) = 2$ $(i = 1, \ldots, k)$.

Thus in case (a) the vertices $v_0, \ldots, v_k$ induce a cycle which is a proper subgraph, while in case (b) the vertices $v_0, \ldots, v_{k+1}$ induce a path and lie in a subgraph isomorphic to the graph $Y_{k+6}$ of Fig. 3.5.

**Theorem 8.1.12** [HofSm]. *Let $G$ be a connected graph with $n$ vertices, and let $G'$ be the graph obtained from $G$ by subdividing the edge $e$ of a connected graph $G$.*

(i) *If $G \neq C_n$ and if $e$ does not belong to an internal path then $\lambda_1(G') > \lambda_1(G)$.*

(ii) *If $G \neq Y_n$ and if $e$ belongs to an internal path of $G$ then $\lambda_1(G') < \lambda_1(G)$.*

*Moreover, if $G = C_n$ $(n \geq 3)$ or $Y_n$ $(n \geq 6)$ then $\lambda_1(G') = \lambda_1(G) = 2$.*

**Proof.** We know that $C_n$ $(n \geq 3)$ and $Y_n$ $(n \geq 6)$ have index 2 (see Theorem 3.11.1), and so we suppose that $G \neq C_n, Y_n$. In case (i), $G'$ has a proper subgraph isomorphic to $G$, and so $\lambda_1(G') > \lambda_1(G)$ by Theorem 8.1.1. It remains to consider case (ii).

Suppose first that $e$ lies on an internal path $v_0, v_1, \ldots v_{k+1}$ of type (a), with the vertices $v_0, v_1, \ldots, v_k$ labelled $0, 1, \ldots, k$. Let $x_0, x_1, \ldots, x_k$ be the corresponding entries of the principal eigenvector $\mathbf{x}$ of $G$, and let $\lambda_1 = \lambda_1(G)$. (It is helpful to picture the components of $\mathbf{x}$ ascribed to the relevant vertices of $G$, as in Fig. 3.5.) We have $\lambda_1 x_i = x_{i-1} + x_{i+1}$ $(i = 1, \ldots, k)$ where $x_{k+1} = x_0$ and (by symmetry) $x_1 = x_k$, $x_2 = x_{k-1}$, and so on. Let $e = uv$. If $k$ is even then without loss of generality we may take $u = \frac{1}{2}k$, $v = \frac{1}{2}k + 1$. Now let $\mathbf{y}$ be obtained from $\mathbf{x}$ by inserting the additional entry $x_w$ equal to $x_u$ and $x_v$. If $A'$ is the adjacency matrix of $G'$ then $A'\mathbf{y}$ and $\lambda_1 \mathbf{y}$ differ only in the $w$-th entries, for which $x_u + x_v = 2x_w < \lambda_1 x_w$. Hence $A'\mathbf{y} \prec \lambda_1 \mathbf{y}$ and we have $\lambda_1(G') < \lambda_1$

by (8.3). If $k$ is odd then we may take $u = \frac{1}{2}(k-1)$, $v = \frac{1}{2}(k+1)$ and take the new entry $x_w$ of $\mathbf{y}$ to be $x_v$. Now $\lambda_1 x_v = 2x_u$ and $\lambda_1 > 2$, whence $x_u > x_v$, $\lambda_1 x_v > x_u + x_w$ and $\lambda_1 x_w > x_u + x_v$. It follows that $A'\mathbf{y} \prec \lambda_1 \mathbf{y}$ and so again $\lambda_1(G') < \lambda_1$ by (8.3).

Secondly, suppose that $uv$ lies on an internal path of type (b), with vertices labelled $0, 1, \ldots, k+1$, and let $x_0, x_1, \ldots, x_{k+1}$ be the corresponding entries of the principal eigenvector $\mathbf{x}$ of $G$. Reversing the path if necessary we may assume that $x_0 \leq x_{k+1}$. Let $t$ be least such that $x_t = \min\{x_0, x_1, \ldots, x_{k+1}\}$ (thus $t < k+1$). Without loss of generality we let $u = t$, $v = t+1$.

Consider first the case $t > 0$: here we take $\mathbf{y}$ to be the vector obtained from $\mathbf{x}$ by inserting an additional component $x_w$ equal to $x_t$. We have $x_{t-1} + x_w \leq x_{t-1} + x_{t+1} = \lambda_1 x_t$ and $x_t + x_{t+1} < x_{t-1} + x_{t+1} = \lambda_1 x_w$, whence $A'\mathbf{y} \prec \lambda_1 \mathbf{y}$ and $\lambda_1(G') < \lambda_1$ by (8.3). Accordingly we suppose that $t = 0$. Let $S$ be the set of neighbours of $0$ other than $1$, and let $s = \sum_{j \in S} x_j$. If $s \geq x_0$ then we construct $\mathbf{y}$ as above with $x_w = x_0$. We have $s + x_w \leq s + x_1 = \lambda_1 x_0$ and $x_0 + x_1 \leq s + x_1 = \lambda_1 x_0 = \lambda_1 x_w$; moreover one of these inequalities is strict for otherwise $\lambda_1 = 2$, contradicting the fact that $Y_{k+6}$ is a proper subgraph of $G$. Hence $A'\mathbf{y} \prec \lambda_1 \mathbf{y}$. Finally suppose that $s < x_0$. In this case we construct $\mathbf{y}$ from $\mathbf{x}$ by replacing $x_0$ with $s$ and inserting $x_w$ equal to $x_0$. Now $x_0 + s < 2s < \lambda_1 s$, and for any $p \in S$ we have $\sum_{q \sim p, q \neq 0} x_q + s < \sum_{q \sim p, q \neq 0} x_q + x_0 = \lambda_1 x_p$. Therefore, it follows again that $A'\mathbf{y} \prec \lambda_1 \mathbf{y}$. This completes the proof. $\qquad\square$

The analogous result holds for the $Q$-index (Exercise 8.4). Hoffman and Smith asked what happens to the index if we subdivide edges repeatedly by inserting arbitrarily many vertices of degree 2; Theorem 2.2.2 answers this question in a particular case (for the adjacency spectrum).

We have seen that we cannot expect our results on $\lambda_1$ and $\xi_1$ to extend to $\nu_1$ in general. We give one theorem which highlights the contrasting situation for the Laplacian index.

**Theorem 8.1.13** [Guo2]. *Let $v$ be a vertex of the connected graph $G$, and let $v_1, v_2, \ldots, v_k$ be the pendant vertices of $G$ adjacent to $v$. Let $G'$ be a graph obtained from $G$ by adding any $q$ edges ($0 \leq q \leq \binom{k}{2}$) between $v_1, v_2, \ldots, v_k$. Then $\nu_1(G') = \nu_1(G)$.*

**Proof.** Let $\hat{G}$ be the graph obtained by adding all $\binom{k}{2}$ edges between $v_1, v_2, \ldots, v_k$. By interlacing, we have $\nu_1(G) \leq \nu_1(G') \leq \nu_1(\hat{G})$, and so it suffices to show that $\nu_1(G) \geq \nu_1(\hat{G})$. We may assume that $G \neq K_{1,k}$ for otherwise $\hat{G} = K_{k+1}$ and we have $\nu_1(G) = k+1 = \nu_1(\hat{G})$.

Let $\mathbf{x}$ be the unit eigenvector of $\hat{G}$ corresponding to $\nu_1(\hat{G})$. The eigenvalue equations for the vertices $v_1, v_2, \ldots, v_k$ of $\hat{G}$ yield:

$$(k - \nu_1(\hat{G}) + 1)x_i = x_v + \sum_{j=1}^{k} x_j \quad (1 \leq i \leq k).$$

Thus, for $1 \leq i < j \leq k$ we have

$$(k + 1 - \nu_1(\hat{G}))(x_i - x_j) = 0.$$

Since $G \neq K_{1,k}$ we have $\Delta(\hat{G}) \geq k + 1$, and by Exercise 7.10, we have $\nu_1(\hat{G}) > \Delta(\hat{G})$. Hence $\nu_1(\hat{G}) > k + 1$ and $x_1 = x_2 = \cdots = x_k$. Now $\nu_1(G) \geq \mathbf{x}^{\top} L_G \mathbf{x} = \mathbf{x}^{\top} L_{\hat{G}} \mathbf{x} = \nu_1(\hat{G})$, as required. $\qquad\square$

Before we introduce further techniques, we consider the behaviour of the least eigenvalue $\lambda_n$ of a graph under a perturbation. We use the following analogue of (8.1):

$$\lambda_n(G') - \lambda_n(G) = \min_{\|\mathbf{y}\|=1} \mathbf{y}^{\top} A\mathbf{y} - \mathbf{x}^{\top} A\mathbf{x} \leq \mathbf{x}^{\top}(A' - A)\mathbf{x}, \qquad (8.4)$$

where $\mathbf{x} = (x_1, x_2, \ldots, x_n)^{\top}$ is an eigenvector of unit length corresponding to $\lambda_n(G)$. We can use (8.4) to prove the following three theorems.

**Theorem 8.1.14** [BelCRS2]. *Let $G'$ be the graph obtained from a graph $G$ by rotating the edge $rs$ to the position of a non-edge $rt$. Then*

(i) $\lambda_n(G') < \lambda_n(G)$ *if $x_r < 0$ and $x_s \leq x_t$, or $x_r = 0$ and $x_s \neq x_t$, or $x_r > 0$ and $x_s \geq x_t$;*
(ii) $\lambda_n(G') \leq \lambda_n(G)$ *if $x_r = 0$ and $x_s = x_t$.*

**Proof.** From (8.4) we have $\lambda_n(G') - \lambda_n(G) \leq 2x_r(x_t - x_s)$. We distinguish two cases.

*Case $x_r = 0$.* Then $\lambda_n(G') \leq \lambda_n(G)$. If $x_s \neq x_t$ then $\lambda_n(G') < \lambda_n(G)$. For otherwise, if $\lambda_n = \lambda_n(G') = \lambda_n(G)$, then $\mathbf{x}$ must be an eigenvector of $G'$ corresponding to its least eigenvalue. Therefore, in $G'$, we must have $\lambda_n x_r = \sum_{v \sim r} x_v$; but this cannot be the case when $x_s \neq x_t$. (Note that if $x_s = x_t$ then $\mathbf{x}$ is an eigenvector of $G'$ corresponding to $\lambda_n(G)$, but $\lambda_n(G)$ is not necessarily the least eigenvalue of $G'$.)

*Case $x_r \neq 0$.* Without loss of generality, $x_r > 0$ (for otherwise we may replace $\mathbf{x}$ by $-\mathbf{x}$). If $x_t < x_s$ then it follows at once that $\lambda_n(G') < \lambda_n(G)$. Assume next that $x_t = x_s$, so that certainly $\lambda_n(G') \leq \lambda_n(G)$. If $\lambda_n = \lambda_n(G') = \lambda_n(G)$ then, as above, $\mathbf{x}$ must be an eigenvector of $G'$ corresponding to $\lambda_n$. This is impossible since, in $G'$, we have $\lambda_n x_u \neq \sum_{v \sim u} x_v$ for $u$ equal to $s$ (or $t$).

This completes the proof. $\qquad\square$

In view of the above arguments, we may safely leave the proof of the following two results as exercises.

**Theorem 8.1.15** [BelCRS2]. *Let $G'$ be the graph obtained from a graph $G$ by relocating the edge st to the position of a non-edge uv, where $\{s, t\} \cap \{u, v\} = \emptyset$. Then*

(i) $\lambda_n(G') < \lambda_n(G)$ *if* $x_u x_v < x_s x_t$;
(ii) $\lambda_n(G') \leq \lambda_n(G)$ *if* $x_u x_v = x_s x_t$, *and in this situation,* $\lambda_n(G') = \lambda_n(G)$ *only if* $x_s = x_t = x_u = x_v = 0$.

**Theorem 8.1.16.** *Let $G'$ be a graph obtained from a connected graph $G$ by the local switching of st, uv to sv, tu. If $(x_s - x_u)(x_v - x_t) \leq 0$ then $\lambda_n(G') \leq \lambda_n(G)$, with $\lambda_n(G') = \lambda_n(G)$ only if $x_s = x_u$ and $x_v = x_t$.*

Analogues of Theorems 8.1.14–8.1.16 hold for the signless Laplacian (Exercise 8.6). They are used in [CarCRS] to show that the connected non-bipartite graph on $n$ vertices ($n > 3$) with minimal least $Q$-eigenvalue is obtained from a triangle by attaching a pendant path of length $n - 3$. Here we show how our results are used to identify the trees and unicyclic graphs with extremal eigenvalues.

**Theorem 8.1.17.** *Among all trees with n vertices ($n > 2$), the path $P_n$ has the smallest index, and the star $K_{1,n-1}$ has the largest index, with respect to the adjacency, or Laplacian, or signless Laplacian spectrum.*

**Proof.** For the adjacency index or $Q$-index, the proof easily follows by a repeated application of Theorem 8.1.7 (either to decrease or to increase the index). For the Laplacian index we may use the corresponding result for the $Q$-index, in view of Proposition 7.8.4. □

**Theorem 8.1.18.** *Among all unicyclic graphs with n vertices ($n > 2$) the cycle $C_n$ has the smallest index, and the graph $K_{1,n-1} + e$ has the largest index, with respect to the adjacency, or Laplacian, or signless Laplacian spectrum.*

**Proof.** Consider first the graphs with smallest index (with respect to the adjacency spectrum). The proof follows by repeated application of Theorem 8.1.1 and Theorem 8.1.12(ii), since the index decreases when an endvertex is deleted and a vertex is inserted in some edge of the cycle. In view of Exercise 8.4, the same argument applies to the $Q$-index. For the Laplacian index $\nu_1$, recall first that $\nu_1 = 4$ for any cycle. Suppose that $U$ is a unicyclic graph on $n$ vertices, other than $C_n$. If $n > 4$ the cycle in $U$ has an edge $e$ such that $U - e$ contains a subtree $T$ with degree sequence $3, 2, 1, 1, 1$. By interlacing (Theorem 7.8.13), we have $\nu_1(U) \geq \nu_1(T) > 4$. If $n \leq 4$ then $U = K_{1,3} + e$, and we are done.

For graphs with largest index or $Q$-index, the result follows immediately from Corollary 8.1.8 and Remark 8.1.9(iii). For the Laplacian index, the required result is a simple consequence of Proposition 7.1.1. $\qquad\square$

For connected graphs with $n$ vertices and $m$ edges ($m \geq n + 2$) we do not have universal results analogous to Theorems 8.1.17 and 8.1.18. The bicyclic graphs for which $\lambda_1$ is smallest or largest are known; see [Sim3] (or [CvDSa] p. 390) and [BruSo]. Most other results refer to graphs with largest index; in particular, the tricyclic graphs with minimal index have not been determined. Although the connected graphs with maximal index or maximal $Q$-index are threshold graphs (see Remark 8.1.9(i)), in the general case it remains for them to be identified within this class of graphs. More details can be found in [ABCHRSS]. To investigate the graphs with minimal least eigenvalue, we examine the sign-pattern of associated eigenvectors as described in Section 8.2.

We note that for the Laplacian spectrum, the graphs (with a fixed number vertices) for which $\nu_1$ is maximal are precisely those whose complements are not connected (see Proposition 7.1.1).

For the remainder of this section, we use characteristic polynomials to compare indices of graphs which differ in the location of an edge. These results (for the adjacency spectrum) are due primarily to Li and Feng, who exploited the following observation:

**Lemma 8.1.19** [LiFe]. *If $H$ is a proper spanning subgraph of the connected graph $G$, then*

$$P_H(x) > P_G(x) \text{ for all } x \geq \lambda_1(G).$$

**Proof.** We first prove by induction on $n = |V(G)|$ that for any spanning subgraph $H$ of an arbitrary graph $G$,

$$P_H(x) \geq P_G(x) \text{ for all } x \geq \lambda_1(G).$$

This clearly holds for $n = 1$; accordingly we suppose that $n > 1$ and the result holds for graphs with $n - 1$ vertices. By Theorem 2.3.1 we have $P_G'(x) = \sum_{j=1}^{n} P_{G-j}(x)$, and using a similar expression for $P_H'(x)$, we find that

$$P_H'(x) - P_G'(x) = \sum_{j=1}^{n}(P_{H-j}(x) - P_{G-j}(x)).$$

For each $j$, $H - j$ is a spanning subgraph of $G - j$ and so, by the induction hypothesis,

$$P_{H-j}(x) \geq P_{G-j}(x) \text{ for all } x \geq \lambda_1(G - j). \qquad (8.5)$$

Since $\lambda_1(G - j) \leq \lambda_1(G)$, it follows that $P'_H(x) - P'_G(x) \geq 0$ for all $x \geq \lambda_1(G)$. Since $\lambda_1(H) \leq \lambda_1(G)$, we have $P_H(x) \geq P_G(x)$ for all $x \geq \lambda_1(G)$. If $G$ is connected and $H$ is a proper spanning subgraph, then $\lambda_1(H) < \lambda_1(G)$ by Proposition 1.3.10, and so $P_H(x) > P_G(x)$ for all $x \geq \lambda_1(G)$.                   $\square$

**Theorem 8.1.20** [LiFe]. *Let $G(k, l)$ $(k, l \geq 0)$ be the graph obtained from a non-trivial connected graph $G$ by attaching pendant paths of length $k$ and $l$ at the same vertex $v$. If $k \geq l \geq 1$ then*

$$\lambda_1(G(k, l)) > \lambda_1(G(k + 1, l - 1)).$$

**Proof.** By Theorem 2.2.1 we have

$$P_{G(k,l)}(x) = x P_{G(k,l-1)}(x) - P_{G(k,l-2)}(x)$$

when $l \geq 2$, and

$$P_{G(k+1,l-1)}(x) = x P_{G(k,l-1)}(x) - P_{G(k-1,l-1)}(x)$$

when $l \geq 1$. It follows that for $k \geq l \geq 2$ we have

$$P_{G(k,l)}(x) - P_{G(k+1,l-1)}(x) = P_{G(k-1,l-1)}(x) - P_{G(k,l-2)}(x). \qquad (8.6)$$

When $k \geq l$, repeated use of (8.6) yields

$$P_{G(k,l)}(x) - P_{G(k+1,l-1)}(x) = P_{G(k-l+1,1)}(x) - P_{G(k-l+2,0)}(x).$$

By Theorem 2.2.1 again.

$$P_{G(k-l+2,0)}(x) = x P_{G(k-l+1,0)}(x) - P_{G(k-l,0)}(x)$$

and

$$P_{G(k-l+1,1)}(x) = x P_{G(k-l+1,0)}(x) - P_H(x),$$

where $H$ is the graph $G(k - l + 1, 0) - v$. Thus

$$P_{G(k+1,l-1)}(x) - P_{G(k,l)}(x) = P_H(x) - P_{G(k-l,0)}(x).$$

Now $H$ is a proper spanning subgraph of $G(k - l, 0)$, and so by Lemma 8.1.19 we have

$$P_{G(k+1,l-1)}(x) - P_{G(k,l)}(x) > 0 \text{ for all } x > \lambda_1(G(k - l, 0)).$$

Since $G(k - l, 0)$ is a proper subgraph of $G(k, l)$ we have $\lambda_1(G(k - l, 0)) < \lambda_1(G(k, l))$. Hence $P_{G(k+1,l-1)}(x)$ is positive at $\lambda_1(G(k, l))$, and the result follows.
□

For the signless Laplacian index we have the same result (Exercise 8.7), while almost the same result holds for Laplacian index:

**Theorem 8.1.21** [Guo2]. *Let $G(k, l)$ $(k, l \geq 0)$ be the graph obtained from a non-trivial connected graph $G$ by attaching pendant paths of length $k$ and $l$ at the same vertex $v$. If $k \geq l \geq 1$ then*

$$\nu_1(G(k, l)) \geq \nu_1(G(k + 1, l - 1)),$$

*with equality if and only if there exists an eigenvector of $G(k, l)$ corresponding to $\nu_1(G(k, l))$ with $v$-th entry 0.*

In general, the inequality in Theorem 8.1.21 is not strict; however if $G$ is bipartite we do indeed have $\nu_1(G(k, l)) > \nu_1(G(k - 1, l + 1))$.

Some extensions of Theorem 8.1.20 can be found in [LiFe]. One of them reads as follows (the proof is left as an exercise).

**Theorem 8.1.22** [LiFe]. *Let $u$, $v$ be adjacent vertices of a connected graph $G$, both of degree at least 2. Let $G(k, l)$ $(k \geq 0, l \geq 0)$ be the graph obtained from $G$ by attaching pendant paths of length $k$ and $l$ at $u$ and $v$. If $k \geq l \geq 1$ then*

$$\lambda_1(G(k, l)) > \lambda_1(G(k + 1, l - 1)).$$

In Theorem 8.1.22, the requirement that neither $u$ nor $v$ is an endvertex is no real restriction, because Theorem 8.1.20 can be applied when $uv$ is a pendant edge.

Our next result concerns graphs of the form $HvwK$ obtained from disjoint graphs $H$, $K$ by adding an edge joining the vertex $v$ of $H$ to the vertex $w$ of $K$. Recall that $H_v$ denotes the graph obtained from $H$ by adding a pendant edge at vertex $v$.

**Theorem 8.1.23** [ZhaZZ]. *If $P_{H_u}(x) < P_{H_v}(x)$ for all $x > \lambda_1(H_v)$ then $\lambda_1(HuwK) > \lambda_1(HvwK)$ for all vertices $w$ of $K$.*

**Proof.** Applying Theorem 2.2.4, we have $P_{HuwK}(x) = P_H(x)P_K(x) - P_{H-u}(x)P_{K-w}(x)$.

Since

$$P_{H_u}(x) = xP_H(x) - P_{H-u}(x),$$

we obtain

$$P_{HuwK}(x) = P_H(x)P_K(x) - P_{K-w}(x)(xP_H(x) - P_{H_u}(x)),$$

and similarly for $P_{HvwK}(x)$. On subtraction we obtain

$$P_{HuwK}(x) - P_{HvwK}(x) = P_{K-w}(x)(P_{H_u}(x) - P_{H_v}(x)).$$

If now $x > \lambda_1(HvwK)$ then $x > \lambda_1(K - w)$ and so $P_{K-w}(x) > 0$. By hypothesis, $P_{H_u}(x) - P_{H_v}(x) < 0$, and the result follows.                    □

### 8.1.2 Bounds on the index

We have already provided some bounds on the eigenvalues of a graph (see Chapters 1 and 3 for the adjacency spectrum, Chapter 7 for the Laplacian and signless Laplacian spectrum). Note that always $\lambda_1 \leq \frac{1}{2}\xi_1$ and $\nu_1 \leq \xi_1$. In this subsection, we discuss further bounds for $\lambda_1$. Some bounds for $\lambda_2$ and $\lambda_n$ can be found in the next section. Throughout we assume that $G$ is a graph with $m$ edges and vertices $1, 2, \ldots, n$. As usual, $d_i = \deg(i)$, and $A$ denotes the adjacency matrix of $G$.

We begin with lower bounds for $\lambda_1$, typically obtained as Rayleigh quotients $\mathbf{x}^\top A\mathbf{x}/\mathbf{x}^\top \mathbf{x}$. If we take $\mathbf{x} = (\sqrt{d_1}, \sqrt{d_2}, \ldots, \sqrt{d_n})^\top$ then we obtain:

**Proposition 8.1.24.** *For any graph $G$ with $m$ edges,*

$$\lambda_1(G) \geq \frac{1}{m} \sum_{i \sim j} \sqrt{d_i d_j}. \tag{8.7}$$

Close to this bound is a bound of Runge [Run], who proved that for a non-trivial connected graph $G$,

$$\lambda_1(G) \geq \sqrt{\frac{m}{\sum_{i \sim j} \frac{1}{d_i d_j}}} \tag{8.8}$$

(see also [Hofm], [HofWH], [SimSte]). In each of (8.7) and (8.8), equality holds if and only if $G$ is regular or semi-regular bipartite (Exercise 8.11).

We saw in Theorem 3.2.1 that $\lambda_1(G)$ is bounded below by the mean degree. The Cauchy–Schwarz inequality shows that the following is a better bound:

**Theorem 8.1.25** [Hof1]. *For any graph $G$ with $n$ vertices,*

$$\lambda_1(G) \geq \sqrt{\frac{d_1^2 + d_2^2 + \cdots + d_n^2}{n}}. \tag{8.9}$$

**Proof.** By Rayleigh's Principle, $\lambda_1(G)^2 \geq \mathbf{j}^\top A^2 \mathbf{j}/\mathbf{j}^\top \mathbf{j}$, where $\mathbf{j}$ is the all-1 vector. Now $\mathbf{j}^\top A^2 \mathbf{j}$ is the number $N_2$ of walks of length 2 in $G$, and

$N_2 = d_1^2 + d_2^2 + \cdots + d_n^2$ because there are $d_i^2$ such walks with $i$ as the mid-vertex. The result follows. $\qquad\square$

For an upper bound (due originally to Hong) we can apply Corollary 1.3.7 to threshold graphs as follows.

**Theorem 8.1.26** [Hon2]. *For any connected graph we have*

$$\lambda_1(G) \le \sqrt{2m - n + 1}, \tag{8.10}$$

*with equality if and only if $G$ is a star $K_{1,n-1}$ or a complete graph $K_n$.*

**Proof.** By Corollary 8.1.8, it suffices to prove (8.10) in the case that $G$ is a threshold graph. Here the stepwise property of $A$ (see Remark 8.1.9(i)) ensures that $(A^2 \mathbf{j})_i \ge (A^2 \mathbf{j})_k$ whenever $d_i \ge d_k$. By Corollary 1.3.7, we have $\lambda_1(G)^2 \le (A^2 \mathbf{j})_1 = 2m - n - 1$, with equality if and only if $\mathbf{j}$ is an eigenvector of $A^2$.

In the case of equality, let $d_n = d$. Then the equation $(A^2 \mathbf{j})_1 = (A^2 \mathbf{j})_n$ becomes $\sum_{i=2}^n (d_i - d) = 0$, whence $d_i = d$ $(i = 2, \ldots, n)$. Since $A$ is a stepwise matrix, this is possible if and only if $d$ is 1 or $n - 1$, that is, if and only if $G$ is $K_{1,n-1}$ or $K_n$. $\qquad\square$

**Remark 8.1.27.** For graphs with $m$ edges (not necessary connected) we have the following result of Stanley [Stan]:

$$\lambda_1(G) \le \frac{1}{2}(\sqrt{8m + 1} - 1), \tag{8.11}$$

with equality if and only if $G$ consists of a complete graph and any number of isolated vertices. When $m = \binom{t}{2} + r$ $(0 < r < t)$, let $G_m$ be the graph with $m$ edges obtained from $K_t$ by adding a vertex of degree $r$. Rowlinson [Row4] showed that among the graphs with $m$ edges, the maximal index is attained solely in those graphs with $G_m$ as the unique non-trivial component. $\qquad\square$

For an upper bound in terms of degrees, we have:

**Proposition 8.1.28** [BerZh]. *For any connected graph $G$,*
$$\lambda_1(G) \le \max_{i \sim j} \sqrt{d_i d_j}. \tag{8.12}$$

**Proof.** Let $(x_1, x_2, \ldots, x_n)^\top$ be the principal eigenvector of $G$, let $x_s = \max_i x_i$, and let $x_t = \max_{i \sim s} x_i$. From the eigenvalue equations for $\lambda_1$, we have

$$\lambda_1 x_s \le d_s x_t \quad \text{and} \quad \lambda_1 x_t \le d_t x_s,$$

whence $\lambda_1^2 x_s x_t \le d_s d_t x_s x_t$. The result follows. $\qquad\square$

Equality holds in (8.12) if and only if $G$ is regular or semi-regular bipartite (Exercise 8.12).

We note some further bounds which involve also the average degrees $m_i = \frac{1}{d_i} \sum_{j\sim i} d_j$ $(i = 1, \ldots, n)$. The first two bounds are due to Favaron *et al.* [FavMS1]. For any graph $G$ without isolated vertices,

$$\min_{1\leq i \leq n} m_i \leq \lambda_1(G) \leq \max_{1\leq i \leq n} m_i, \qquad (8.13)$$

and for any graph $G$,

$$\min_{1\leq i \leq n} \sqrt{m_i d_i} \leq \lambda_1(G) \leq \max_{1\leq i \leq n} \sqrt{m_i d_i}. \qquad (8.14)$$

The bounds (8.13) are established by taking $\mathbf{y} = (d_1, d_2, \ldots, d_n)^\top$ in Corollary 1.3.7. Note that equality holds throughout if and only if $G$ is harmonic. For the bounds (8.14) we may apply Corollary 1.3.7 to the matrix $A^2$, with $\mathbf{y} = \mathbf{j}$.

The upper bound in (8.13) was improved slightly by Das and Kumar [DasKu]: for any connected graph $G$ we have

$$\lambda_1(G) \leq \max_{i\sim j} \sqrt{m_i m_j}, \qquad (8.15)$$

with equality if and only if $G$ is either a regular graph or a bipartite graph for which $m_i$ is constant on each colour class.

Finally, we mention some intriguing questions relating to the bounds from Theorem 3.2.1: it is of interest to establish how far $\lambda_1(G)$ can be from $\bar{d}$, or from $\Delta$, when $G$ is non-regular. Both $\lambda_1(G) - \bar{d}$ and $\Delta - \lambda_1(G)$ have been considered as a measure of irregularity.

A lower bound for $\lambda_1(G) - \bar{d}$ was recently obtained by Cioabă and Gregory [CiGr]: for a non-regular graph $G$,

$$\lambda_1(G) - \bar{d} > \frac{1}{\Delta + 1}. \qquad (8.16)$$

A sharp upper bound was obtained by Bell [Bel2]: for any graph $G$,

$$\lambda_1(G) - \bar{d} \leq \frac{n}{4} - \frac{1}{2} + \epsilon(n), \qquad (8.17)$$

where $\epsilon(n) = 1/4n$ if $n$ is odd, and 0 if $n$ is even.

For similar bounds on $\Delta - \lambda_1(G)$, see [Ste3] and [CiGN].

## 8.2 Eigenvectors and structure

We have already seen in Chapter 1 that a graph is connected if and only if it has a simple largest eigenvalue with a corresponding eigenvector in which all components are of the same (non-zero) sign (Corollary 1.3.8). Here we examine how the sign pattern, or even zero-non-zero pattern, of other eigenvectors influences connectedness.

We start by fixing some notation. For any vector $\mathbf{x} = (x_1, x_2, \ldots, x_n)^\top \in \mathbb{R}^n$, let

$$P(\mathbf{x}) = \{i : x_i > 0\}, \quad N(\mathbf{x}) = \{i : x_i < 0\}, \quad Z(\mathbf{x}) = \{i : x_i = 0\}.$$

When $G$ is a graph with $V(G) = \{1, 2, \ldots, n\}$, we shall say that the vertex $i$ is *positive, negative,* or *null* (with respect to $\mathbf{x}$) according as $i$ belongs to $P(\mathbf{x})$, $N(\mathbf{x})$ or $Z(\mathbf{x})$, respectively. If $U \subseteq V(G)$, then $\langle U \rangle$ denotes the subgraph of $G$ induced by the vertices in $U$. For any graph $H$, comp($H$) denotes the number of components of $H$.

The following lemma is a direct consequence of the Interlacing Theorem; as usual, the eigenvalues $\lambda_1, \lambda_2, \ldots, \lambda_n$ are assumed to be in non-increasing order.

**Lemma 8.2.1** [Pow2]. *Let $B$ be a principal submatrix of the real symmetric matrix $A$, and suppose that $B = \mathrm{diag}(B_1, B_2, \ldots, B_k)$, where $B_i$ ($i = 1, 2, \ldots, k$) are irreducible matrices. If $\lambda_1(B_i) \geq \beta$ for each $i$, then $\lambda_k(A) \geq \beta$, with equality only if $\lambda_1(B_i) = \beta$ for each $i$.*

Our first theorem bounds, in terms of eigenvalues, the number of components in a subgraph induced by the non-negative vertices associated with an appropriate vector.

**Theorem 8.2.2** [Pow2]. *Let $A$ be the adjacency matrix of a non-trivial connected graph. If $\mathbf{x}$ is a vector such that for some real $\alpha$*

$$A\mathbf{x} \geq \alpha \mathbf{x}, \tag{8.18}$$

*then*

$$\mathrm{comp}(\langle P(\mathbf{x}) \cup Z(\mathbf{x}) \rangle) \leq \max\{i : \lambda_i(A) > \alpha\}.$$

**Proof.** Suppose that

$$A = \begin{pmatrix} B & C \\ C^\top & D \end{pmatrix}, \quad \mathbf{x} = \begin{pmatrix} \mathbf{y} \\ -\mathbf{z} \end{pmatrix},$$

where the partitions are determined by the non-negative and negative vertices. Next, let

$$B = \begin{pmatrix} B_1 & & O \\ & \ddots & \\ O & & B_k \end{pmatrix}, \quad \mathbf{y} = \begin{pmatrix} \mathbf{y_1} \\ \vdots \\ \mathbf{y_k} \end{pmatrix}, \quad C = \begin{pmatrix} C_1 \\ \vdots \\ C_k \end{pmatrix}$$

where $B_1, \ldots, B_k$ are irreducible. The hypothesis $A\mathbf{x} \geq \alpha\mathbf{x}$ implies that $B_i\mathbf{y_i} - C_i\mathbf{z} \geq \alpha\mathbf{y_i}$ for each $i$. Since $A$ is irreducible, no $C_i$ is a zero matrix. Thus $C_i\mathbf{z} \geq \mathbf{0}$ with strict inequality for some entry, and hence $B_i\mathbf{y_i} \geq \alpha\mathbf{y_i}$ with strict inequality for some entry. Therefore,

$$\mathbf{y_i}^\top B_i\mathbf{y_i} > \alpha\mathbf{y_i}^\top\mathbf{y_i} \tag{8.19}$$

and hence $\lambda_1(B_i) > \alpha$ for each $i \in \{1, 2, \ldots, k\}$. By Lemma 8.2.1 we have $\lambda_k(A) > \alpha$, and so (by interlacing) $k \leq \max\{i : \lambda_i(A) > \alpha\}$, as required. $\quad\square$

When the scalar $\alpha$ is positive, we can deduce a little more from the above proof:

**Corollary 8.2.3.** *If* $\alpha > 0$ *in* (8.18) *then*
(i) *no component of* $\langle P(\mathbf{x}) \cup Z(\mathbf{x})\rangle$ *is trivial,*
(ii) *no component of* $\langle P(\mathbf{x}) \cup Z(\mathbf{x})\rangle$ *contains only vertices from* $Z(\mathbf{x})$.

**Proof.** If the $i$-th component is trivial, then $B_i = 0$ and (8.19) is contradicted. If the $i$-th component contains vertices from $Z(\mathbf{x})$ alone, then $\mathbf{y_i} = \mathbf{0}$, and again (8.19) is contradicted. $\quad\square$

Some other corollaries can be deduced by taking $\alpha$ from different ranges, for example $0 \leq \alpha < 1$ or $1 \leq \alpha < \sqrt{2}$ (see Exercise 8.14). Theorem 8.2.2 is essentially a graph-theoretical version of a theorem of Fiedler on irreducible symmetric matrices, namely Theorem 2.1 of [Fie2], where the result is proved for $\alpha = \lambda_s$, $s \geq 2$. Note that for $\alpha < \lambda_1(A)$, $\langle P(\mathbf{x}) \cup Z(\mathbf{x})\rangle$ is non-empty because otherwise $\mathbf{x} < \mathbf{0}$, $A(-\mathbf{x}) \leq \alpha(-\mathbf{x})$ and we have $\lambda_1(A) \leq \alpha$ by Rayleigh's Principle.

The following variant of Theorem 8.2.2 provides an upper bound for the number of components in $\langle P(\mathbf{x})\rangle$; in the case that $\alpha$ is not an eigenvalue of $A$ the bound is the same as for the number of components in $\langle P(\mathbf{x}) \cup Z(\mathbf{x})\rangle$. The result demonstrates the role of null vertices in determining the components of the subgraph induced by non-negative vertices.

**Theorem 8.2.4** [Pow2]. *In the notation of Theorem 8.2.2, with* $\alpha$ *as in* (8.18), *we have*

$$\mathrm{comp}(\langle P(\mathbf{x})\rangle) \leq \max\{i : \lambda_i(A) \geq \alpha\}.$$

**Proof.** We may repeat the arguments for Theorem 8.2.2 with $\mathbf{y} > \mathbf{0}$, $\mathbf{z} \geq \mathbf{0}$ to show that $B_i \mathbf{y_i} - C_i \mathbf{z} \geq \alpha \mathbf{y_i}$. In this situation, $C_i \mathbf{z}$ may be zero and so we have only

$$B_i \mathbf{y_i} \geq \alpha \mathbf{y_i} \qquad (8.20)$$

for each $i$. Thus $\lambda_1(B_i) \geq \alpha$ for each $i$, and the conclusion follows from Lemma 8.2.1. $\qquad \square$

The next result provides a spectral bound on the number of components of any induced subgraph of a connected graph.

**Theorem 8.2.5** [Pow2]. *Let $G$ be a connected graph, $U$ a proper subset of $V(G)$, and $\delta$ the minimum vertex degree in $\langle U \rangle$. Then*

$$\mathrm{comp}(\langle U \rangle) \leq \max\{i : \lambda_i(A) \geq \delta\}.$$

**Proof.** Let $A$ be the adjacency matrix of a graph $G$ and let $B$ be the principal submatrix of $A$ corresponding to $U$. If $B = \mathrm{diag}(B_1, B_2, \ldots, B_k)$, where the $B_i$ are irreducible, then $B_i \mathbf{j} \geq \delta \mathbf{j}$, where $\mathbf{j}$ is the all-1 vector of appropriate size. Thus $\lambda_1(B_i) \geq \delta$ for each $i$. By Lemma 8.2.1, $\lambda_k(A) \geq \delta$, which is equivalent to the assertion of the theorem. $\qquad \square$

**Remark 8.2.6.** From the above theorem it follows immediately that

$$\mathrm{comp}(H) \leq \max\{i : \lambda_i(G) \geq 0\},$$

for any induced subgraph $H$ of $G$. Similar results are due to Cvetković (see [CvDSa, pp. 88–9]). $\qquad \square$

In what follows we assume that equality holds in (8.18). In this case, both $\mathbf{x}$ and $-\mathbf{x}$ are eigenvectors of $A$, and if we apply Theorem 8.2.2 to both of these vectors when $\alpha = \lambda_2$, we obtain the following result.

**Theorem 8.2.7** [Fie2]. *Let $G$ be a connected graph, and let $\mathbf{x}$ be an eigenvector corresponding to the second largest eigenvalue. Then both of the subgraphs $\langle P(\mathbf{x}) \cup Z(\mathbf{x}) \rangle$, $\langle N(\mathbf{x}) \cup Z(\mathbf{x}) \rangle$ are connected.*

In the case that $\mathbf{x}$ is an eigenvector $(x_1, x_2, \ldots, x_n)^\top$ corresponding to an eigenvalue $\alpha < \lambda_1$ one can obtain the following basic upper bound for the number of null vertices. This in turn gives an upper bound for the multiplicity of $\alpha$ (see Corollary 8.2.9).

**Theorem 8.2.8** [Pow2]. *Let A be the adjacency matrix of a connected graph with n vertices, $n > 2$. If $A\mathbf{x} = \alpha\mathbf{x}$, then*

$$|Z(\mathbf{x})| \leq \begin{cases} n - 2 - 2\alpha & if\ \alpha > 0, \\ n - 2 & if\ -1 < \alpha \leq 0, \\ n + 2 - 2|\alpha| & if\ \alpha \leq -1. \end{cases}$$

**Proof.** Let $|Z(\mathbf{x})| = a$, $|P(\mathbf{x})| = b$, $|N(\mathbf{x})| = c$. Note that since $\mathbf{x}$ is orthogonal to the principal eigenvector, we have $b \geq 1$ and $c \geq 1$, whence $a \leq n - 2$ whatever the value of $\alpha$.

Let $h$ and $k$ be such that $x_h \geq x_i \geq x_k$ for all $i$. On comparing the $h$-th entries of $\alpha\mathbf{x}$ and $A\mathbf{x}$ we obtain $\alpha x_h \leq (b - 1)x_h$, and similarly, $\alpha|x_k| \leq (c - 1)|x_k|$. It follows that

$$\alpha \leq \min\{b, c\} - 1, \tag{8.21}$$

and hence that $\alpha \leq \frac{1}{2}(b + c) - 1 = \frac{1}{2}(n - a) - 1$. This gives the required bound in the case that $\alpha > 0$. From the equation $A\mathbf{x} = \alpha\mathbf{x}$ we also deduce that $|\alpha|x_h \leq c|x_k|$ and $|\alpha||x_k| \leq bx_h$. Now we have

$$|\alpha| \leq \sqrt{bc} \leq \frac{1}{2}(b + c) = \frac{1}{2}(n - a), \tag{8.22}$$

and this gives the required bound in the case $\alpha \leq -1$. $\qquad\square$

**Corollary 8.2.9.** *Let $\alpha$ be an eigenvalue of a connected graph on n vertices. If $\alpha$ has multiplicity $m > 1$ then*

$$m \leq \begin{cases} n - 1 - 2\alpha & if\ \alpha > 0, \\ n - 1 & if\ -1 < \alpha \leq 0, \\ n + 1 - 2|\alpha| & if\ \alpha \leq -1. \end{cases}$$

**Proof.** This bound follows from Theorem 8.2.8 because there exists an eigenvector $\mathbf{x}$ corresponding to $\alpha$ with $|Z(\mathbf{x})| = m - 1$. $\qquad\square$

We may use the foregoing arguments to establish bounds on $\lambda_2$ and $\lambda_n$.

**Proposition 8.2.10.** *If $G$ is a connected graph on n vertices then*

$$\lambda_2(G) \leq \left\lfloor \frac{n}{2} \right\rfloor - 1.$$

*The bound is attained for all odd $n > 1$, and is asymptotically sharp for even n.*

**Proof.** The bound follows from (8.21). If $n = 2s + 1$ then the bound is attained in the connected graph constructed from $2K_s \cup K_1$ by adding two edges incident with an isolated vertex. If $n = 2s$ and $G_s$ is the graph constructed from

$2K_s$ by adding an edge, then $\lambda_2(G_s) = \frac{1}{2}(s - 3 + \sqrt{s^2 + 2s - 3})$, which tends to $s - 1$ as $s \to \infty$. $\quad\square$

Secondly, (8.22) yields:

**Proposition 8.2.11** [Con2]. *If G is a connected graph with n vertices then*

$$\lambda_n(G) \geq -\sqrt{\left\lfloor \frac{n}{2} \right\rfloor \left\lceil \frac{n}{2} \right\rceil}.$$

*Equality holds for* $G = K_{\lfloor \frac{n}{2} \rfloor, \lceil \frac{n}{2} \rceil}$.

The least eigenvalue of any non-trivial complete graph is equal to $-1$. For other graphs we have the following upper bound of Yong, stated without proof:

**Theorem 8.2.12** [Yong]. *If G is a non-complete connected graph with n vertices* $(n \geq 4)$ *then*

$$\lambda_n(G) < -\frac{1}{2}\left(1 + \sqrt{1 + 4\frac{n-3}{n-1}}\right).$$

The following lemma will help us to describe the non-complete connected graphs $G$ with $n$ vertices and $m$ edges for which $\lambda_n$ is minimal. For the remainder of this section, $G$ is such a graph, and $\mathbf{x}$ is any eigenvector $(x_1, x_2, \ldots, x_n)^\top$ of $G$ corresponding to $\lambda_n$.

**Lemma 8.2.13.** *If* $Z(\mathbf{x}) \neq \emptyset$ *then* $\deg(u) = n - 1$ *for every vertex* $u \in Z(\mathbf{x})$.

**Proof.** Assume the contrary, and let $r$ be a vertex in $Z(\mathbf{x})$ such that $\deg(r) < n - 1$. Let $S_r = \{s \in V(G) : s \sim r\}$, and $T_r = \{t \in V(G) : t \not\sim r, t \neq r\}$. Note that $S_r \neq \emptyset$ because $G$ is connected and non-trivial. Now choose a vertex $s$ from $S_r$ and a vertex $t$ from $T_r$. Let $G'$ be the graph obtained from $G$ by rotating the edge $rs$ to the position $rt$.

Assume first that $G'$ is connected for any choice of $s$ and $t$. If $x_s \neq x_t$ for some $s$ and $t$ then $\lambda_n(G') < \lambda_n(G)$ by Theorem 8.1.14. This contradicts the choice of $G$, and so $x_s = x_t$ for any choice of $s$ and $t$. But then $x_v = c$ for any $v \neq r$, where $c$ is a real constant. Now $\lambda_n(G)x_r = \sum_{v \in S_r} x_v = \deg(r)c$. Since $\deg(r) \neq 0$ and $x_r = 0$, we conclude that $c = 0$ and hence $\mathbf{x} = \mathbf{0}$, a contradiction.

Accordingly, suppose that, for some choice of $s$ and $t$, the graph $G'$ is not connected. Then $rs$ must be a bridge in $G$, and $s, t$ lie in different components $G_s, G_t$ of $G'$, respectively. Let $t'$ be a vertex (if any) in $G_s$ different from $s$. Note that $t' \in T_r$, for otherwise there exists an $r$-$s$ path in $G$ avoiding the bridge $rs$. If $x_s \neq x_{t'}$, then we obtain a contradiction by applying the argument above to $t'$ instead of $t$ (note that the corresponding graph $G'$ is now connected). Consequently, $x_u = x_s$ for every $u \in V(G_s)$. By the eigenvalue equation for the vertex $s$, applied in $G$, we obtain $\lambda_n(G)x_s = (\deg(s) - 1)x_s$, whence $x_s = 0$. Therefore, $G_t$ contains a vertex $u$ such that $x_u \neq 0$. Now the graph $G''$ obtained from $G$ by rotating $sr$ to $su$ is connected, and $\lambda_n(G'') < \lambda_n(G)$ by Theorem 8.1.14.

This final contradiction completes the proof.                                    □

**Theorem 8.2.14** [BelCRS2]. *Let $G$ be a connected graph whose least eigenvalue $\lambda_n(G)$ is minimal among the connected graphs with $n$ vertices and $m$ edges, where $m < \binom{n}{2}$. Then $\lambda_n(G)$ is a simple eigenvalue of $G$.*

**Proof.** Suppose that $\lambda_n(G)$ has multiplicity at least two. Then, for any vertex $u \in V(G)$, there exists an eigenvector $\mathbf{x}$ whose $u$-th entry is equal to zero (so that $u \in Z(\mathbf{x})$ and $Z(\mathbf{x}) \neq \emptyset$). Since $G$ is not complete, we may choose $u$ to be a vertex such that $\deg(u) < n - 1$. Now we have a contradiction to Lemma 8.2.13, and the proof follows.                                                      □

As an immediate consequence of Theorem 8.2.14 we see that if $G$ is not complete then the partition of $V(G)$ induced by the sign pattern of any eigenvector corresponding to $\lambda_n(G)$ is unique (since only the role of negative and positive vertices can be exchanged). Accordingly, in what follows we assume that $m < \binom{n}{2}$, and write $P, N, Z$ for $P(\mathbf{x}), N(\mathbf{x}), Z(\mathbf{x})$. If $H = \langle P \cup N \rangle$ and $K = \langle Z \rangle$ then by Lemma 8.2.13, $K$ is a complete graph and $G = K \triangledown H$. To describe $H$, let $H^+ = \langle P \rangle$, $H^- = \langle N \rangle$. Note that the subsets $P$ and $N$ are non-empty since the eigenspaces of $\lambda_n(G)$ and $\lambda_1(G)$ are orthogonal and the latter is spanned by a positive eigenvector; in contrast, $Z$ may be an empty set.

**Proposition 8.2.15.** *Both $H^+$ and $H^-$ are threshold graphs.*

**Proof.** Let $P = \{1, 2, \ldots, k\}$ where $x_1 \leq x_2 \leq \cdots \leq x_k$. We shall prove that $jq \in E(G)$ implies that $ip \in E(G)$ whenever $1 \leq i \leq j \leq k$ and $1 \leq p \leq q \leq k$. Suppose by way of contradiction that

$$1 \leq i \leq j \leq k, \quad 1 \leq p \leq q \leq k, \quad jq \in E(G), \quad ip \notin E(G).$$

We delete $jq$ and add $ip$, to obtain the graph $G'$. Taking $\|\mathbf{x}\| = 1$, we have $\lambda_n(G) = 2\sum_{uv \in E(G)} x_u x_v$, and so

$$0 \leq \lambda_n(G') - \lambda_n(G) = 2(x_i - x_j)x_p + 2(x_p - x_q)x_j \leq 0,$$

whence $x_i = x_j$, $x_p = x_q$. Moreover $\mathbf{x}$ is an eigenvector corresponding to $\lambda_n(G') = \lambda_n(G)$. This is a contradiction, since $q$ has lost a neighbour from $P$. Hence $H^+$ is a threshold graph. In a similar way we can derive the same conclusion for $H^-$. $\qquad\square$

**Lemma 8.2.16.** *If $P$ or $N$ induces an edge $ij$, then $pq \in E(G)$ for all $p \in N$, $q \in P$.*

**Proof.** If the assertion is false, we can remove $ij$ and add an edge between $V^-$ and $V^+$ to reduce $2\Sigma_{uv \in E(G)} x_u x_v$. $\qquad\square$

Accordingly, we arrive at the following conclusion:

**Proposition 8.2.17.** *If at least one of the graphs $H^-$ or $H^+$ is not totally disconnected then $H = H^- \triangledown H^+$; otherwise, $H$ is a bipartite graph (not necessarily a complete bipartite graph).*

In addition, we have:

**Lemma 8.2.18.** *If $Z \neq \emptyset$ then $H = H^- \triangledown H^+$.*

**Proof.** If $H \neq H^- \triangledown H^+$ then we obtain a contradiction by applying Theorem 8.1.15 to four vertices chosen as follows. First, let $p$ and $q$ be two non-adjacent vertices taken from $N$ and $P$, respectively; secondly, choose $i$ from $Z$ and $j$ from $N \cup P$. By Lemma 8.2.13, $i$ is adjacent to $j$, and $ij$ is not a bridge. Moreover, $x_p x_q < x_i x_j$. If we replace the edge $ij$ with $pq$ then we obtain a connected graph $G'$ for which $\lambda_n(G') < \lambda_n(G)$ by Theorem 8.1.16(i). This contradicts the minimality of $\lambda_n(G)$, and so every vertex of $H^-$ is adjacent to every vertex of $H^+$. $\qquad\square$

It follows from Lemmas 8.2.13 and 8.2.18 that when $Z \neq \emptyset$, $G$ has the form $K \triangledown L$, where $K$ and $L$ are threshold graphs, the vertices of $K$ are non-negative, and those of $L$ are non-positive. Here $V(K) = P \cup X$ and $V(L) = N \cup Y$, where $X \cup Y$ is an arbitrary bipartition of $Z$. Combining this observation with Proposition 8.2.17, we obtain the following general description of $G$:

**Theorem 8.2.19** [BelCRS2]. *Let $G$ be a connected graph whose least eigenvalue is minimal among the connected graphs with $n$ vertices and $m$ edges $(n - 1 \leq m < \binom{n}{2})$. Then $G$ is either*

(a) *a bipartite graph, or*
(b) *a join of two threshold graphs (not both totally disconnected).*

## 8.3 Reconstructing the characteristic polynomial

Let $G$ be a graph with vertices $1, 2, \ldots, n$. The family of $n$ vertex-deleted subgraphs $G - 1, G - 2, \ldots, G - n$ is called the *deck* of $G$, denoted by $\mathcal{D}(G)$. Ulam famously asked whether every graph with at least three vertices is reconstructible (up to isomorphism) from its deck. The question remains unanswered.

In spectral graph theory, we can consider also the *polynomial deck* (or *p-deck*)

$$\mathcal{P}(G) = \{P_{G-1}(x), P_{G-2}(x), \ldots, P_{G-n}(x)\}.$$

Now four questions arise when we ask whether $\mathcal{D}(G)$ or $\mathcal{P}(G)$ determine $G$ or $P_G(x)$. These four reconstruction problems were discussed by Schwenk in [Sch3] (see also [CvDGT, Section 3.5] and [SchWi, Section 12]). In fact, Tutte [Tut2] showed that $P_G(x)$ is reconstructible from $\mathcal{D}(G)$; an alternative proof may be found in [LauSc, Chapter 10]. On the other hand, $\mathcal{P}(G)$ does not determine $G$: the Clebsch graph and $L(K_{4,4})$ have the same eigenvalues and the same angles, hence (by Proposition 2.2.6) the same polynomial deck. (For another instance of this phenomenon, see Example 4.3.1.) Here we discuss the problem of reconstructing $P_G(x)$ from $\mathcal{P}(G)$. The problem was posed by Cvetković in 1973 at the 18th International Scientific Colloquium in Ilmenau, and it has not yet been resolved. The Polynomial Reconstruction Conjecture, denoted by (P), states that for any graph $G$ with at least three vertices, $\mathcal{P}(G)$ determines $P_G(x)$.

It has been shown that the conjecture holds for graphs with up to 10 vertices (a result attributed to B. McKay, obtained by a computer search).

We know from Theorem 2.3.1 that

$$P'_G(x) = P_{G-1}(x) + P_{G-2}(x) + \cdots + P_{G-n}(x), \tag{8.23}$$

and so $\mathcal{P}(G)$ determines $P_G(x)$ to within an additive constant. If we know just one eigenvalue of $G$, then the constant term can be calculated. In particular, this is the case if some $P_{G-i}(x)$ has a repeated root $\lambda$, for then (by interlacing) $\lambda$ is an eigenvalue of $G$.

The following invariants (1)–(6) and properties (7)–(8) are reconstructible from the polynomial deck:

(1) the numbers of vertices and edges;
(2) the vertex degree sequence;
(3) the length of the shortest odd cycle, and the number of such cycles;
(4) the number of closed walks of length $k$ starting and terminating at the $i$-th vertex;

(5) the numbers of triangles, quadrangles and pentagons;
(6) the spectral moments $s_0, s_1, \ldots, s_{n-2}$;
(7) regularity (and strong regularity);
(8) bipartiteness.

The proofs of (1)–(8) are left to the reader (Exercise 8.15; see also [Cve12]).

For some classes of graphs, one of (1)–(8) suffices to reconstruct the characteristic polynomial (or even the graph). For example, it was shown in [GuCv] that (P) holds for regular graphs (Exercise 8.16) and various classes of bipartite graphs such as trees without a perfect matching. For trees in general, the problem was open for many years; here we show how the earlier result was extended to the remaining trees (see [Cve12] and [CvLe2]). We denote by $\sigma(G)$ the set of distinct eigenvalues of a graph $G$.

**Theorem 8.3.1** [CvLe2]. *Let $H$ be a graph with at least three vertices and exactly two components. If these components have different numbers of vertices, then the characteristic polynomial of $H$ is determined by $\mathcal{P}(H)$.*

**Proof.** Suppose by way of contradiction that there exists at least one graph $G \neq H$ such that $P_G(x) = P_H(x) + c$ ($c \neq 0$) and $P_{G-i}(x) = P_{H-i}(x)$ ($i = 1, 2, \ldots, n$). Let $H_1, H_2$ be the two components of $H$, with $n_1, n_2$ vertices respectively, where $n_1 > n_2$. Clearly,

$$\sigma(H - u) = \sigma(H_1) \cup \sigma(H_2 - u) \quad \left(u \in V(H_2)\right). \tag{8.24}$$

Since $H - u$ has no multiple eigenvalues, the same is true of $H_1$, and we let $\sigma(H_1) = \{\lambda_1^*, \lambda_2^*, \ldots, \lambda_{n_1}^*\}$, where $\lambda_1^* > \lambda_2^* > \cdots > \lambda_{n_1}^*$.

Let $v$ be a fixed vertex of $H_2$. Since $|\sigma(H_2 - v)| < n_1 - 1$, there exists at least one index $i = i_0$ ($1 \leq i_0 \leq n_1 - 1$) such that no eigenvalue of $H_2 - v$ lies in the open interval $(\lambda_{i_0+1}^*, \lambda_{i_0}^*)$. Therefore,

$$(\lambda_{i_0+1}^*, \lambda_{i_0}^*) \cap \sigma(H - v) = \emptyset. \tag{8.25}$$

Since $P_{G-v}(x) = P_{H-v}(x)$, we know from (8.24) that $\lambda_{i_0}^*$ and $\lambda_{i_0+1}^*$ lie in $\sigma(G - v)$. By the Interlacing Theorem, there exists at least one eigenvalue $\alpha$ of $G$ in the interval $(\lambda_{i_0+1}^*, \lambda_{i_0}^*)$. Since $P_G(\alpha) = 0$ and $P_G(\lambda_{i_0+1}^*) = P_G(\lambda_{i_0}^*) = c$, there exist at least two eigenvalues $\alpha, \beta$ of $G$ (not necessarily distinct) in $(\lambda_{i_0+1}^*, \lambda_{i_0}^*)$.

Finally, using the interlacing theorem again, we see that $G - v$ has at least one eigenvalue $\gamma \in [\alpha, \beta] \subseteq (\lambda_{i_0+1}^*, \lambda_{i_0}^*)$. Since $\sigma(G - v) = \sigma(H - v)$, this is a contradiction to (8.25). $\qquad \square$

We can deal with more than two components in similar fashion:

**Theorem 8.3.2** [CvLe2]. *If $H$ is a graph with at least three components, then the characteristic polynomial of $H$ is determined by $\mathcal{P}(H)$.*

**Proof.** Suppose that $H$ has components $H_1, H_2, \ldots, H_k$ ($k > 2$). We may assume that all eigenvalues of $H$ are simple, for otherwise some $H - v$ has a multiple eigenvalue. Accordingly we take $\lambda_1(H_1) > \lambda_1(H_2) > \cdots > \lambda_1(H_k)$. It follows that

$$\big(\lambda_1(H_1), \lambda_1(H_2)\big) \cap \sigma(H_k - v) = \emptyset \quad \big(v \in V(H_k)\big),$$

and the proof now follows as before.                                    □

In proving the next result, we make use of the fact (Exercise 2.14) that the characteristic polynomial of a tree with $n$ vertices has constant term $(-1)^{n/2}$ or 0 (according as it does or does not have a perfect matching). We write $e(G)$ for the number of edges of the graph $G$.

**Theorem 8.3.3** [CvLe2]. *If $T$ is a tree, then its characteristic polynomial is determined by $\mathcal{P}(T)$.*

**Proof.** Let $n = |V(T)|$ and suppose that $H$ is a graph such that $\mathcal{P}(H) = \mathcal{P}(T)$.

Consider first the case that $n$ is odd. Applying Theorem 3.2.3 to $P_{H-i}(x)$ ($= P_{T-i}(x)$), we see that each $H - i$ is bipartite. Hence either $H$ is an odd cycle or $H$ is bipartite. In the former case, $H$ is regular and so $P_H(x) = P_T(x)$ by Exercise 8.16. If $H$ is bipartite with parts of size $n_1, n_2$ then we may suppose that $n_1 > n_2$. Now for some vertex $u$ of $H$, the graph $H - u$ is bipartite with parts of size $n_1, n_2 - 1$. Inspecting the rank of $A(H - u)$, we see that $H - u$ has 0 as an eigenvalue of multiplicity at least $n_1 - (n_2 - 1)$. Hence $H$ has 0 as an eigenvalue, and so again $P_H(x) = P_T(x)$.

Now suppose that $n$ is even, say $n = 2k$, and that $P_H(x) \neq P_T(x)$. We shall obtain a contradiction. If $H$ is connected, it is a tree because $e(H) = e(T)$. Since $P_H(0) \neq P_T(0)$, we have $\{P_H(0), P_T(0)\} = \{(-1)^k, 0\}$. We may assume that $P_H(0) = 0$ without loss of generality. By Exercise 2.6, if $u$ is the neighbour of an end vertex of $H$, then 0 is a multiple eigenvalue of $H - u$. Then $H$ has 0 as an eigenvalue and $P_H(x) = P_T(x)$, contrary to assumption. Hence $H$ is not connected.

By Theorems 8.3.1 and 8.3.2, $H$ has exactly two connected components $H_1$ and $H_2$, each with exactly $k$ vertices. Since $e(H) = e(T) = 2k - 1$, we know that one component, say $H_1$, is a unicyclic graph and the other component is a tree. Let $\lambda_1^* > \lambda_2^* > \cdots > \lambda_k^*$ be the eigenvalues of the tree $H_2$. If there exist a vertex $v$ of $H_1$ and an index $i_0$ such that $(\lambda_{i_0+1}^*, \lambda_{i_0}^*) \cap \sigma(H_1 - v) = \emptyset$, then the proof proceeds as in Theorem 8.3.1. Otherwise, for any vertex $v$ of $H_1$, the eigenvalues $\gamma_1, \ldots, \gamma_{k-1}$ of $H_1 - v$ interlace those of $H_2$ – that is,

$$\gamma_i \in \left(\lambda_{i+1}^*, \lambda_i^*\right) \quad (i = 1, 2, \ldots, k - 1). \tag{8.26}$$

Now, because it has no multiple eigenvalues, the unicyclic graph $H_1$ is not a cycle, and so we may choose $v$ to be an endvertex. Then $H_1 - v$ is unicyclic and we have from (8.26) the contradiction

$$2(k - 1) = 2e(H_1 - v) = \sum_{i=1}^{k-1} \gamma_i^2 < \sum_{i=1}^{k} \left(\lambda_i^*\right)^2 = 2e(H_2) = 2(k - 1).$$

This completes the proof. □

Simić and Stanić [SimSta1] have shown that (P) holds for unicyclic graphs. Further results on graphs with pendant vertices are proved by Sciriha *et al.* [Sci1, SciFo]. The main feature of these results is that (P) holds if the number of terminal vertices of the graph is sufficiently high; for example, (P) holds for coronas of the form $G \circ K_1$.

The results of Section 3.4 can be used to show that (P) holds for graphs whose vertex-deleted subgraphs have least eigenvalue $\geq -2$ (see [Sim4, Sim5] for connected graphs, and [SimSta1, SimSta2] for others). In particular, the characteristic polynomial of any line graph is reconstructible. As a consequence of the latter observation, it was shown in [CvSi5] that the $Q$-polynomial can be reconstructed from the polynomial deck determined by edge-deleted subgraphs. We mention in passing that the $Q$-polynomial of a graph is reconstructible from the deck of vertex-deleted subgraphs (see [Ded]), a result analogous to Tutte's result for adjacency spectrum.

A graph without $P_4$ as an induced subgraph is called a *cograph*. It was proved in [Roy] that 0 is an eigenvalue of a cograph $G$ only if $G$ contains duplicate vertices, and $-1$ is an eigenvalue of $G$ only if $G$ contains co-duplicate vertices. We know from Section 3.11 that graphs whose second largest eigenvalue is less than $\frac{\sqrt{5}-1}{2}$ are cographs. This observation was used in [BiySS] to show that (P) holds for graphs whose vertex-deleted subgraphs have second largest eigenvalue less or equal to $\frac{\sqrt{5}-1}{2}$.

We note that Hagos [Hag] proved that for any graph $G$, $P_G(x)$ is reconstructible from $\mathcal{P}(G)$ and $\mathcal{P}(\overline{G})$; then the same is true of $P_{\overline{G}}(x)$. In view of Proposition 2.1.3, this means that the eigenvalues and main angles of a graph are determined by the eigenvalues and main angles of its vertex-deleted subgraphs.

Finally, let $c = a + b\sqrt{m}$ and $\bar{c} = a - b\sqrt{m}$, where $a$ and $b$ are non-zero integers and $m$ is a positive integer which is not a perfect square. The *conjugate* adjacency matrix of a graph $G$ has entries $c$ and $\bar{c}$ for adjacent and non-adjacent vertices, respectively, while diagonal entries are equal to 0. The characteristic polynomial of the conjugate adjacency matrix is called the *conjugate* characteristic polynomial of $G$. It is proved in [Lep3] that the conjugate characteristic polynomial of a graph is determined by the conjugate characteristic polynomials of its vertex-deleted subgraphs.

## 8.4 Integral graphs

A graph is said to be *integral* if all the eigenvalues of an adjacency matrix are integers. Since the eigenvalues of any graph are algebraic integers, an eigenvalue is an integer if and only if it is rational. Attractive examples of integral graphs include strongly regular graphs (other than the conference graphs) and the skeleta of the Platonic solids. By Theorem 2.1.2, the complement of any regular integral graph is also integral.

The quest for integral graphs began in 1974 with a paper by Harary and Schwenk [HarSc]. They identified some large collections of integral graphs, and observed that various graph operations can be used to construct new integral graphs from old (cf. Chapter 2). As noted in [SteAFD], an application of integral graphs was recently found in the context of quantum spin networks (see [CDDEKL]).

In what follows, we consider only connected integral graphs, since the spectrum of a disconnected graph consists of the spectra of its components.

**Remark 8.4.1** [Cve5]. There are only finitely many connected integral graphs whose vertices have bounded degree. For then the number of distinct eigenvalues is bounded (by Proposition 1.1.1), and this in turn bounds the diameter (by Theorem 3.3.5). ☐

The connected integral non-regular graphs with maximum degree at most 3 were identified by Cvetković *et al.* [CvGT]; there are only 7 such graphs. The cubic integral graphs were found (in part using a computer search) by Bussemaker and Cvetković [BuCv], and independently (by hand) by Schwenk [Sch2]. There are just 13 connected cubic integral graphs. Schwenk made the important observation that a search for integral graphs can be restricted to bipartite graphs as follows. If $G$ is connected, non-bipartite and integral, then $G \times K_2$ is connected, bipartite and integral, since the eigenvalues of $K_2$ are 1 and $-1$. Accordingly, if we know all connected bipartite integral graphs, the non-bipartite integral graphs can be extracted from those which are decomposable with respect to the above product.

Simić and Radosavljević [SimRa] determined all 13 connected non-regular non-bipartite integral graphs with maximum degree 4. Such graphs have least eigenvalue $-2$ and so the results of Section 3.4 can be exploited. The corresponding problem for non-regular bipartite graphs, with some values avoided in the spectrum, was investigated by a mixture of theoretical arguments and computer search by Balińska *et al.* [BaSi1, BaSi2, BaSZ1]. The question was finally resolved in full generality by Lepović *et al.* [LepSBZ]: using brute force and a computer search which lasted more than a year, they showed that there are exactly 93 connected non-regular bipartite integral graphs with maximum degree 4.

There have been some attempts to find all the connected regular bipartite graphs of degree 4. In such a graph, let $2k, q, h$ be the numbers of vertices, quadrangles and hexagons. Considering the spectral moments $s_0, s_2, s_4, s_6$, we see that the spectrum has the form

$$4, 3^x, 2^y, 1^z, 0^{2w}, -1^z, -2^y, -3^x, -4,$$

where (by Theorem 3.1.1):

$$1 + x + y + z + w = k$$
$$16 + 9x + 4y + z = 4k$$
$$256 + 81x + 16y + z = 28k + 4q$$
$$4096 + 729x + 64y + z = 232k + 72q + 6h.$$

The solutions of these diophantine equations have been obtained by computer, and some are reproduced in [CvSiS]; the largest putative graph that appears has 5040 vertices. The non-existence of graphs with some of the feasible spectra from [CvSiS] was established in [Ste1, Ste2, SteAFD] (in [Ste2], graph angles were used). In [CvSiS] just 65 graphs were identified; all those with at most 19 vertices have been generated by Balińska *et al.* [BaSZ2]. Substantial progress was made in [SteAFD], where only 12 spectra remained unresolved for graphs with more than 360 vertices (the largest having 560 vertices).

We now turn to small integral graphs. For $1 \leq n \leq 12$ the number $i_n$ of connected integral graphs with $n$ vertices is given in the following table.

| $n$ | 1 | 2 | 3 | 4 | 5 | 6 | 7 | 8 | 9 | 10 | 11 | 12 |
|-----|---|---|---|---|---|---|---|---|---|----|----|----|
| $i_n$ | 1 | 1 | 1 | 2 | 3 | 6 | 7 | 22 | 24 | 83 | 113 | 236 |

These results may be found in [BaCLS] for $1 \leq n \leq 10$, and in [BaKSZ1, BaKSZ2] for $n = 11, 12$. At the time of writing, the search for all connected integral graphs with 13 vertices continues; to date, 547 such graphs have been generated by a probabilistic algorithm of Balińska *et al.* [BaKSZ3].

From Tables A1 and A3 in the Appendix, we see that the 14 connected integral graphs with at most 6 vertices are $K_1$, $K_2$, $K_3$, $K_4$, $C_4$, $K_5$, $\overline{2K_1 \cup K_3}$, $K_{1,4}$, $K_6$, $CP(3)$, $C_3 + K_2$, $K_{3,3}$, $C_6$ and the unique tree with degree sequence $3, 3, 1, 1, 1, 1$ (the corona $K_2 \circ 2K_1$). The 7 connected integral graphs with 7 vertices are $S(K_{1,3})$, $K_1 \triangledown 3K_2$, $L(K_{1,2} + K_2)$, $L(\overline{K_{3,3} - e})$, $\overline{C_3 \cup C_4}$, $\overline{C_4 \cup 3K_1}$ and $K_7$.

There are no cospectral (non-isomorphic) connected integral graphs with fewer than 8 vertices. There is just one triplet of connected integral graphs on 8 vertices; there are three pairs on 9 vertices and ten pairs, one triplet, two quadruplets and one quintuplet on 10 vertices. None of the graphs in these sets is regular, and in all cases the cospectral graphs can be distingushed by angles. We have already noted in Section 4.3 that the smallest cospectral graphs with the same angles have 10 vertices and we find that there are no integral graphs among the 58 pairs of such cospectral graphs.

Other results on integral graphs concern specific classes of graphs; for example, there are many results on integral trees. One of the first is due to Watanabe:

**Theorem 8.4.2** [Wat]. *The only integral tree with a perfect matching is $K_2$.*

Most of the early results on integral trees may be found in [BaCRSS], while further results were collected by Wang [Wan1, Wan2]. Nearly all trees from the literature have diameter at most 10; no construction of integral trees with arbitrarily large diameter is known. Other results concern 3-partite graphs and complete split graphs [HanMS]. Typically, the problem of integrality of graphs is addressed by considering diophantine equations such as those mentioned above.

We say that a graph $G$ is *L-integral* (respectively, *Q-integral*) if all eigenvalues of the Laplacian (respectively, signless Laplacian) are integers. Note that for the spectrum, $L$-spectrum and $Q$-spectrum of a regular graph, integrality of one spectrum implies integrality of the other two. By (7.30), the graph $G$ is $Q$-integral if and only if its line graph $L(G)$ is integral; such graphs are investigated in [Sta] and [SimSta3].

A general observation from the literature on $L$-integral graphs is that they appear to be more common than integral graphs. For example, all the complete bipartite graphs $K_{m,n}$ are $L$-integral (and indeed $Q$-integral). Moreover, the class of $L$-integral graphs is closed with respect to the operations of complementation (Proposition 7.1.1), sum (Theorem 7.1.12) and join (Exercise 8.16).

Finally we mention two interesting results proved by Merris [Mer2]. Recall first that a graph is *degree maximal* if its degree sequence $(d_1 \geq d_2 \geq \cdots \geq d_n)$ cannot be majorized by any other graphic sequence. Then we have: any degree maximal graph is $L$-integral. Merris also showed that if $u$, $v$ are co-duplicate vertices of the $L$-integral graph $G$ then $G - uv$ is $L$-integral.

# Exercises

**8.1** Prove Theorem 8.1.1.

**8.2** Prove Theorem 8.1.5, and also an analogue for the $Q$-index.

**8.3** Show that a graph with maximal $Q$-index among the connected graphs with a fixed number of vertices and a fixed number of edges is a threshold graph [CvRS11].

**8.4** State and prove analogues of Theorems 8.1.11 and 8.1.12 for the signless Laplacian. [Hint: Use formula (7.30) for the first and Theorem 2.4.4 for the second.]

**8.5** Let $G'$ be the graph obtained from a graph $G$ by relocating edges $st$ and $uv$ to the positions of non-edges $sv$ and $tu$ ('local switching'). Let $(x_1, x_2, \ldots, x_n)^\top$ be the principal $Q$-eigenvector of $G$. Prove that if $(x_s - x_u)(x_v - x_t) \geq 0$ then $\xi_1(G') \geq \xi_1(G)$, with equality if and only if $x_s = x_u$ and $x_t = x_v$.

**8.6** State and prove analogues of Theorems 8.1.14–8.1.16 for the signless Laplacian [CarCRS].

**8.7** Let $G(k, l)$ ($k \geq 2, l \geq 2$) be the graph obtained from a non-trivial connected graph $G$ by attaching pendant paths of length $k$ and $l$ at the same vertex $v$. Show that if $k \geq l \geq 3$ then $\xi_1(G(k, l)) > \xi_1(G(k + 1, l - 1))$ [CvSi5].

**8.8** Prove Theorem 8.1.22.

**8.9** Use Corollary 1.3.5 to prove that

$$\lambda_1(G)^r \geq \frac{N_{2q+r}}{N_{2q}},$$

where $q \geq 0$, $r > 0$ and $N_k$ is the number of walks of length $k$ in $G$ [Nik2].

**8.10** Prove that for any graph $G$, $\lambda_1(G) \geq \sqrt{\Delta(G)}$.

**8.11** Show that for each of (8.7), (8.8), equality holds if and only if $G$ is regular or semi-regular bipartite.

**8.12** Show that equality holds in (8.12) if and only if $G$ is regular or semi-regular bipartite [BerZh].

**8.13** Prove Lemma 8.2.1.

**8.14** With the notation of Section 8.2, suppose that $A\mathbf{x} \geq \alpha\mathbf{x}$. Show that the number of components in $(\langle P(\mathbf{x}) \cup Z(\mathbf{x})\rangle)$ is at most $\frac{1}{2}(n-1)$ when $0 \leq \alpha < 1$, and at most $\frac{1}{3}(n-1)$ when $1 \leq \alpha < \sqrt{2}$ [Pow2].

**8.15** Prove that the invariants (1)–(6) and properties (7)–(8) of a graph $G$, listed in Section 8.3, are determined by the polynomial deck of $G$.

**8.16** Show that if $G$ is a regular graph then $P_G(x)$ is reconstructible from $\mathcal{P}(G)$.

**8.17** Show that the join of two $L$-integral graphs is $L$-integral.

# Notes

A survey of graph perturbations appears in [Row6], and the characteristic polynomials of modified graphs are reviewed in [Row11]. The subdivision of an edge always results in a topologically equivalent (or *homeomorphic*) graph, and such graphs are discussed in [HofSm] and [SimKo]. The effect on the Laplacian index $\nu_1$ of adding or deleting an edge is investigated in [Guo2]. Further results on the behaviour of eigenvalues under graph modifications may be obtained by applying the analytical theory of matrix perturbations to the adjacency matrix of $G$; see [CvRS2, Chapter 6].

A survey of results on the index $\lambda_1$ of a graph may be found in [CvRo3], and a survey concerning $\lambda_2$ appears in [CvSi3]. Some bounds on $\lambda_k$ are discussed in [Hon1] and [Pow3]. For the largest eigenvalue of the Laplacian and of the signless Laplacian, see [BrHS], [HonZh] and [CvRS11], [OLAH], respectively. For connected graphs with prescribed numbers of vertices and edges, the maximal index is investigated in [Bel1], [BruSo], [CvRo2], [SimMB], and the minimal least eigenvalue is discussed in [BelCRS2], [BelCRS3].

Fiedler [Fie2] was the first to show that, for a connected graph, information can be extracted from an eigenvector corresponding to the second largest eigenvalue. Subsequent observations are due to Powers [Pow1, Pow2], and more of his results appear in [CvRS2, Chapter 9].

Sections 8.3 and 8.4 bring up to date Sections 3.6 and 3.8 of [CvRo4]. For an introduction to polynomial reconstruction, see [Sci2]; and for a survey of integral graphs, see [BaCRSS]. The integral trees with at most 50 vertices are identified in [Bro], and the integral trees with index 3 are determined in [BroHae]. The $L$-integral graphs with maximal degree 3 are identified in [Kir3], while addition of a vertex to preserve $L$-integrality is considered in [Kir1].

# 9

# Applications

In this chapter we present a small selection of applications of the theory of graph spectra. We limit ourselves to applications in physics, chemistry, computer science and mathematics itself; although we devote a section to each of these four subjects, the topics covered are not as compartmentalized as this might suggest. The recurring themes of approximation and optimization are found also in applications to many other scientific areas, including biology, geography, economics and the social sciences.

## 9.1 Physics

We explain how the theory of graph spectra is used in treating the vibration of a membrane and in a combinatorial enumeration problem which arises in chemical physics.

### 9.1.1 Vibration of a membrane

In the approximate numerical solution of certain partial differential equations, graphs and their spectra arise quite naturally. Consider, for example, the partial differential equation

$$\frac{\partial^2 z}{\partial x^2} + \frac{\partial^2 z}{\partial y^2} + \lambda z = 0, \tag{9.1}$$

that is, $\nabla z + \lambda z = 0$, where $\nabla$ denotes the Laplacian operator. We seek solutions $z = z(x, y)$ subject to the boundary condition $z(x, y) = 0$ on a simple closed curve $\Gamma$ lying in the $(x, y)$-plane. The non-zero solutions are called *eigenfunctions*, and they correspond to an infinite sequence of discrete values of $\lambda$ called *eigenvalues*. For example, if $\Gamma$ is the rectangle with vertices

259

$(0, 0)$, $(a, 0)$, $(0, b)$, $(a, b)$, the eigenvalues and corresponding eigenfunctions (to within a scalar multiple) are given by

$$\lambda_{ij} = \pi^2 \left( \frac{i^2}{a^2} + \frac{j^2}{b^2} \right), \quad z_{ij} = \sin \frac{i\pi}{a} x \, \sin \frac{j\pi}{b} y. \tag{9.2}$$

To approximate $z$ we consider the values only for a set of points $(x_i, y_i)$ which form a regular lattice (square, triangular or hexagonal) in the $xy$-plane. The corresponding (infinite) graph has the points $(x_i, y_i)$ as vertices, with edges joining points at minimal distance. The points (or vertices) lying in the interior of $\Gamma$ are called *internal* points (or vertices) and the other points (or vertices) are called *external*. Let $z_i = z(x_i, y_i)$. In view of the boundary condition, we take $z_i = 0$ for all external points.

We consider the case of a square lattice aligned with the co ordinate axes. Let $(x_0, y_0)$ be a fixed point of the lattice, let $z_0 = z(x_0, y_0)$, and let the values of $z$ at the neighbouring points (labelled 1, 2, 3, 4 in Fig. 9.1) be $z_1 = z(x_0+h, y_0)$, $z_2 = z(x_0 - h, y_0)$, $z_3 = z(x_0, y_0 + h)$, $z_4 = z(x_0, y_0 - h)$. The value of $\partial^2 z/\partial x^2 + \partial^2 z/\partial y^2$ at the point $(x_0, y_0)$ can, as usual, be approximated by

$$\frac{1}{h^2}(z_1 + z_2 + z_3 + z_4 - 4z_0).$$

Equation (9.1) then becomes

$$(4 - \lambda h^2)z_0 = z_1 + z_2 + z_3 + z_4. \tag{9.3}$$

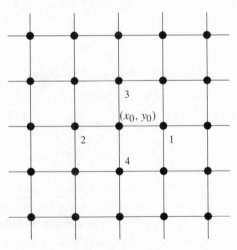

Figure 9.1 Vertices of a membrane graph.

Now we write $\zeta = 4 - \lambda h^2$ and label the internal points $1, 2, \ldots, n$. In view of (9.3) we have

$$\zeta z_i = \sum_{j \sim i} z_j \quad (i = 1, 2, \ldots, n), \tag{9.4}$$

where the summation is taken over all indices $j$ corresponding to internal points $(x_j, y_j)$ neighbouring $(x_i, y_i)$. (External points neighbouring $(x_i, y_i)$ are excluded since the value of $z$ at such points is zero.) If $G$ is the subgraph of the lattice graph induced by the internal vertices, the equations (9.4) are just the eigenvalue equations for $G$. If $\zeta_1, \ldots, \zeta_n$ are the eigenvalues of $G$, then the numbers

$$\lambda_i' = \frac{4 - \zeta_i}{h^2} \quad (i = 1, \ldots, n)$$

are approximate eigenvalues of Equation (9.1).

This procedure is often used in practical problems (see, for example, [Col]) to obtain approximate solutions of partial differential equations. We deal with a vibrating membrane $\Omega$ held fixed along its boundary $\Gamma$. Its displacement $F(x, y, t)$ orthogonal to the $(x, y)$-plane at time $t$ is given by the wave equation

$$\frac{\partial^2 F}{\partial t^2} = c^2 \left( \frac{\partial^2 F}{\partial x^2} + \frac{\partial^2 F}{\partial y^2} \right), \tag{9.5}$$

where $c$ is a constant depending on the physical properties of the membrane and of the tension under which the membrane is held. The harmonic vibrations are given by solutions of the form $F(x, y, t) = z(x, y)e^{i\omega t}$, where $i = \sqrt{-1}$. If we substitute this expression in (9.5), we obtain

$$-\omega^2 z(x, y) = c^2 \left( \frac{\partial^2 z(x, y)}{\partial x^2} + \frac{\partial^2 z(x, y)}{\partial y^2} \right),$$

and this is just Equation (9.1) with $\lambda = \omega^2/c^2$. In this situation, $G$ is called the *membrane graph*.

Essentially the same graph $G$ arises if, instead of a discrete approximation to a contnuous model, we start with a discrete model. Here the membrane consists of a set of atoms which in the equilibrium state lie at the points of a lattice, and each atom acts on its neighbouring atoms by elastic forces. It is assumed that all atoms have the same mass and that elastic forces are of the same intensity for all neighbouring pairs of atoms. If $z_i(t)$ and $z_j(t)$ are displacements of neighbouring atoms $i$ and $j$ at time $t$, the elastic force tending to reduce the relative displacement between these atoms is

$$F_{ij} = -K \left( z_i(t) - z_j(t) \right),$$

where $K$ is a constant determined by the elastic properties of the membrane.

The equation of motion of the $k$-th atom is

$$m\frac{d^2 z_k(t)}{dt^2} = -K \sum_{j \sim k} \left( z_k(t) - z_j(t) \right) \qquad (9.6)$$

where $m$ is the mass of an atom, and the summation is taken over the neighbours of the $k$-th atom. For an external vertex $j$ of the lattice graph, at which there is no atom of the membrane, we have $z_j(t) = 0$.

We can again consider pure harmonic oscillations and take $z_k(t) = z_k e^{i\omega t}$, where $i = \sqrt{-1}$. If we insert this expression into (9.6) and do so for each atom $k$, then we obtain the eigenvalue equations for the Laplacian matrix $L_G$. Since the lattice is 4-regular, the eigenvalues of $L_G$ are approximately $4 - \zeta_i$ ($i = 1, \ldots, n$).

**Example 9.1.1.** We consider the vibrations of a membrane whose perimeter is the rectangle with vertices $(0, 0)$, $(a, 0)$, $(0, b)$, $(a, b)$. We take the points of our lattice to be the points $(ph, qh)$ $(p, q \in \mathbb{Z})$. Then $G = P_m + P_n$, where $m = \lceil \frac{a}{h} - 1 \rceil$ and $n = \lceil \frac{b}{h} - 1 \rceil$. From Section 2.6 we know that the eigenvalues of $G$ are

$$\zeta_{ij} = 2\cos\frac{\pi}{m+1}i + 2\cos\frac{\pi}{n+1}j, \quad (i = 1, 2, \ldots, m; \ j = 1, 2, \ldots, n).$$
$$(9.7)$$

To within a scalar multiple, a corresponding eigenvector has coordinates

$$\sin\frac{\pi}{m+1}ip \, \sin\frac{\pi}{n+1}jq \quad (p = 1, 2, \ldots, m; \ q = 1, 2, \ldots, n). \qquad (9.8)$$

The approximate eigenvalues $\lambda'_{ij}$ of Equation (9.1) are therefore:

$$\lambda'_{ij} = \frac{4 - \zeta_{ij}}{h^2} = \frac{2}{h^2}\left(1 - \cos\frac{\pi}{m+1}i + 1 - \cos\frac{\pi}{n+1}j\right)$$
$$= \frac{4}{h^2}\left(\sin^2\frac{\pi}{2(m+1)}i + \sin^2\frac{\pi}{2(n+1)}j\right).$$

To compare $\lambda'_{ij}$ with the eigenvalue $\lambda_{ij}$ of Equation (9.2), note that for sufficiently large $m$ and $n$ (and fixed values of $i$ and $j$) we may use the approximation $\sin x \approx x$ to obtain

$$\lambda'_{ij} \approx \frac{4}{h^2}\left(\frac{\pi^2 i^2}{4(m+1)^2} + \frac{\pi^2 j^2}{4(n+1)^2}\right) = \pi^2\left(\frac{i^2}{((m+1)h)^2} + \frac{j^2}{((n+1)h)^2}\right)$$
$$\approx \pi^2\left(\frac{i^2}{a^2} + \frac{j^2}{b^2}\right) = \lambda_{ij}.$$

Hence $\lambda'_{ij}$ approximates $\lambda_{ij}$ well if the distance $h$ between the neighbouring points of the lattice is small enough. We can see similarly that (9.8) gives a good approximation to the corresponding eigenfunction of (9.2): at the point $(x, y) = (ph, qh)$ we have

$$\sin \frac{\pi}{m+1} ip \sin \frac{\pi}{n+1} jq = \sin \frac{i\pi x}{(m+1)h} \sin \frac{j\pi y}{(n+1)h} \approx \sin \frac{i\pi x}{a} \sin \frac{j\pi y}{b}.$$

$\square$

### 9.1.2 The dimer problem

The spectra of graphs, or the spectra of certain matrices which are closely related to adjacency matrices, appear in a number of problems in statistical physics (see, for example, [Kast], [Mon], [Per]). We shall describe the so-called *dimer problem*, which arises in the investigation of the thermodynamic properties of a system of diatomic molecules ('dimers') adsorbed on the surface of a crystal. The most favourable points for the adsorption of atoms on such a surface form a two-dimensional lattice, and a dimer can occupy two neighbouring points. It is required to count all the ways in which dimers can be arranged on the lattice without overlapping each other, so that every lattice point is occupied. In other words, the task is to determine the number $k(m, n)$ of 1-factors in the graph $G_{m,n} = P_m + P_n$ (Fig. 9.2). Since $k(m, n) = 0$ when $mn$ is odd, we assume without loss of generality that $n$ is even. When $m = n$ the problem is equivalent to that of enumerating the ways in which an $n \times n$ chess-board can be completely covered by $\frac{1}{2}n^2$ dominoes.

For the enumeration of 1-factors one can make use of the *permanent* of a square matrix, defined as follows. If $A = (a_{ij})$ is an $n \times n$ matrix then

$$\text{per } A = \sum_{\sigma \in S_n} a_{1\sigma(1)} a_{2\sigma(2)} \cdots a_{n\sigma(n)}.$$

For properties of the permanent, see [BruRy, Chapter 7]. Note that $\text{per}(A^\top) = \text{per } A$, and that the value of the permanent is unchanged when rows or columns are permuted. However, the elementary row operation of adding a multiple of one row to another row can change the value of the permanent, and this accounts for the general computational difficulty in evaluating per $A$. Indeed, the problem of computing the permanent is NP-complete (in fact, #P-complete; see [BruRy, Chapter 7]). In Section 9.4 we discuss an efficient means of approximating permanents of certain $(0, 1)$-matrices.

Given an $n \times n$ matrix $A$ with non-negative entries, consider the weighted bipartite graph $G(A)$ which has $n$ black vertices corresponding to the rows of $A$, $n$ white vertices corresponding to the columns of $A$, and an edge of weight

$a_{ij}$ between the $i$-th black vertex and the $j$-th white vertex $(i, j = 1, \ldots, n)$. If we define the weight $w(F)$ of a perfect matching $F$ to be the product of the weights of its edges then

$$\text{per } A = \sum_{F \in \mathcal{F}(A)} w(F),$$

where $\mathcal{F}(A)$ is the set of all 1-factors of $G(A)$. In particular, if $A$ is a $(0, 1)$-matrix then per $A$ is the number of 1-factors of $G(A)$. In the case that $A$ is the adjacency matrix of a bipartite graph $G$, say

$$A = \begin{pmatrix} O & B^\top \\ B & O \end{pmatrix},$$

we have per $A = (\text{per } B)^2$, while per $B$ is the number of 1-factors of $G$. Hence we have the following:

**Theorem 9.1.2.** *For a bipartite graph $G$ with adjacency matrix $A$, we have*

$$\text{per } A = k^2,$$

*where $k$ is the number of 1-factors of $G$.*

We shall now count the 1-factors in $P_m + P_n$ using one of several possible variants for transforming the permanent into a determinant. Let $G_{m,n} = P_m + P_n$, and let $H_{m,n}$ denote the digraph obtained from $G_{m,n}$ by replacing every edge by a corresponding pair of arcs of opposite orientation. In accordance with Fig. 9.2, arcs may be described as horizontal or vertical. A *circuit* in a digraph is a directed cycle of length $\geq 2$. The following lemma, stated without proof, can be established using arguments from [Per].

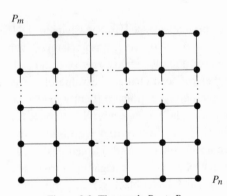

Figure 9.2 The graph $P_m + P_n$.

**Lemma 9.1.3.** *For every spanning collection L of circuits in $H_{m,n}$ (n even) we have*

$$2c(L) \equiv h(L) \mod 4,$$

*where $c(L)$ is the number of circuits in L, and $h(L)$ is the number of horizontal arcs in L.*

**Theorem 9.1.4.** *The number k of 1-factors in $G_{m,n}$ (n even) is given by*

$$k^2 = \det(A_m \otimes I_n + i I_m \otimes A_n),$$

*where $i = \sqrt{-1}$ and $A_s$ is the adjacency matrix of a path with s vertices.*

**Proof.** From Section 2.5 we know that $A(G_{m,n}) = A_m \otimes I_n + I_m \otimes A_n$. Clearly, 1s from $A_m \otimes I_n$ correspond to vertical edges and 1s from $I_m \otimes A_n$ correspond to horizontal edges of $G_{m,n}$. The matrix

$$A^*_{m,n} = A_m \otimes I_n + i I_m \otimes A_n$$

differs from $A_{m,n}$ in that 1s corresponding to horizontal edges are multiplied by $i$. Now

$$\det A_{m,n} = \sum_L (-1)^{c(L)},$$

where the summation runs over all spanning collections $L$ of circuits in $H_{m,n}$. Hence

$$\det A^*_{m,n} = \sum_L (-1)^{c(L)} i^{h(L)}.$$

By Lemma 9.1.3, we have $i^{h(L)} = (-1)^{c(L)}$, and so $\det A^*_{m,n} = \operatorname{per} A_{m,n}$. The result now follows from Theorem 9.1.2 $\qquad\square$

Since $A^*_{m,n}(\mathbf{x} \otimes \mathbf{y}) = A_m \mathbf{x} \otimes \mathbf{y} + i \mathbf{x} \otimes A_n \mathbf{y}$, the eigenvalues of $A^*_{m,n}$ are

$$2\cos \frac{\pi}{m+1} j + 2i \cos \frac{\pi}{n+1} l \quad (j = 1, \ldots, m; \; l = 1, \ldots, n),$$

and so

$$k^2 = \prod_{j=1}^{m} \prod_{l=1}^{n} \left( 2\cos \frac{\pi}{m+1} j + 2i \cos \frac{\pi}{n+1} l \right)$$

$$= 2^{mn} \prod_{j=1}^{m} \prod_{l=1}^{n/2} \left( \cos^2 \frac{\pi}{m+1} j + \cos^2 \frac{\pi}{n+1} l \right).$$

For $n \times n$ lattices with $n = 2, 4, 6, 8$ we have $k = 2, 36, 6728, 12\,988\,816$, respectively. The last number is $2^4 \cdot 901^2$ and this is the number of ways in which an $8 \times 8$ chess-board can be covered by 32 dominoes.

In the general case, we have also (cf. [Per])

$$k \sim e^{\frac{mn}{\pi}C} \quad (m \to +\infty, n \to +\infty),$$

where $C$ $(= 0.91596...)$ is Catalan's constant.

## 9.2 Chemistry

One of the most important applications of the theory of graph spectra is in chemistry, in treating unsaturated hydrocarbons by an approximating technique called the Hückel molecular orbital theory. We first describe this technique and then discuss the mathematical notion of graph energy which arises naturally from the Hückel theory.

### 9.2.1 The Hückel molecular orbital theory

The equation of motion of a particle of mass $m$ in a potential field of force $V = V(x, y, z)$ is

$$ih\frac{\partial \psi}{\partial t} = -\frac{h^2}{m}\left(\frac{\partial^2 \psi}{\partial x^2} + \frac{\partial^2 \psi}{\partial y^2} + \frac{\partial^2 \psi}{\partial z^2}\right) + V\psi, \tag{9.9}$$

where $i = \sqrt{-1}$, $h$ is Planck's constant and $\psi$ is a complex-valued function of $x, y, z, t$ describing the state of the system. If the stationary states of the system are $\Psi_k = \Psi_k(x, y, z)$ with energy levels $E_k$ then the general solution of (9.9) is

$$\psi = \sum_k c_k \Psi_k e^{-iE_k t/h}.$$

Here the functions $\Psi_k$ satisfy the time-independent Schrödinger equation

$$\hat{H}\Psi_k = E_k\Psi_k, \tag{9.10}$$

where $\hat{H}$ is the Hamiltonian operator defined by

$$\hat{H}\psi = -\frac{h^2}{m}\left(\frac{\partial^2 \psi}{\partial x^2} + \frac{\partial^2 \psi}{\partial y^2} + \frac{\partial^2 \psi}{\partial z^2}\right) + V\psi.$$

The complex-value functions $\Psi_k(x, y, z)$ are called *wave functions* and in the context of molecules, they are known as *molecular orbitals*.

One of the basic goals of quantum chemistry is to describe the electronic structure of a system of molecules. This requires the solution of (9.10) for complicated molecular systems with many electrons, and various approximations are used. The Hückel theory applies to *conjugated hydrocarbons*, which we now describe. A hydrocarbon is a chemical compound composed of only two elements – carbon (C) and hydrogen (H); there are single bonds between a hydrogen atom and a carbon atom, while two carbon atoms may have single or double bonds between them. We assume that in a hydrocarbon molecule all carbon atoms have valency 4 (i.e. feature in 4 bonds) and all hydrogen atoms have valency 1. Associated with each carbon atom are three $\sigma$-electrons local to the atom and one $\pi$-electron. If a function $\Psi_k$ satisfying (9.10) is normalized so that $\int\int\int |\Psi_k(x, y, z)|^2 dx\, dy\, dz = 1$ then $|\Psi_k(x, y, z)|^2$ is a probability distribution for $\pi$-electrons in the molecule.

In a conjugated hydrocarbon, each carbon atom features in exactly one double bond and two single bonds. The corresponding *Hückel graph* [GuPo] (or *carbon skeleton*) has the carbon atoms as its vertices, with an edge between atoms precisely when there is a single or double bond between them. Figure 9.3 shows the molecular structure of Styrene and the associated Hückel graph.

The Hückel theory assumes that the energy of electrons is determined simply by the adjacencies in the Hückel graph. Then (9.10) can be expressed in matrix form

$$Hw = \lambda w, \quad H = \alpha I + \beta A, \qquad (9.11)$$

where $\alpha, \beta$ are constants and $A$ is the adjacency matrix of the Hückel graph. (The parameters $\alpha$ and $\beta$ are called the *Coulomb integral* and the *resonance integral*.) If $\lambda_1, \ldots, \lambda_n$ are the eigenvalues of $A$, then the eigenvalues of $H$ are

$$\epsilon_j = \alpha + \beta\lambda_j \ (j = 1, \ldots, n),$$

and $\epsilon_j$ is a measure of $\pi$-electron energy in the $j$-th quantum state. Let $Hw_j = \epsilon_j w_j \ (j = 1, \ldots, n)$, where $\epsilon_1 \geq \cdots \geq \epsilon_n$. The eigenvectors $w_1, \ldots, w_n$ are taken as discrete approximations to molecular orbitals $\Psi_1, \ldots, \Psi_n$. It is assumed that $n$ is even and that each of $\Psi_1, \ldots, \Psi_{n/2}$

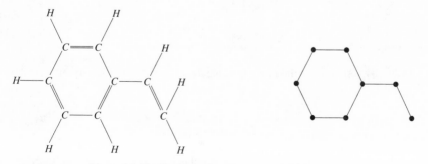

Figure 9.3 The Styrene molecule and its Hückel graph.

determines the distribution of two $\pi$-electrons (an assumption supported by Pauli's Principle, which implies that at most two $\pi$-electrons are associated with any orbital). Accordingly the total $\pi$-electron energy is calculated as

$$\epsilon = 2\sum_{j=1}^{\frac{n}{2}} \epsilon_j = 2\sum_{j=1}^{\frac{n}{2}} (\alpha + \beta\lambda_j) = n\alpha + 2\beta\sum_{j=1}^{\frac{n}{2}} \lambda_j,$$

where $\lambda_1 \geq \cdots \geq \lambda_n$. The significant part of this expression is the sum $2\sum_{j=1}^{\frac{n}{2}} \lambda_j$, which we denote by $E$. Since $\sum_{j=1}^{n} \lambda_j = 0$, we have the following important observation:

**Remark 9.2.1.** If $\lambda_{n/2} \geq 0 \geq \lambda_{n/2+1}$, then $E = \sum_{j=1}^{n} |\lambda_j|$. □

By Theorem 3.2.3 the hypotheses of Remark 9.2.1 are satisfied when the molecular multigraph, and hence also the Hückel graph $G$, is bipartite. (Then our conjugated hydrocarbon is said to be *alternant*.) In this situation, the spectrum of $G$ is symmetric about 0 and the eigenspaces $\mathcal{E}_H(\alpha + \beta\lambda) = \mathcal{E}_A(\lambda)$, $\mathcal{E}_H(\alpha - \beta\lambda) = \mathcal{E}_A(-\lambda)$ are paired in the following sense (see Exercise 1.6): with appropriate labelling,

$$\begin{pmatrix} \mathbf{x} \\ \mathbf{y} \end{pmatrix} \in \mathcal{E}_A(\lambda) \Longleftrightarrow \begin{pmatrix} \mathbf{x} \\ -\mathbf{y} \end{pmatrix} \in \mathcal{E}_A(-\lambda). \tag{9.12}$$

In quantum chemistry the corresponding pairing of molecular orbitals is known as the 'Pairing Theorem'.

We shall explain the role of graph angles (cf. [Cve11]). As usual, we take the distinct eigenvalues of $G$ to be $\mu_1, \ldots, \mu_m$ in decreasing order. Let $\mathbf{x}_i = (x_{i1}, \ldots, x_{in})^\top$ $(i = 1, \ldots, n)$ be orthonormal eigenvectors of $G$, and define $M_p = \{j : A\mathbf{x}_j = \mu_p\mathbf{x}_j\}$. By Equation (1.4), the angles $\alpha_{pq}$ satisfy

$$\alpha_{pq}^2 = \sum_{j \in M_p} x_{jq}^2 \quad (p = 1, \ldots, m).$$

In general, the $\pi$-electron *charges* are the numbers $c_q^2 = \sum_{j=1}^{n} g_j x_{jq}^2$ $(q = 1, \ldots, n)$, where $g_j$ is the number of electrons ascribed to the $j$-th orbital. We can write

$$c_q^2 = \sum_{p=1}^{m} \sum_{j \in M_p} g_j x_{jq}^2 = \sum_{p=1}^{m} c_{pq}^2$$

where $c_{pq}^2 = \sum_{j \in M_p} g_j x_{jq}^2$. We call the numbers $c_{pq}^2$ *partial electron charges*.

In our situation, we have $g_j = 2$ for $j \leq n/2$ and $g_j = 0$ for $j > n/2$. Moreover, it follows from (9.12) that if $\lambda_j = -\lambda_i$ then $x_{iq}^2 = x_{jq}^2$ ($q = 1, \ldots, n$). Now, using also Propositon 1.3.3, we have

$$c_q^2 = \sum_{j=1}^{n/2} g_j x_{jq}^2 = 2 \sum_{j=1}^{n/2} x_{jq}^2 = \sum_{j=1}^{n} x_{jq}^2 = \sum_{p=1}^{m} \alpha_{pq}^2 = 1.$$

This is a result of Coulson and Rushbrooke [CouRu] from 1940. Although this result is of great importance in chemistry, electron charges in bipartite graphs are of little mathematical interest since, unlike partial electron charges, they do not provide any structural information.

For *stable* molecules, 0 is not an eigenvalue of $G$, and so $m$ is even. In this situation, the connection between partial charges and angles is given by

$$c_{pq}^2 = \sum_{j \in M_p} 2x_{jq}^2 = 2\alpha_{pq}^2 \quad (p = 1, \ldots, m/2; q = 1, \ldots, n).$$

Since paired eigenvalues have the same angle sequence, knowledge of partial charges is equivalent to the knowledge of graph angles. The relation between graph structure and graph angles has been explored in Chapters 1–4. For example, Equation (2.21) tells us that the number of $q$-$q$ walks in $G$ of length $s$ is $\sum_{p=1}^{m} \alpha_{pq}^2 \mu_p^s$. Taking $s = 2$, we see that the degree of vertex $q$ is given by

$$d_q = \sum_{p=1}^{m/2} c_{pq}^2 \mu_p^2.$$

In chemical terms, the partial $\pi$-electron charges for an atom and $\pi$-electron energy levels determine the number of neighbouring carbon atoms. We can also determine the number of quadrangles in the Hückel graph (Theorem 3.1.5), the characteristic polynomials of vertex-deleted subgraphs (Proposition 2.2.6) and other graph invariants. However, beyond the class of bipartite graphs under consideration, partial charges appear less important from a mathematical point of view.

## 9.2.2 Graph energy

Let $G$ be a graph with $n$ vertices and $m$ edges ($m > 0$). The energy $E_G$ of $G$ is defined by

$$E_G = \sum_{j=1}^{n} |\lambda_j|,$$

where $\lambda_1, \lambda_2, \ldots, \lambda_n$ are the eigenvalues of $G$. The energy of a graph was defined by Gutman [Gut3] and has attracted much attention from researchers in the last few years. We have seen in the previous subsection that in some cases the energy defined in this way corresponds to the energy of a molecule (cf. Remark 9.2.1). However, $E_G$ can be studied for any graph $G$ independently of the chemical context.

Since $\sum_{j=1}^{n} \lambda_j^2 = 2m$, we have (e.g. by the Cauchy–Schwarz inequality):

**Proposition 9.2.2** [McC]. *For a graph $G$ with $n$ vertices and $m$ edges, $E_G \leq \sqrt{2mn}$.*

Numerous relations are known for $E_G$, and we mention two without proof (see [Gut1] and [Cou]). If $G$ has adjacency matrix $A$ then

$$2m - n(\det A)^{2/n} \leq 2mn - E_G^2 \leq (n-1)\left(2m - n(\det A)^{2/n}\right),$$

and if $G$ has characteristic polynomial $P_G(x)$ then (with $i = \sqrt{-1}$):

$$E_G = \frac{1}{\pi} \int_{-\infty}^{+\infty} t^{-2} \log \left| t^n P_G\left(\frac{i}{t}\right) \right| dt.$$

It is known that for $n \leq 7$, the graphs with maximal energy are the complete graphs $K_n$, $n = 1, 2, \ldots, 7$. The maximal values of energy for graphs with $n$ vertices have been determined heuristically by the system Auto-GraphiX [CapCGH, CvGr] for $n \leq 12$. The graph with maximal energy among 10-vertex graphs is $L(K_5)$. The $n$-vertex graphs with maximal energy, for an infinite sequence of values of $n$, are determined in [KooMo1]; like $L(K_5)$, these graphs are strongly regular, and the smallest such graph is the Clebsch graph (Example 1.2.4). Indeed it is proved in [KooMo1] that for a graph $G$ on $n$ vertices,

$$E_G \leq \frac{n}{2}(1 + \sqrt{n}),$$

with equality if and only if $G$ is a strongly regular graph with parameters $\left(n, \frac{1}{2}n + \sqrt{\frac{1}{4}n}, \frac{1}{4}n + \sqrt{\frac{1}{4}n}, \frac{1}{4}n + \sqrt{\frac{1}{4}n}\right)$. Such strongly regular graphs exist for $n = 4\tau^2$ with $\tau = 2^m$ ($m \in \mathbb{N}$). By Theorem 3.6.5, the distinct eigenvalues of such a graph are $\tau(2\tau + 1)$, $\pm\tau$. The conjecture that, for any $\epsilon > 0$, for almost all $n$ there exists a graph $G$ on $n$ vertices such that

$$E_G \geq (1 - \epsilon)\frac{n}{2}(1 + \sqrt{n})$$

has been confirmed in a slightly improved form in [Nik3]. These results suggest that graphs with maximal energy have a small number of distinct eigenvalues, but there are significant exceptions which seem to make the maximal energy problem very difficult.

In what follows, we use a calculus approach to search for graphs with maximal energy in the class $\mathcal{G}_{m,n}$ of graphs with $n$ vertices and $m$ edges ($m > 0$). Although our procedure provides a good heuristic, it is limited by the fact that the maximum value of the continuous variable under consideration is not necessarily attained in a graph.

We define

$$I = \{1, \ldots, n\}, \quad I_+ = \{i \in I : \lambda_i \geq 0\}, \quad I_- = \{i \in I : \lambda_i < 0\}.$$

Since $m > 0$, both $I_+$ and $I_-$ are non-empty. Moreover the energy can be represented in the form

$$E = \sum_{i \in I_+} \lambda_i - \sum_{i \in I_-} \lambda_i,$$

while the eigenvalues satisfy the relations

$$\sum_{i \in I} \lambda_i = 0, \quad \sum_{i \in I} \lambda_i^2 = 2m. \tag{9.13}$$

We consider an auxiliary function of $x_1, \ldots, x_n$ involving these constraints:

$$F = \sum_{i \in I_+} x_i - \sum_{i \in I_-} x_i + \alpha \sum_{i \in I} x_i + \beta \left( \sum_{i \in I} x_i^2 - 2m \right),$$

where $\alpha, \beta$ are Lagrange multipliers. The extremal values of the function $E$ satisfying (9.13) are found from:

$$\frac{\partial F}{\partial x_j} = \pm 1 + \alpha + 2\beta x_j = 0 \ (j \in I).$$

Here the first term in the sum is equal to $+1$ if $j \in I_+$ and is equal to $-1$ if $j \in I_-$. We obtain

$$\lambda_j = \frac{-\alpha \mp 1}{2\beta} \ (j \in I).$$

Now a graph has just two distinct eigenvalues if and only if it has the form $cK_t$ ($c \in \mathbb{N}$), and the maximal value of $E$ is attained when $n = ct$ and $m = \frac{1}{2}ct(t-1)$. If $m$ and $n$ are not of this form then a graph in $\mathcal{G}_{m,n}$ with maximal energy has at least three distinct eigenvalues. In this situation, we can extend our procedure if we assume that some of the eigenvalues are prescribed, say the eigenvalues $\lambda_i$ ($i \in K$). Let $\mathcal{H}$ be the set of graphs in $\mathcal{G}_{m,n}$ whose spectrum includes these eigenvalues. We write $J = I \setminus K$ and extend our notation so that

$$E = \sum_{i \in J_+} \lambda_i - \sum_{i \in J_-} \lambda_i + \sum_{i \in K_+} \lambda_i - \sum_{i \in K_-} \lambda_i$$

and

$$\sum_{i \in J} \lambda_i + \sum_{i \in K} \lambda_i = 0, \quad \sum_{i \in J} \lambda_i^2 + \sum_{i \in K} \lambda_i^2 = 2m.$$

Let

$$C_+ = \sum_{i \in K_+} \lambda_i, \quad C_- = \sum_{i \in K_-} \lambda_i, \quad C = \sum_{i \in K} \lambda_i, \quad D = \sum_{i \in K} \lambda_i^2.$$

We can write

$$F = \sum_{i \in J_+} x_i - \sum_{i \in J_-} x_i + C_+ - C_- + \alpha \left( \sum_{i \in J} x_i + C \right) + \beta \left( \sum_{i \in J} x_i^2 + D - 2m \right).$$

As before, the equations $\frac{\partial F}{\partial x_j} = 0$ $(j \in J)$ yield

$$\lambda_j = \frac{-\alpha \mp 1}{2\beta} \quad (j \in J).$$

Assuming that both sets $J_+$ and $J_-$ are non-empty, we conclude that the unknown eigenvalues take just two values ($\xi, \eta$ say) in a graph with extremal energy. (If $J_+$ or $J_-$ is empty, our approach does not give a solution.) The multiplicities $p, q$ of $\xi, \eta$ are such that $p + q = |J| = n - |K|$, and we may formulate a Lagrange multiplier problem for each possible pair $(p, q)$. In this way we obtain $|J| - 1$ problems, one for each situation $(|J_+|, |J_-|) = (i, |J| - i)$ $(i = 1, \ldots, |J| - 1)$. For a given distribution of unknown positive and negative eigenvalues, the solution of the corresponding Lagrange multiplier problem yields an upper bound on the maximal energy of graphs in $\mathcal{G}_{m,n}$ with a corresponding distribution of eigenvalues. If we take the maximal value of $E$ over all such solutions, and that energy value is realized by a graph, then we know we have a maximal energy graph in $\mathcal{G}_{m,n}$. We denote by $\mathcal{L}$ this procedure for extending a partial spectrum and maximizing $E$. Now we have:

**Theorem 9.2.3** [CvGr]. *let $\mathcal{K}$ be a family of real numbers, let $G$ be a graph with maximal energy in $\mathcal{G}_{m,n}$, and suppose that the spectrum of $G$ has all elements of $\mathcal{K}$ as eigenvalues. Let $S(\mathcal{K})$ be the spectrum obtained from $\mathcal{K}$ by the procedure $\mathcal{L}$. Then every graph with spectrum $S(\mathcal{K})$ is a graph in $\mathcal{G}_{m,n}$ with maximal energy.*

An analogous result holds for minimal energy graphs. In practice, it is often convenient to avoid explicit use of the procedure $\mathcal{L}$ by exploiting the fact that there are just two distinct unknown eigenvalues. Indeed, in view of (9.13) we have (in the notation above):

$$p + q = |J|, \quad p\xi + q\eta = -C, \quad p\xi^2 + q\eta^2 = 2m - D, \tag{9.14}$$

where, without loss of generality, $p \leq q$. For each solution $p, q, \xi, \eta$ of (9.14) with $p \leq q$, we calculate $E = p|\xi| + q|\eta| + C_+ - C_-$, and then ask whether there exists a graph with spectrum $\mathcal{K} \cup \{\xi^p, \eta^q\}$ when $p, q, \xi, \eta$ determine the maximum (or minimum) value of $E$.

**Example 9.2.4** [CvGr]. To investigate the graphs with maximal energy among the $r$-regular graphs with $n$ vertices, we take $m = \frac{1}{2}rn$ and $\mathcal{K} = \{r\}$. Note that if $E$ is maximized for values $p, q, \xi, \eta$ that are attained in a graph $G$ then $G$ is strongly regular by Theorems 3.2.1 and 3.6.4. For example, when $n = 16$ and $r = 10$ we have $m = 80$, $C = 10$, $D = 100$, and there are 14 solutions of (9.14) with $p \leq q$. We find that the largest value of $E$ is 40, which arises when $p = 5, q = 10, \xi = 2$ and $\eta = -2$. The corresponding spectrum arises only in the Clebsch graph (cf. Theorem 5.2.8). We conclude that the Clebsch graph has maximal energy among the 10-regular graphs on 16 vertices. We do not obtain the graph(s) with minimal energy because the least value of $E$ arises for values of $p, q, \xi, \eta$ that are not attained in a graph (e.g. because $\frac{1}{6} \sum_i \lambda_i^3$ is not an integer). $\quad\square$

# 9.3 Computer science

We have already noted in Chapter 7 how graph spectra determine expansion properties of a communication network, and we shall see in Section 9.4 how graph spectra are related to the complexity of certain mathematical problems. In the last decade there has been growing recognition that graph spectra have further important applications in computer science, for example in internet technologies, pattern recognition and computer vision. Here we describe an elementary application in so-called interconnection topologies for multiprocessors, and a connection between generalized line graphs and the security of statistical databases.

## 9.3.1 Load balancing

Let $G$ be a connected graph with $n$ vertices and Laplacian matrix $L$. Let $\zeta_1, \ldots, \zeta_m$ be the distinct eigenvalues of $L$; we take $\zeta_m = 0$ but we do not assume any ordering of the positive eigenvalues $\zeta_1, \ldots, \zeta_{m-1}$. We define

$$f_k(x) = \prod_{i=1}^{k} \left(1 - \frac{x}{\zeta_i}\right) \quad (k = 1, \ldots, m - 1).$$

Then for any vector $\mathbf{x} \in \mathbb{R}^n$, we have

$$f_k(L)\mathbf{x} \in \mathcal{E}_L(\zeta_{k+1}) \oplus \cdots \oplus \mathcal{E}_L(\zeta_m) \quad (k = 1, \ldots, m - 1). \tag{9.15}$$

Since $L\mathbf{j} = \mathbf{0}$, we also have

$$\mathbf{j}^\top \mathbf{x} = \mathbf{j}^\top f_k(L)\mathbf{x} \quad (k = 1, \ldots, m - 1). \tag{9.16}$$

These simple mathematical facts have an interesting application in the design of multiprocessor computer networks. Such a network is modelled by a graph $G$ in which vertices denote processors and edges represent direct communication links between processors. A job which is to be executed is divided into elementary tasks assigned to particular processors. With the notation above, we take $\mathbf{x} = (x_1, \ldots, x_n)^\top$, where $x_i$ is the number of tasks initially allocated to the $i$-th processor. The idea of load balancing is to reallocate the tasks in $m - 1$ steps to obtain a uniform distribution among the processors. We define

$$\mathbf{x}^{(0)} = \mathbf{x}, \quad \mathbf{x}^{(k)} = \left(I - \frac{1}{\zeta_k}L\right)\mathbf{x}^{(k-1)} \quad (k = 1, \ldots, m - 1).$$

Thus $\mathbf{x}^{(k)} = f_k(L)\mathbf{x}$ $(k = 1, \ldots, m - 1)$. If $\mathbf{x}^{(k)} = (x_1^{(k)}, \ldots, x_n^{(k)})^\top$ then

$$x_i^{(k)} = x_i^{(k-1)} - \frac{1}{\zeta_k} \sum_{j \sim i} (x_i - x_j) \quad (k = 1, \ldots, m - 1). \tag{9.17}$$

Thus at the $k$-th step, the net flow of tasks from a processor $i$ to a neighbouring processor $j$ is $\frac{1}{\zeta_k}(x_i^{(k-1)} - x_j^{(k-1)})$, the direction of flow determined by sign. By (9.16) (or (9.17)), the total number of tasks is unchanged. By (9.15), $\mathbf{x}^{(m-1)} \in \mathcal{E}_L(\zeta_m)$, and since $\mathcal{E}_L(\zeta_m)$ is spanned by $\mathbf{j}$, the tasks are indeed uniformly distributed after $m - 1$ steps. In practice, the numbers $x_i^{(k)}$ should be integers; more importantly, the whole process is feasible only if, for some ordering of $\zeta_1, \ldots, \zeta_{m-1}$, all $x_i^{(k)}$ are non-negative. Nevertheless, this use of the Laplacian underpins an optimal scheme for load balancing described in [ElKM]. Complexity considerations show that efficiency depends essentially on the parameter $m\Delta$, where $m$ is the number of distinct eigenvalues of $L$ and $\Delta$ is the maximum degree of $G$. The hypercubes $Q_k$ are used in [ElKM] to construct an infinite family of graphs $G_{(n)}$ $(n > 2)$ such that $G_{(n)}$ has $n$ vertices,

$O(\log^2 n)$ distinct eigenvalues and maximum degree at most $3 \log n + o(\log n)$. For further details the reader is referred to [DieFM] and [ElKM].

## 9.3.2  A problem in the security of statistical databases

We saw in Section 5.5 how certain subgraphs can be used to construct a basis for the eigenspace of $-2$ in generalized line graphs. Here we explain how essentially the same result emerged independently in the context of database security. We may think of a database as an array in which rows (or *records*) are indexed by individuals (say, the employees of a company) and columns are indexed by attributes (such as salary, gender and address). To fix ideas, suppose that individual salaries are treated as confidential, and that in accessing salary data, users of the database are restricted to types of queries such as the sum, average, maximum and minimum salary over a set of individuals. This set is called the *query set*, and is specified in terms of attributes; for example, the average salary of female employees might be requested. It is clear that in some circumstances, individual data can be extracted from statistical data obtained in this way. For example, the salary of a sole male employee can be calculated from the average salary of female employees, the average salary of all employees and the number of employees of each gender. The general problem is to identify sequences of query sets which do not allow confidential data to be revealed. Such a sequence is said to be *compromise-free*, and a database is *secure* if queries are restricted to compromise-free sequences. In practice there is a trade-off between the number of query types and the proportion of possible sequences that are compromise-free.

Several security mechanisms are described in the literature, but most of them are either insecure or overly restrictive. One exception is the so-called 'Audit Expert' first proposed in [ChOz] to deal principally with sums of quantifiable attributes. The database security system keeps track of all previously answered queries and each new query is answered only if the database remains secure. For a mathematical formulation of Audit Expert, consider a database consisting of $n$ records. An answered request for a sum can be thought of as a linear equation

$$\beta_1 x_1 + \beta_2 x_2 + \cdots + \beta_n x_n = r$$

where $\beta_i = 1$ if $i$ is in the query set and $\beta_i = 0$ otherwise; $x_i$ is the value of the confidential attribute of the $i$-th individual, and $r$ is the answer to the query. Then a sequence of $k$ answered queries can be viewed as a system of $k$ linear equations in $n$ variables:

$$Q\mathbf{x} = \mathbf{r}$$

where $Q = (\beta_{ij})$, $\mathbf{x} = (x_1, x_2, \ldots, x_n)^\top$ and $\mathbf{r} = (r_1, r_2, \ldots, r_k)^\top$. The matrix $Q$ is called the *query matrix*, and $Q$ is said to be compromise-free if and only if the corresponding sequence of query sets is compromise-free. As observed in [ChOz], $Q$ is compromise-free if and only if for each $i \in \{1, 2, \ldots, n\}$, there exists a vector $\mathbf{v} = (v_1, v_2, \ldots, v_n)^\top$ with $v_i \neq 0$ such that $Q\mathbf{v} = \mathbf{0}$.

Now consider the case in which each individual features in at most two queries; then $Q$ is said to be *restricted*. In this situation, $Q$ is the incidence matrix of a graph $G[Q]$, possibly with multiple edges, where queries correspond to vertices and individuals correspond to edges. To cater for the situation in which an individual features in only one query, we allow semi-edges in our graph (cf. Fig. 9.4); semi-edges correspond to columns of $Q$ having exactly one non-zero entry. Thus $G[Q]$ is obtained from a $B$-graph $\hat{H}$ by replacing petals with semi-edges and then repeating edges if necessary. We modify the definition of an odd dumbbell in $G[Q]$ accordingly, replacing 'petal' with 'semi-edge'. Note also that in $G[Q]$ an even cycle may be a 2-cycle. Now recall that if $\hat{H}$ has incidence matrix $C$ then the eigenvectors of $L(\hat{H})$ corresponding to $-2$ are the non-zero vectors $\mathbf{x}$ such that $C\mathbf{x} = \mathbf{0}$ (see Section 1.2). Moreover, a basis of $\mathcal{E}_{L(\hat{H})}(-2)$ can be constructed from odd dumbbells and even cycles (see Section 5.5), and in this situation, the non-zero entries of a vector $\mathbf{x}$ are the weights shown in Fig. 5.7. In the case of $G[Q]$, where an odd dumbbell may have a semi-edge, and an even cycle may be a 2-cycle, a non-zero vector $\mathbf{x}$ such that $Q\mathbf{x} = \mathbf{0}$ is constructed as illustrated in Fig. 9.4. In the example shown, we assign the weight of $-2$ to a semi-edge instead of assigning weights of $-1$ to each edge of a petal.

With the interpretations above, we may now apply Corollary 5.5.9 to obtain the following result.

**Theorem 9.3.4** [Bra, BraMS]. *The restricted query matrix $Q$ is compromise-free if and only if each edge of the graph $G[Q]$ is contained in an even cycle or an odd dumbbell.*

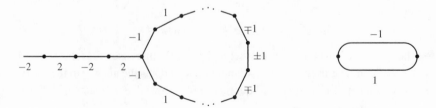

Figure 9.4 Constructing vectors from $G[Q]$.

## 9.4 Mathematics

There are many interactions between the theory of graph spectra and other branches of mathematics. Here we illustrate one application in combinatorial optimization and another in combinatorial enumeration. In both cases, the complexity considerations demonstrate the relevance to computer science.

### 9.4.1 The travelling salesperson problem

A salesperson wishes to pay one visit to each city on a given list, starting and finishing at the same city; the problem is to minimize the cost of travel (in time or money). The *travelling salesperson problem* (TSP) is therefore a combinatorial optimization problem for a weighted complete digraph $D$, where the weight $d_{ij}$ of arc $ij$ measures the cost of travelling from $i$ to $j$. Here we discuss only the *symmetric* travelling salesperson problem (STSP), where it is assumed that $d_{ij} = d_{ji}$ for all $i$, $j$. Then the problem is to find a Hamiltonian cycle of least weight in a weighted complete graph. Since weights can be made arbitrarily large, the problem embraces that of determining whether a given graph is Hamiltonian.

The travelling salesperson problem is one of the best-known NP-hard combinatorial optimization problems, and there is an extensive literature on both its theoretical and practical aspects. The most important theoretical results on TSP can be found in [LawLRS] (see also [CoCPS], [CvDM]). Many algorithms and heuristics for TSP have been proposed, and for a review we refer the reader to Laporte [Lap]. We shall mention here only one approach, which applies semi-definite programming (SDP) to the problem of minimizing the weight of a Hamiltonian cycle. An upper bound for the least weight is obtained by relaxing the STSP and exploiting a property of algebraic connectivity (cf. [CvCK1, CvCK2]). The method can be used in an algorithm of branch-and-bound type as first proposed by Christofides [Chr].

The crucial observation is the following, where $h_n = 2 - 2\cos(2\pi/n)$, the algebraic connectivity of an $n$-cycle.

**Theorem 9.4.1.** *Let $G$ be a graph with $n$ vertices, and let $H$ be a 2-regular spanning subgraph of $G$ with Laplacian matrix $L_H$. Let $X = L_H + \alpha J - \beta I$, where $\alpha$ and $\beta$ are real parameters such that $\alpha > h_n/n, 0 < \beta \leq h_n$. Then $H$ is a Hamiltonian cycle in $G$ if and only if the matrix $X$ is positive semi-definite.*

**Proof.** Let $v_1 \geq v_2 \geq \cdots \geq v_n = 0$ be the eigenvalues of $L_H$ and let $\mathbf{x}_1, \mathbf{x}_2, \ldots, \mathbf{x}_n$ be corresponding eigenvectors which are pairwise orthogonal, with $\mathbf{x}_n = \mathbf{j}$. Then $\mathbf{x}_1, \mathbf{x}_2, \ldots, \mathbf{x}_n$ are linearly independent eigenvectors of $X$ with corresponding eigenvalues $\nu_1 - \beta$, $\nu_2 - \beta, \ldots$, $\nu_{n-1} - \beta$, $n\alpha - \beta$. Since $H$ is a union of disjoint cycles, either $H$ is a Hamiltonian cycle (with $\nu_{n-1} = h_n$) or $H$ is disconnected (with $\nu_{n-1} = 0$). In either case, $\nu_{n-1} - \beta$ is the smallest eigenvalue because $n\alpha > h_n \geq \nu_{n-1}$.

If $H$ is a Hamiltonian cycle then $\nu_{n-1} - \beta = h_n - \beta \geq 0$ and $X$ is semi-definite. Conversely, if $X$ is semi-definite then $\nu_{n-1} \geq \beta > 0$ and so $H$ is connected by Theorem 7.1.2; therefore $H$ is a Hamiltonian cycle.    □

It follows from Theorem 9.4.1 that a spanning subgraph $H$ of $G$ is a Hamiltonian cycle if and only if its Laplacian $L_H = (l_{ij})$ satisfies the following conditions:

$$l_{ii} = 2 \ (i = 1, \ldots, n), \tag{9.18}$$

$$L_H + \alpha J - \beta I \geq 0 \quad \text{when} \quad \alpha > h_n/n \quad \text{and} \quad 0 < \beta \leq h_n. \tag{9.19}$$

If we introduce the matrix $X = (x_{ij}) = L_H + \alpha J - \beta I$, we can define a discrete SDP model of STSP as follows:

$$\text{minimize } F(X) = \sum_{i=1}^{n} \sum_{j=1}^{n} \left( -\frac{1}{2} d_{ij} \right) x_{ij} + \frac{\alpha}{2} \sum_{i=1}^{n} \sum_{j=1}^{n} d_{ij} \tag{9.20}$$

subject to

$$x_{ii} = 2 + \alpha - \beta \ (i = 1, \ldots, n), \tag{9.21}$$

$$\sum_{j=1}^{n} x_{ij} = n\alpha - \beta, \ (i = 1, \ldots, n), \tag{9.22}$$

$$x_{ij} \in \{\alpha - 1, \alpha\} \ (j = 1, \ldots, n : i < j), \tag{9.23}$$

$$X \geq 0. \tag{9.24}$$

Here $X \geq 0$ means that the matrix $X$ is symmetric and positive semi-definite, while $\alpha$ and $\beta$ are chosen as in Theorem 9.4.1. The matrix $L = X + \beta I - \alpha J$ represents the Laplacian of a Hamiltonian cycle if and only if $X$ satisfies (9.21)–(9.24). Indeed, the constraints (9.21)–(9.23) ensure that $L$ has the form of a Laplacian with diagonal entries equal to 2, while condition (9.24) guarantees that $L$ corresponds to a Hamiltonian cycle. Therefore, if $X^*$ is an optimal solution of problem (9.20)–(9.24) then $L^* = X^* + \beta I - \alpha J$ is the Laplacian

$(l_{ij}^*)$ of an optimal Hamiltonian cycle of $G$ with the objective function value

$$\sum_{i=1}^{n}\sum_{j=1}^{n}\left(-\frac{1}{2}d_{ij}\right)l_{ij}^* = F(X^*).$$

A natural semi-definite relaxation of the travelling salesperson problem is obtained when the discrete condition (9.23) is replaced by inequalities:

$$\text{minimize } F(X) \tag{9.25}$$

subject to

$$x_{ii} = 2 + \alpha - \beta \ (i = 1, \ldots, n), \tag{9.26}$$

$$\sum_{j=1}^{n} x_{ij} = n\alpha - \beta \ (i = 1, \ldots, n), \tag{9.27}$$

$$\alpha - 1 \leq x_{ij} \leq \alpha \ (i, j = 1, \ldots, n; \ i < j), \tag{9.28}$$

$$X \geq 0. \tag{9.29}$$

It is easy to see that the relaxation (9.25)–(9.29) can be expressed in the standard form of an SDP problem. Indeed, the constraint (9.26) can be written as $A_i \cdot X = 2 + \alpha - \beta$, where $A_i \cdot X$ denotes the Frobenius inner product $\mathbf{j}^\top(A_i \circ X)\mathbf{j}$, and $A_i$ is a symmetric $n \times n$ matrix with 1 at position $(i, i)$ and all other entries equal to 0. Similarly, condition (9.27) is equivalent to $B_i \cdot X = 2(n\alpha - \beta)$, where $B_i$ has 2 at position $(i, i)$, all remaining elements of the $i$-th row and the $i$-th column are equal to 1, and all the other entries are zero. Finally, condition (9.28) can be expressed as $2(\alpha - 1) \leq C_{ij} \cdot X \leq 2\alpha$, where $C_{ij}$ has 1 at the positions $(i, j)$ and $(j, i)$ and zero elewhere. Since the SDP problem (9.25)–(9.29) depends on parameters $\alpha$ and $\beta$ it represents a class of semi-definite relaxations of the STSP.

The SDP model (9.25)–(9.29) has an equivalent formulation in terms of Laplacians as follows, where $L$ is a symmetric matrix $(l_{ij})$ with second smallest eigenvalue $\nu_{n-1}(L)$.

$$\text{minimize } \Phi(L) = \sum_{i=1}^{n}\sum_{j=1}^{n}\left(-\frac{1}{2}d_{ij}\right)l_{ij} \tag{9.30}$$

subject to

$$l_{ii} = 2 \ (i = 1, \ldots, n), \tag{9.31}$$

$$\sum_{j=1}^{n} l_{ij} = 0 \quad (i = 1, \ldots, n), \tag{9.32}$$

$$-1 \le l_{ij} \le 0 \quad (i, j = 1, \ldots, n; \; i < j), \tag{9.33}$$

$$v_{n-1}(L) \ge \beta. \tag{9.34}$$

This continuous SDP model (9.30)–(9.34) yields a solution in polynomial time, and provides a good approximation to an optimum solution for the STSP.

### 9.4.2 Markov chains

Consider an infinite sequence $X_0, X_1, X_2, \ldots$ of random variables in a time-homogeneous Markov chain on a finite state space $\{1, \ldots, n\}$ with transition matrix $(p_{ij})$. Thus $p_{ij} = \Pr(X_{t+1} = j | X_t = i)$, independent of time $t$. Moreover, $P\mathbf{j} = \mathbf{j}$ and $P^s = (p_{ij}^{(s)})$, where $p_{ij}^{(s)} = \Pr(X_{t+s} = j | X_t = i)$. The distribution of $X_t$ is given by the vector $\mathbf{p}^{(t)} = (p_1^{(t)}, \ldots, p_n^{(t)})^\top$, where $p_i^{(t)} = \Pr(X_t = i) \; (i = 1, \ldots, n)$.

The chain is *ergodic* if there exists a distribution $\mathbf{p} = (p_1, \ldots, p_n)^\top$ such that

$$\lim_{s \to \infty} p_{ij}^{(s)} = p_j > 0 \quad \text{for all} \quad i, j \in \{1, \ldots, n\},$$

equivalently the chain is *irreducible* (any state can be reached from any other) and *aperiodic* ($\gcd\{s : p_{ij}^{(s)} > 0\} = 1$). In this situation, $\mathbf{p}^{(t)} \to \mathbf{p}$ as $t \to \infty$ whatever the initial state $\mathbf{p}^{(0)}$. The distribution $\mathbf{p}$ is stationary since $\mathbf{p}^\top P = \mathbf{p}^\top$; indeed, $\mathbf{p}$ is the unique vector such that $\mathbf{p}^\top P = \mathbf{p}^\top$ and $\mathbf{p}^\top \mathbf{j} = 1$. The chain is *time-reversible* if

$$p_{ij} p_i = p_{ji} p_j \quad \text{for all} \quad i, j \in \{1, \ldots, n\}.$$

To investigate the convergence of $\mathbf{p}^{(t)}$ to $\mathbf{p}$ we define the *relative pointwise distance* after $t$ steps by:

$$\Delta(t) = \max_{i,j} \frac{|p_{ij}^{(t)} - p_j|}{p_j}.$$

**Theorem 9.4.2** [SinJe]. *Let $P$ be the transition matrix of a time-reversible ergodic Markov chain, with eigenvalues $1 = \rho_1 > \rho_2 \ge \cdots \ge \rho_n$. If the probabilities in the stationary state are $p_1, \ldots, p_n$ then*

$$\Delta(t) \le \rho^t / p_{\min},$$

where $\rho = \max\{|\rho_2|, \ldots, |\rho_n|\}$ and $p_{\min} = \min\{p_1, \ldots, p_n\}$.

**Proof.** Let $D = \mathrm{diag}(p_1, \ldots, p_n)$ and $Q = D^{\frac{1}{2}} P^{-\frac{1}{2}}$. Since our chain is time-reversible, $Q$ is symmetric. Let $\mathbf{x}_1, \ldots, \mathbf{x}_n$ be orthonormal eigenvectors of $Q$ such that $Q\mathbf{x}_i = \rho_i \mathbf{x}_i$ ($i = 1, \ldots, n$) and $\mathbf{x}_1^\top = \mathbf{p}^\top D^{-\frac{1}{2}}$. The spectral decomposition of $Q$ may be written

$$Q = \sum_{i=1}^n \rho_i \mathbf{x}_i \mathbf{x}_i^\top,$$

and so we have

$$P^t = D^{-\frac{1}{2}} Q^t D^{\frac{1}{2}} = \sum_{i=1}^n \rho_i^t (D^{-\frac{1}{2}} \mathbf{x}_i)(D^{\frac{1}{2}} \mathbf{x}_i)^\top$$

$$= \mathbf{j}\mathbf{p}^\top + \sum_{i=2}^n \rho_i^t (D^{-\frac{1}{2}} \mathbf{x}_i)(D^{\frac{1}{2}} \mathbf{x}_i)^\top.$$

It follows that if $\mathbf{x}_i = (x_{i1}, \ldots, x_{in})^\top$ ($i = 2, \ldots, n$) then

$$p_{jk}^{(t)} = p_k + \sqrt{\frac{p_k}{p_j}} \sum_{i=2}^n \rho_i^t x_{ij} x_{ik}.$$

Using the Cauchy–Schwarz inequality, we deduce that

$$\Delta(t) = \max_{j,k} \frac{|\sum_{i=2}^n \rho_i^t x_{ij} x_{ik}|}{\sqrt{p_j p_k}} \le \rho^t \frac{\sum_{i=2}^n |x_{ij}||x_{ik}|}{p_{\min}} \le \frac{\rho^t}{p_{\min}}.$$

$\square$

To specify the number of steps required to ensure that $\mathbf{p}^{(t)}$ is close to $\mathbf{p}$, we define $\tau : \mathbb{R}^+ \to \mathbb{N}$ by:

$$\tau(\epsilon) = \min\{t \in \mathbb{N} : \Delta(t') \le \epsilon \text{ for all } t' \ge t\}.$$

**Corollary 9.4.3.** *With the above notation, $\tau(\epsilon) \le (\ln p_{\min}^{-1} + \ln \epsilon^{-1})/(1 - \rho)$.*

**Proof.** If $\rho^s / p_{\min} = \epsilon$ then $s = -\ln(p_{\min}\epsilon)/\ln \rho^{-1} \le (\ln p_{\min}^{-1} + \ln \epsilon^{-1})/(1 - \rho)$. $\square$

In practice, we may usually replace $P$ with $\frac{1}{2}(I + P)$ because $P$ and $\frac{1}{2}(I + P)$ have the same stationary distribution. Since the eigenvalues of $\frac{1}{2}(I + P)$ are non-negative, we shall now assume that $\rho_n \ge 0$, so that $\rho = \rho_2$.

Let $H$ be the undirected weighted graph (in general with loops) having adjacency matrix $A = DP$, i.e. $(a_{ij}) = (p_{ij}p_i)$ (cf. Remark 7.7.1). By analogy with (7.18) the conductance of $H$ is defined by

$$\Phi(H) = \min \left\{ \frac{\sum_{i \in S, j \notin S} a_{ij}}{\sum_{i \in S} p_i} : 0 < \sum_{i \in S} p_i < \frac{1}{2} \right\}.$$

Note that if $P$ is replaced with $\frac{1}{2}(I + P)$ then the conductance of $H$ is halved.

Now let $\tilde{\nu}_1 \geq \cdots \geq \tilde{\nu}_n$ be the eigenvalues of the normalized Laplacian $D^{-\frac{1}{2}}(D - A)D^{-\frac{1}{2}}$ $(= I - D^{\frac{1}{2}}PD^{-\frac{1}{2}})$. As in Theorem 7.7.5, we have (cf. [SinJe, Lemma 3.3]):

$$2\Phi(H) \geq \tilde{\nu}_{n-1} \geq \frac{1}{2}\Phi(H)^2.$$

Since $\rho = \rho_2 = 1 - \tilde{\nu}_{n-1}$, it follows from Corollary 9.4.3 that

$$\tau(\epsilon) \leq \frac{2}{\Phi(H)^2}(\ln p_{\min}^{-1} + \ln \epsilon^{-1}). \tag{9.35}$$

In the mathematical modelling of a physical system, we often have a very large state space, consisting of configurations of the system, and we want a sample of the space distributed according to **p**. Here we are interested in the situation where the states can be identified with combinatorial structures. We can simulate the Markov chain from an initial state if the probabilities $p_{ij}$ are computable locally, that is, if we can calculate $p_{ij}$ for any given states $i$ and $j$. If **p** is uniform, i.e. $\mathbf{p} = \frac{1}{n}\mathbf{j}$, we can then obtain an approximation to the number $n$ of structures. For $0 < \epsilon < 1$, an approximation of $\frac{1}{n}$ to within a ratio of $1 + \epsilon$ is guaranteed after at least $\tau\left(\frac{1}{2}\epsilon\right)$ steps, since for $t \geq \tau\left(\frac{1}{2}\epsilon\right)$ we have

$$(1 + \epsilon)^{-1} < 1 - \frac{1}{2}\epsilon \leq \frac{p_{ij}^{(t)}}{\left(\frac{1}{n}\right)} \leq 1 + \frac{1}{2}\epsilon < 1 + \epsilon.$$

**Example 9.4.4.** We describe how the above technique is applied in [JeSi] to approximate the number $m(G)$ of perfect matchings in a *dense* bipartite graph $G$: such a graph has $2k$ vertices, with colour classes of size $k$ and minimum degree $\delta(G) \geq \frac{1}{2}k$. We have seen in Subsection 9.1.2 that the problem of determining $m(G)$ is NP-hard, and equivalent to calculating the permanent of a $(0, 1)$-matrix.

Let $\mathcal{M}_r(G)$ denote the set of matchings of size $r$ in $G$. We take as state space the set $\mathcal{N} = \mathcal{M}_k(G) \cup \mathcal{M}_{k-1}(G)$, and specify transitions as follows. For any $M \in \mathcal{N}$, choose an edge $e = uv \in E(G)$ uniformly at random and move to state $M'$, where

(i)   if $M \in \mathcal{M}_k(G)$ and $e \in M$ then $M' = M - e$,

(ii)  if $M \in \mathcal{M}_{k-1}(G)$ and $u, v$ are not matched in $M$ then $M' = M + e$,

(iii) if $M \in \mathcal{M}_{k-1}(G)$, $u$ is matched to $w$ in $M$ and $v$ is not matched in $M$, then $M' = (M - uw) + e$; and if $M \in \mathcal{M}_{k-1}(G)$, $v$ is matched to $w$ in $M$ and $u$ is not matched in $M$, then $M' = (M - vw) + e$,

(iv)  otherwise, $M' = M$.

Note that these transitions determine a time-reversible ergodic Markov chain. For the reasons explained above, we replace the transition matrix $P = (p_{ij})$ with $\frac{1}{2}(I + P)$. A major result of [JeSi] asserts that the underlying graph $H$ has conductance $\Phi(H) \geq 1/12k^6$. Since also $p_{\min}^{-1} = n \leq 2^{k^2}$, it follows from (9.35) that $\tau(\frac{1}{2}\epsilon)$ is bounded above by a polynomial function $f_\epsilon(k)$. (In these circumstances the Markov chain is said to be *rapidly mixing*.) One perfect matching in $G$ can be found in polynomial time, and taken as an initial state; moreover the $p_{ij}$ are locally computable in polynomial time. Accordingly a simulation of the Markov chain yields an approximately uniform distribution in polynomial time. This distribution provides estimates for $|\mathcal{M}_k(G)| + |\mathcal{M}_{k-1}(G)|$, $|\mathcal{M}_k(G)|/|\mathcal{M}_{k-1}(G)|$ and hence also for $|\mathcal{M}_k(G)|$. Further details may be found in [JeSi]. $\qquad\square$

# Notes

The motivation for founding the theory of graph spectra came from applications in chemistry and physics. The paper [Huc] is considered to be the first paper where graph spectra appear, though in an implicit form. The first mathematical paper on graph spectra [ColSi] was motivated by the membrane vibration problem and similar problems concerning oscillations (see [Col],[Kac], [Rut]). More details on Hückel's molecular orbital theory may be found in the books [Bal], [CouLM], [Dia], [GrGT], [Gut7], [GuTr], [Tri].

The dimer problem is not the only problem that can be reduced to the enumeration of 1-factors. Others include the famous *Ising problem* that arises in the theory of ferromagnetism; see, for example, the books [Kast] and [Mon]. These texts include a discussion of the enumeration of walks of various kinds in a lattice graph.

For the construction of a matrix $A^*$ from a matrix $A$ such that per $A = \det A^*$, see [BruRy, Section 7.5]. A treatment of permanents and determinants using digraphs can be found in [BruCv]. Various means of calculating the number of 1-factors in $P_{2m} + P_{2n}$ are described in [Per].

Our technique for finding graphs with maximal energy in $\mathcal{G}_{m,n}$ was used in [Gut4] to obtain an alternative proof of Proposition 9.2.2. The problem of finding graphs with minimal energy appears to be easier, and there are several recent results in this direction; see, for example, [Yan] and [Hua]. The paper [Gut2] is a seminal article on graphs with extremal energy. The survey papers [Gut5] and [Gut6] on graph energy are written for mathematicians and chemists respectively.

Doob's original description [Doo7] of the eigenspace of $-2$ for line graphs, in terms of even cycles and odd dumbbells, appeared in 1973. For generalized line graphs, a description of $\mathcal{E}(-2)$ in terms of chain groups was given by Cvetković, Doob and Simić [CvDS2] in 1981. In 1996, with their observations on even cycles and odd dumbbells in the context of database security, Branković, Miller and Širáň [BraMS] implicitly shed further light on the extension of Doob's description to generalized line graphs. This was achieved independently by Cvetković, Rowlinson and Simić in a paper [CvRS4] submitted in 1998, in the context of graph foundations. The link between [BraMS] and [CvRS4] was noted in [BraCv]. A refinement of Audit Expert (called 'Hybrid Audit Expert') is considered in [Bra] and [BraMS]. Further combinatorial questions relating to Audit Expert are investigated in [DemKM]

Semi-definite programming has many applications to various classes of optimization problems (see e.g. [VanBo]); in particular, there is a growing interest in the application of SDP to combinatorial optimization, where it is used to obtain satisfactory bounds on an optimal objective function value; see [Goe] for a survey. Semi-definite relaxations have recently been introduced for the max-cut problem [GoeWi] and the graph colouring problem [KaMS].

# Appendix

This Appendix contains the following graph tables:

A1. The spectra and characteristic polynomials of the adjacency matrix, Seidel matrix, Laplacian and signless Laplacian for connected graphs with at most 5 vertices;

A2. The eigenvalues, angles and main angles of connected graphs with 2 to 5 vertices;

A3. The spectra and characteristic polynomials of the adjacency matrix for connected graphs with 6 vertices;

A4. The spectra and characteristic polynomials of the adjacency matrix for trees with at most 9 vertices;

A5. The spectra and characteristic polynomials of the adjacency matrix for cubic graphs with at most 12 vertices.

In Tables A1 and A2, the graphs are given in the same order as in Table 1 in the Appendix of [CvDSa]. In Table A1, the spectra and coefficients for the characteristic polynomials with respect to the adjacency matrix, Laplacian, signless Laplacian and Seidel matrix, appear in consecutive lines. Table A2, which is taken from [CvPe2], was also published in [CvRS3]. This table contains, for each graph, the eigenvalues (first line), the main angles (second line) and the vertex angle sequences, with vertices labelled as in the diagrams alongside. Vertices of graphs in Table A2 are ordered in such a way that the corresponding vertex angle sequences are in lexicographical order. Since similar vertices have the same angle sequence, just one sequence is given for each orbit.

In Tables A3, A4 and A5, the the spectra and coefficients for the characteristic polynomials are listed in consecutive lines. Table A3 comes from the paper [CvPe1], and here graphs are ordered lexicographically by spectral moments. In Table A4, the trees with up to 9 vertices are also ordered by spectral

285

moments. The corresponding data for trees with up to 10 vertices appear in [CvDSa, Table 2]; there the trees are ordered by characteristic polynomials. In Table A5, taken from [BuČCS], the graphs are ordered lexicographically by spectrum. The same information appears in [CvDSa, Table 3], but with the graphs in a different order.

## A.1 TABLE A1

The spectra and characteristic polynomials of the adjacency matrix, Seidel matrix, Laplacian and signless Laplacian for connected graphs with at most 5 vertices

```
01    1.0000 -1.0000      1      0    -1
      2.0000  0.0000      1     -2     0
      2.0000  0.0000      1     -2     0
      1.0000 -1.0000      1      0    -1

02    2.0000 -1.0000 -1.0000      1     0    -3    -2
      3.0000  3.0000  0.0000      1    -6     9     0
      4.0000  1.0000  1.0000      1    -6     9    -4
      1.0000  1.0000 -2.0000      1     0    -3     2

03    1.4142  0.0000 -1.4142      1     0    -2     0
      3.0000  1.0000  0.0000      1    -4     3     0
      3.0000  1.0000  0.0000      1    -4     3     0
      2.0000 -1.0000 -1.0000      1     0    -3    -2

04    3.0000 -1.0000 -1.0000 -1.0000      1     0    -6    -8    -3
      4.0000  4.0000  4.0000  0.0000      1   -12    48   -64     0
      6.0000  2.0000  2.0000  2.0000      1   -12    48   -80    48
      1.0000  1.0000  1.0000 -3.0000      1     0    -6     8    -3

05    2.5616  0.0000 -1.0000 -1.5616      1     0    -5    -4     0
      4.0000  4.0000  2.0000  0.0000      1   -10    32   -32     0
      5.2361  2.0000  2.0000  0.7639      1   -10    32   -40    16
      2.2361  1.0000 -1.0000 -2.2361      1     0    -6     0     5

06    2.1701  0.3111 -1.0000 -1.4812      1     0    -4    -2     1
      4.0000  3.0000  1.0000  0.0000      1    -8    19   -12     0
      4.5616  2.0000  1.0000  0.4384      1    -8    19   -16     4
      2.2361  1.0000 -1.0000 -2.2361      1     0    -6     0     5

07    2.0000  0.0000  0.0000 -2.0000      1     0    -4     0     0
      4.0000  2.0000  2.0000  0.0000      1    -8    20   -16     0
      4.0000  2.0000  2.0000  0.0000      1    -8    20   -16     0
      3.0000 -1.0000 -1.0000 -1.0000      1     0    -6    -8    -3

08    1.7321  0.0000  0.0000 -1.7321      1     0    -3     0     0
      4.0000  1.0000  1.0000  0.0000      1    -6     9    -4     0
      4.0000  1.0000  1.0000  0.0000      1    -6     9    -4     0
      3.0000 -1.0000 -1.0000 -1.0000      1     0    -6    -8    -3

09    1.6180  0.6180 -0.6180 -1.6180      1     0    -3     0     1
      3.4142  2.0000  0.5858  0.0000      1    -6    10    -4     0
      3.4142  2.0000  0.5858  0.0000      1    -6    10    -4     0
      2.2361  1.0000 -1.0000 -2.2361      1     0    -6     0     5
```

| | | | | | | | | | |
|---|---|---|---|---|---|---|---|---|---|
| 10 | 4.0000 | -1.0000 | -1.0000 | -1.0000 | -1.0000 | 1 | 0 | -10 | -20 | -15 | -4 |
| | 5.0000 | 5.0000 | 5.0000 | 5.0000 | 0.0000 | 1 | -20 | 150 | -500 | 625 | 0 |
| | 8.0000 | 3.0000 | 3.0000 | 3.0000 | 3.0000 | 1 | -20 | 150 | -540 | 945 | -648 |
| | 1.0000 | 1.0000 | 1.0000 | 1.0000 | -4.0000 | 1 | 0 | -10 | 20 | -15 | 4 |
| | | | | | | | | | | | |
| 11 | 3.6458 | 0.0000 | -1.0000 | -1.0000 | -1.6458 | 1 | 0 | -9 | -14 | -6 | 0 |
| | 5.0000 | 5.0000 | 5.0000 | 3.0000 | 0.0000 | 1 | -18 | 120 | -350 | 375 | 0 |
| | 7.3723 | 3.0000 | 3.0000 | 3.0000 | 1.6277 | 1 | -18 | 120 | -378 | 567 | -324 |
| | 2.3723 | 1.0000 | 1.0000 | -1.0000 | -3.3723 | 1 | 0 | -10 | 8 | 9 | -8 |
| | | | | | | | | | | | |
| 12 | 3.3234 | 0.3579 | -1.0000 | -1.0000 | -1.6813 | 1 | 0 | -8 | -10 | -1 | 2 |
| | 5.0000 | 5.0000 | 4.0000 | 2.0000 | 0.0000 | 1 | -16 | 93 | -230 | 200 | 0 |
| | 6.8284 | 3.0000 | 3.0000 | 2.0000 | 1.1716 | 1 | -16 | 93 | -250 | 312 | -144 |
| | 2.5616 | 1.0000 | 1.0000 | -1.5616 | -3.0000 | 1 | 0 | -10 | 4 | 17 | -12 |
| | | | | | | | | | | | |
| 13 | 3.2361 | 0.0000 | 0.0000 | -1.2361 | -2.0000 | 1 | 0 | -8 | -8 | 0 | 0 |
| | 5.0000 | 5.0000 | 3.0000 | 3.0000 | 0.0000 | 1 | -16 | 94 | -240 | 225 | 0 |
| | 6.5616 | 3.0000 | 3.0000 | 2.4384 | 1.0000 | 1 | -16 | 94 | -256 | 321 | -144 |
| | 3.0000 | 1.5616 | -1.0000 | -1.0000 | -2.5616 | 1 | 0 | -10 | -4 | 17 | 12 |
| | | | | | | | | | | | |
| 14 | 3.0861 | 0.4280 | -1.0000 | -1.0000 | -1.5141 | 1 | 0 | -7 | -8 | 0 | 2 |
| | 5.0000 | 4.0000 | 4.0000 | 1.0000 | 0.0000 | 1 | -14 | 69 | -136 | 80 | 0 |
| | 6.3723 | 3.0000 | 2.0000 | 2.0000 | 0.6277 | 1 | -14 | 69 | -152 | 148 | -48 |
| | 2.3723 | 1.0000 | 1.0000 | -1.0000 | -3.3723 | 1 | 0 | -10 | 8 | 9 | -8 |
| | | | | | | | | | | | |
| 15 | 3.0000 | 0.0000 | 0.0000 | -1.0000 | -2.0000 | 1 | 0 | -7 | -6 | 0 | 0 |
| | 5.0000 | 5.0000 | 2.0000 | 2.0000 | 0.0000 | 1 | -14 | 69 | -140 | 100 | 0 |
| | 6.3723 | 3.0000 | 2.0000 | 2.0000 | 0.6277 | 1 | -14 | 69 | -152 | 148 | -48 |
| | 3.3723 | 1.0000 | -1.0000 | -1.0000 | -2.3723 | 1 | 0 | -10 | -8 | 9 | 8 |
| | | | | | | | | | | | |
| 16 | 2.9354 | 0.6180 | -0.4626 | -1.4728 | -1.6180 | 1 | 0 | -7 | -6 | 3 | 2 |
| | 5.0000 | 4.4142 | 3.0000 | 1.5858 | 0.0000 | 1 | -14 | 70 | -146 | 105 | 0 |
| | 6.1249 | 3.0000 | 2.6367 | 1.2384 | 1.0000 | 1 | -14 | 70 | -158 | 161 | -60 |
| | 2.2361 | 2.2361 | 0.0000 | -2.2361 | -2.2361 | 1 | 0 | -10 | 0 | 25 | 0 |
| | | | | | | | | | | | |
| 17 | 2.8558 | 0.3216 | 0.0000 | -1.0000 | -2.1774 | 1 | 0 | -7 | -4 | 2 | 0 |
| | 5.0000 | 4.0000 | 3.0000 | 2.0000 | 0.0000 | 1 | -14 | 71 | -154 | 120 | 0 |
| | 5.7785 | 3.0000 | 2.7108 | 2.0000 | 0.5107 | 1 | -14 | 71 | -162 | 160 | -48 |
| | 3.3723 | 1.0000 | -1.0000 | -1.0000 | -2.3723 | 1 | 0 | -10 | -8 | 9 | 8 |
| | | | | | | | | | | | |
| 18 | 2.6855 | 0.3349 | 0.0000 | -1.2713 | -1.7491 | 1 | 0 | -6 | -4 | 2 | 0 |
| | 5.0000 | 4.0000 | 2.0000 | 1.0000 | 0.0000 | 1 | -12 | 49 | -78 | 40 | 0 |
| | 5.7785 | 2.7108 | 2.0000 | 1.0000 | 0.5107 | 1 | -12 | 49 | -86 | 64 | -16 |
| | 3.0000 | 1.5616 | -1.0000 | -1.0000 | -2.5616 | 1 | 0 | -10 | -4 | 17 | 12 |
| | | | | | | | | | | | |
| 19 | 2.6412 | 0.7237 | -0.5892 | -1.0000 | -1.7757 | 1 | 0 | -6 | -4 | 3 | 2 |
| | 4.4812 | 4.0000 | 2.6889 | 0.8299 | 0.0000 | 1 | -12 | 50 | -82 | 40 | 0 |
| | 5.4679 | 2.9128 | 2.0000 | 1.2011 | 0.4182 | 1 | -12 | 50 | -90 | 68 | -16 |
| | 2.5616 | 1.0000 | 1.0000 | -1.5616 | -3.0000 | 1 | 0 | -10 | 4 | 17 | -12 |
| | | | | | | | | | | | |
| 20 | 2.5616 | 1.0000 | -1.0000 | -1.0000 | -1.5616 | 1 | 0 | -6 | -4 | 5 | 4 |
| | 5.0000 | 3.0000 | 3.0000 | 1.0000 | 0.0000 | 1 | -12 | 50 | -84 | 45 | 0 |
| | 5.5616 | 3.0000 | 1.4384 | 1.0000 | 1.0000 | 1 | -12 | 50 | -92 | 77 | -24 |
| | 2.5616 | 1.0000 | 1.0000 | -1.5616 | -3.0000 | 1 | 0 | -10 | 4 | 17 | -12 |
| | | | | | | | | | | | |
| 21 | 2.4812 | 0.6889 | 0.0000 | -1.1701 | -2.0000 | 1 | 0 | -6 | -2 | 4 | 0 |
| | 4.6180 | 3.6180 | 2.3820 | 1.3820 | 0.0000 | 1 | -12 | 51 | -90 | 55 | 0 |
| | 5.1149 | 2.7459 | 2.6180 | 1.1392 | 0.3820 | 1 | -12 | 51 | -94 | 71 | -16 |
| | 3.0000 | 1.5616 | -1.0000 | -1.0000 | -2.5616 | 1 | 0 | -10 | -4 | 17 | 12 |
| | | | | | | | | | | | |
| 22 | 2.4495 | 0.0000 | 0.0000 | 0.0000 | -2.4495 | 1 | 0 | -6 | 0 | 0 | 0 |
| | 5.0000 | 3.0000 | 2.0000 | 2.0000 | 0.0000 | 1 | -12 | 51 | -92 | 60 | 0 |
| | 5.0000 | 3.0000 | 2.0000 | 2.0000 | 0.0000 | 1 | -12 | 51 | -92 | 60 | 0 |
| | 4.0000 | -1.0000 | -1.0000 | -1.0000 | -1.0000 | 1 | 0 | -10 | -20 | -15 | -4 |

| | | | | | | | | | | | |
|---|---|---|---|---|---|---|---|---|---|---|---|
| 23 | 2.3429 | 0.4707 | 0.0000 | -1.0000 | -1.8136 | 1 | 0 | -5 | -2 | 2 | 0 |
| | 5.0000 | 3.0000 | 1.0000 | 1.0000 | 0.0000 | 1 | -10 | 32 | -38 | 15 | 0 |
| | 5.3234 | 2.3579 | 1.0000 | 1.0000 | 0.3187 | 1 | -10 | 32 | -42 | 23 | -4 |
| | 3.3723 | 1.0000 | -1.0000 | -1.0000 | -2.3723 | 1 | 0 | -10 | -8 | 9 | 8 |
| | | | | | | | | | | | |
| 24 | 2.3028 | 0.6180 | 0.0000 | -1.3028 | -1.6180 | 1 | 0 | -5 | -2 | 3 | 0 |
| | 4.3028 | 3.6180 | 1.3820 | 0.6972 | 0.0000 | 1 | -10 | 33 | -40 | 15 | 0 |
| | 4.9354 | 2.6180 | 1.5374 | 0.5272 | 0.3820 | 1 | -10 | 33 | -44 | 23 | -4 |
| | 2.2361 | 2.2361 | 0.0000 | -2.2361 | -2.2361 | 1 | 0 | -10 | 0 | 25 | 0 |
| | | | | | | | | | | | |
| 25 | 2.2143 | 1.0000 | -0.5392 | -1.0000 | -1.6751 | 1 | 0 | -5 | -2 | 4 | 2 |
| | 4.1701 | 3.0000 | 2.3111 | 0.5188 | 0.0000 | 1 | -10 | 34 | -44 | 15 | 0 |
| | 4.6412 | 2.7237 | 1.4108 | 1.0000 | 0.2243 | 1 | -10 | 34 | -48 | 27 | -4 |
| | 2.3723 | 1.0000 | 1.0000 | -1.0000 | -3.3723 | 1 | 0 | -10 | 8 | 9 | -8 |
| | | | | | | | | | | | |
| 26 | 2.1358 | 0.6622 | 0.0000 | -0.6622 | -2.1358 | 1 | 0 | -5 | 0 | 2 | 0 |
| | 4.4812 | 2.6889 | 2.0000 | 0.8299 | 0.0000 | 1 | -10 | 34 | -46 | 20 | 0 |
| | 4.4812 | 2.6889 | 2.0000 | 0.8299 | 0.0000 | 1 | -10 | 34 | -46 | 20 | 0 |
| | 3.3723 | 1.0000 | -1.0000 | -1.0000 | -2.3723 | 1 | 0 | -10 | -8 | 9 | 8 |
| | | | | | | | | | | | |
| 27 | 2.0000 | 0.6180 | 0.6180 | -1.6180 | -1.6180 | 1 | 0 | -5 | 0 | 5 | -2 |
| | 3.6180 | 3.6180 | 1.3820 | 1.3820 | 0.0000 | 1 | -10 | 35 | -50 | 25 | 0 |
| | 4.0000 | 2.6180 | 2.6180 | 0.3820 | 0.3820 | 1 | -10 | 35 | -50 | 25 | -4 |
| | 2.2361 | 2.2361 | 0.0000 | -2.2361 | -2.2361 | 1 | 0 | -10 | 0 | 25 | 0 |
| | | | | | | | | | | | |
| 28 | 2.0000 | 0.0000 | 0.0000 | 0.0000 | -2.0000 | 1 | 0 | -4 | 0 | 0 | 0 |
| | 5.0000 | 1.0000 | 1.0000 | 1.0000 | 0.0000 | 1 | -8 | 18 | -16 | 5 | 0 |
| | 5.0000 | 1.0000 | 1.0000 | 1.0000 | 0.0000 | 1 | -8 | 18 | -16 | 5 | 0 |
| | 4.0000 | -1.0000 | -1.0000 | -1.0000 | -1.0000 | 1 | 0 | -10 | -20 | -15 | -4 |
| | | | | | | | | | | | |
| 29 | 1.8478 | 0.7654 | 0.0000 | -0.7654 | -1.8478 | 1 | 0 | -4 | 0 | 2 | 0 |
| | 4.1701 | 2.3111 | 1.0000 | 0.5188 | 0.0000 | 1 | -8 | 20 | -18 | 5 | 0 |
| | 4.1701 | 2.3111 | 1.0000 | 0.5188 | 0.0000 | 1 | -8 | 20 | -18 | 5 | 0 |
| | 3.0000 | 1.5616 | -1.0000 | -1.0000 | -2.5616 | 1 | 0 | -10 | -4 | 17 | 12 |
| | | | | | | | | | | | |
| 30 | 1.7321 | 1.0000 | 0.0000 | -1.0000 | -1.7321 | 1 | 0 | -4 | 0 | 3 | 0 |
| | 3.6180 | 2.6180 | 1.3820 | 0.3820 | 0.0000 | 1 | -8 | 21 | -20 | 5 | 0 |
| | 3.6180 | 2.6180 | 1.3820 | 0.3820 | 0.0000 | 1 | -8 | 21 | -20 | 5 | 0 |
| | 2.5616 | 1.0000 | 1.0000 | -1.5616 | -3.0000 | 1 | 0 | -10 | 4 | 17 | -12 |

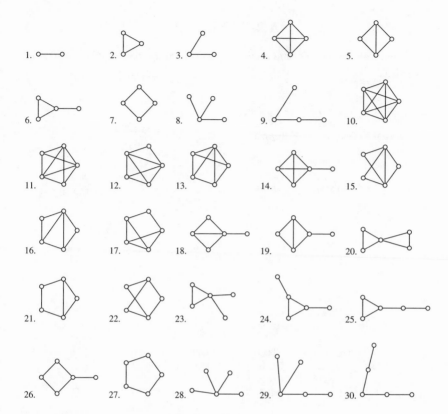

# A.2  TABLE A2

The eigenvalues, angles and main angles of connected graphs with 2 to 5 vertices

**1.**

| | | |
|---|---|---|
| | 1.0000 | −1.0000 |
| | 1.0000 | 0.0000 |
| 1,2. | 0.7071 | 0.7071 |

**2.**

| | | |
|---|---|---|
| | 2.0000 | $-1.0000^2$ |
| | 1.0000 | 0.0000 |
| 1,2,3. | 0.5774 | 0.8165 |

**3.**

| | | | |
|---|---|---|---|
| | 1.4142 | 0.0000 | −1.4142 |
| | 0.9856 | 0.0000 | 0.1691 |
| 1. | 0.7071 | 0.0000 | 0.7071 |
| 2,3. | 0.5000 | 0.7071 | 0.5000 |

**4.**

| | | |
|---|---|---|
| | 3.0000 | $-1.0000^3$ |
| | 1.0000 | 0.0000 |
| 1,2,3,4. | 0.5000 | 0.8660 |

**5.**

| | | | | |
|---|---|---|---|---|
| | 2.5616 | 0.0000 | −1.0000 | −1.5616 |
| | 0.9925 | 0.0000 | 0.0000 | 0.1222 |
| 1,2. | 0.5573 | 0.0000 | 0.7071 | 0.4352 |
| 3,4. | 0.4352 | 0.7071 | 0.0000 | 0.5573 |

**6.**

| | | | | |
|---|---|---|---|---|
| | 2.1701 | 0.3111 | −1.0000 | −1.4812 |
| | 0.9695 | 0.1663 | 0.0000 | 0.1803 |
| 1. | 0.6116 | 0.2536 | 0.0000 | 0.7494 |
| 2,3. | 0.5227 | 0.3682 | 0.7071 | 0.3020 |
| 4. | 0.2818 | 0.8152 | 0.0000 | 0.5059 |

**7.**

| | | | |
|---|---|---|---|
| | 2.0000 | $0.0000^2$ | −2.0000 |
| | 1.0000 | 0.0000 | 0.0000 |
| 1,2,3,4. | 0.5000 | 0.7071 | 0.5000 |

**8.**

| | | | |
|---|---|---|---|
| | 1.7321 | $0.0000^2$ | −1.7321 |
| | 0.9659 | 0.0000 | 0.2588 |
| 1. | 0.7071 | 0.0000 | 0.7071 |
| 2,3,4. | 0.4082 | 0.8165 | 0.4082 |

**9.**

| | | | | |
|---|---|---|---|---|
| | 1.6180 | 0.6180 | −0.6180 | −1.6180 |
| | 0.9732 | 0.0000 | 0.2298 | 0.0000 |
| 1,2. | 0.6015 | 0.3717 | 0.3717 | 0.6015 |
| 3,4. | 0.3717 | 0.6015 | 0.6015 | 0.3717 |

**10.**

|  | 4.0000 | $-1.0000^4$ |  |  |
|---|---|---|---|---|
|  | 1.0000 | 0.0000 |  |  |
| 1,2,3,4,5. | 0.4472 | 0.8944 |  |  |

**11.**

|  | 3.6458 | 0.0000 | $-1.0000^2$ | $-1.6458$ |
|---|---|---|---|---|
|  | 0.9957 | 0.0000 | 0.0000 | 0.0930 |
| 1,2,3. | 0.4792 | 0.0000 | 0.8165 | 0.3220 |
| 4,5. | 0.3943 | 0.7071 | 0.0000 | 0.5869 |

**12.**

|  | 3.3234 | 0.3579 | $-1.0000^2$ | $-1.6813$ |
|---|---|---|---|---|
|  | 0.9861 | 0.0837 | 0.0000 | 0.1432 |
| 1,2. | 0.5100 | 0.1378 | 0.7071 | 0.4700 |
| 3,4. | 0.4390 | 0.4294 | 0.7071 | 0.3505 |
| 5. | 0.3069 | 0.7702 | 0.0000 | 0.5590 |

**13.**

|  | 3.2361 | $0.0000^2$ | $-1.2361$ | $-2.0000$ |
|---|---|---|---|---|
|  | 0.9960 | 0.0000 | 0.0898 | 0.0000 |
| 1. | 0.5257 | 0.0000 | 0.8507 | 0.0000 |
| 2,3,4,5. | 0.4253 | 0.7071 | 0.2629 | 0.5000 |

**14.**

|  | 3.0861 | 0.4280 | $-1.0000^2$ | $-1.5141$ |
|---|---|---|---|---|
|  | 0.9567 | 0.2306 | 0.0000 | 0.1774 |
| 1. | 0.5236 | 0.3610 | 0.0000 | 0.7717 |
| 2,3,4. | 0.4820 | 0.2297 | 0.8165 | 0.2196 |
| 5. | 0.1697 | 0.8435 | 0.0000 | 0.5097 |

**15.**

|  | 3.0000 | $0.0000^2$ | $-1.0000$ | $-2.0000$ |
|---|---|---|---|---|
|  | 0.9798 | 0.0000 | 0.0000 | 0.2000 |
| 1,2. | 0.5477 | 0.0000 | 0.7071 | 0.4472 |
| 3,4,5. | 0.3651 | 0.8165 | 0.0000 | 0.4472 |

**16.**

|  | 2.9354 | 0.6180 | $-0.4626$ | $-1.4728$ | $-1.6180$ |
|---|---|---|---|---|---|
|  | 0.9839 | 0.0000 | 0.0738 | 0.1629 | 0.0000 |
| 1. | 0.5590 | 0.0000 | 0.3069 | 0.7702 | 0.0000 |
| 2,3. | 0.4700 | 0.3717 | 0.5100 | 0.1378 | 0.6015 |
| 4,5. | 0.3505 | 0.6015 | 0.4390 | 0.4294 | 0.3717 |

**17.**

|  | 2.8558 | 0.3216 | 0.0000 | $-1.0000$ | $-2.1774$ |
|---|---|---|---|---|---|
|  | 0.9898 | 0.1363 | 0.0000 | 0.0000 | 0.0416 |
| 1,2. | 0.4912 | 0.3870 | 0.0000 | 0.7071 | 0.3301 |
| 3,4. | 0.4558 | 0.1312 | 0.7071 | 0.0000 | 0.5244 |
| 5. | 0.3192 | 0.8161 | 0.0000 | 0.0000 | 0.4817 |

| | | | | | | |
|---|---|---|---|---|---|---|
| **18.** | | 2.6855 | 0.3349 | 0.0000 | −1.2713 | −1.7491 |
| | | 0.9602 | 0.1692 | 0.0000 | 0.0486 | 0.2170 |
| | 1. | 0.5825 | 0.2835 | 0.0000 | 0.4008 | 0.6478 |
| | 2. | 0.5237 | 0.3506 | 0.0000 | 0.7611 | 0.1534 |
| | 3,4. | 0.4119 | 0.2004 | 0.7071 | 0.2834 | 0.4581 |
| | 5. | 0.2169 | 0.8464 | 0.0000 | 0.3153 | 0.3704 |

| | | | | | | |
|---|---|---|---|---|---|---|
| **19.** | | 2.6412 | 0.7237 | −0.5892 | −1.0000 | −1.7757 |
| | | 0.9550 | 0.1833 | 0.2319 | 0.0000 | 0.0262 |
| | 1,2. | 0.5371 | 0.1655 | 0.1955 | 0.7071 | 0.3820 |
| | 3. | 0.4747 | 0.5030 | 0.3529 | 0.0000 | 0.6301 |
| | 4. | 0.4067 | 0.4573 | 0.6636 | 0.0000 | 0.4303 |
| | 5. | 0.1797 | 0.6950 | 0.5989 | 0.0000 | 0.3549 |

| | | | | | |
|---|---|---|---|---|---|
| **20.** | | 2.5616 | 1.0000 | −1.0000[2] | −1.5616 |
| | | 0.9802 | 0.0000 | 0.0000 | 0.1979 |
| | 1. | 0.6154 | 0.0000 | 0.0000 | 0.7882 |
| | 2,3,4,5. | 0.3941 | 0.5000 | 0.7071 | 0.3077 |

| | | | | | | |
|---|---|---|---|---|---|---|
| **21.** | | 2.4812 | 0.6889 | 0.0000 | −1.1701 | −2.0000 |
| | | 0.9850 | 0.1223 | 0.0000 | 0.1220 | 0.0000 |
| | 1,2. | 0.5299 | 0.1793 | 0.5000 | 0.4325 | 0.5000 |
| | 3. | 0.4271 | 0.5207 | 0.0000 | 0.7392 | 0.0000 |
| | 4,5. | 0.3578 | 0.5765 | 0.5000 | 0.1993 | 0.5000 |

| | | | | | |
|---|---|---|---|---|---|
| **22.** | | 2.4495 | 0.0000[3] | −2.4495 | |
| | | 0.9949 | 0.0000 | 0.1005 | |
| | 1,2. | 0.5000 | 0.7071 | 0.5000 | |
| | 3,4,5. | 0.4082 | 0.8165 | 0.4082 | |

| | | | | | | |
|---|---|---|---|---|---|---|
| **23.** | | 2.3429 | 0.4707 | 0.0000 | −1.0000 | −1.8136 |
| | | 0.9506 | 0.1587 | 0.0000 | 0.0000 | 0.2667 |
| | 1. | 0.6359 | 0.2414 | 0.0000 | 0.0000 | 0.7331 |
| | 2,3. | 0.4735 | 0.4560 | 0.0000 | 0.7071 | 0.2606 |
| | 4,5. | 0.2714 | 0.5128 | 0.7071 | 0.0000 | 0.4042 |

| | | | | | | |
|---|---|---|---|---|---|---|
| **24.** | | 2.3028 | 0.6180 | 0.0000 | −1.3028 | −1.6180 |
| | | 0.9444 | 0.0000 | 0.2582 | 0.2035 | 0.0000 |
| | 1,2. | 0.5651 | 0.3717 | 0.0000 | 0.4250 | 0.6015 |
| | 3. | 0.4908 | 0.0000 | 0.5774 | 0.6525 | 0.0000 |
| | 4,5. | 0.2454 | 0.6015 | 0.5774 | 0.3263 | 0.3717 |

| | | | | | | |
|---|---|---|---|---|---|---|
| **25.** | | 2.2143 | 1.0000 | −0.5392 | −1.0000 | −1.6751 |
| | | 0.9370 | 0.2828 | 0.2021 | 0.0000 | 0.0347 |
| | 1. | 0.6037 | 0.0000 | 0.4762 | 0.0000 | 0.6394 |
| | 2,3. | 0.4972 | 0.3162 | 0.3094 | 0.7071 | 0.2390 |
| | 4. | 0.3425 | 0.6325 | 0.3620 | 0.0000 | 0.5930 |
| | 5. | 0.1547 | 0.6325 | 0.6714 | 0.0000 | 0.3540 |

| | | | | | | |
|---|---|---|---|---|---|---|
| **26.** | | 2.1358 | 0.6622 | 0.0000 | −0.6622 | −2.1358 |
| | | 0.9762 | 0.0742 | 0.0000 | 0.1835 | 0.0885 |
| | 1. | 0.5573 | 0.4352 | 0.0000 | 0.4352 | 0.5573 |
| | 2,3. | 0.4647 | 0.1845 | 0.7071 | 0.1845 | 0.4647 |
| | 4. | 0.4352 | 0.5573 | 0.0000 | 0.5573 | 0.4352 |
| | 5. | 0.2610 | 0.6572 | 0.0000 | 0.6572 | 0.2610 |

**27.**

|  | 2.0000 | 0.6180² | −1.6180² |  |  |
|---|---|---|---|---|---|
|  | 1.0000 | 0.0000 | 0.0000 |  |  |
| 1,2,3,4,5. | 0.4472 | 0.6325 | 0.6325 |  |  |

**28.**

|  | 2.0000 | 0.0000³ | −2.0000 |  |  |
|---|---|---|---|---|---|
|  | 0.9487 | 0.0000 | 0.3162 |  |  |
| 1. | 0.7071 | 0.0000 | 0.7071 |  |  |
| 2,3,4,5. | 0.3536 | 0.8660 | 0.3536 |  |  |

**29.**

|  | 1.8478 | 0.7654 | 0.0000 | −0.7654 | −1.8478 |
|---|---|---|---|---|---|
|  | 0.9530 | 0.0785 | 0.0000 | 0.2638 | 0.1267 |
| 1. | 0.6533 | 0.2706 | 0.0000 | 0.2706 | 0.6533 |
| 2. | 0.5000 | 0.5000 | 0.0000 | 0.5000 | 0.5000 |
| 3,4. | 0.3536 | 0.3536 | 0.7071 | 0.3536 | 0.3536 |
| 5. | 0.2706 | 0.6533 | 0.0000 | 0.6533 | 0.2706 |

**30.**

|  | 1.7321 | 1.0000 | 0.0000 | −1.0000 | −1.7321 |
|---|---|---|---|---|---|
|  | 0.9636 | 0.0000 | 0.2582 | 0.0000 | 0.0692 |
| 1. | 0.5774 | 0.0000 | 0.5774 | 0.0000 | 0.5774 |
| 2,3. | 0.5000 | 0.5000 | 0.0000 | 0.5000 | 0.5000 |
| 4,5. | 0.2887 | 0.5000 | 0.5774 | 0.5000 | 0.2887 |

## A.3  TABLE A3

The spectra and characteristic polynomials of the adjacency matrix for connected graphs with 6 vertices

```
001    5.0000  -1.0000  -1.0000  -1.0000  -1.0000  -1.0000
          1        0      -15      -40      -45      -24      -5

002    4.7016   0.0000  -1.0000  -1.0000  -1.0000  -1.7016
          1        0      -14      -32      -27       -8       0

003    4.4279   0.3757  -1.0000  -1.0000  -1.0000  -1.8035
          1        0      -13      -26      -15        2       3

004    4.3723   0.0000   0.0000  -1.0000  -1.3723  -2.0000
          1        0      -13      -24      -12        0       0

005    4.2015   0.5451  -1.0000  -1.0000  -1.0000  -1.7466
          1        0      -12      -22       -9        6       4

006    4.1623   0.0000   0.0000  -1.0000  -1.0000  -2.1623
          1        0      -12      -20       -9        0       0

007    4.1190   0.6180  -0.4316  -1.0000  -1.6180  -1.6874
          1        0      -12      -20       -4        8       3

008    4.0678   0.3616   0.0000  -1.0000  -1.2446  -2.1848
          1        0      -12      -18       -3        4       0

009    4.0000   0.0000   0.0000   0.0000  -2.0000  -2.0000
          1        0      -12      -16        0        0       0

010    4.0514   0.4827  -1.0000  -1.0000  -1.0000  -1.5341
          1        0      -11      -20       -9        4       3

011    3.8951   0.3973   0.0000  -1.0000  -1.2924  -2.0000
          1        0      -11      -16       -2        4       0

012    3.8590   0.7792  -0.3791  -1.0000  -1.4758  -1.7832
          1        0      -11      -16        1       10       3

013    3.8284   1.0000  -1.0000  -1.0000  -1.0000  -1.8284
          1        0      -11      -16        3       16       7

014    3.8201   0.4594   0.0000  -1.0000  -1.0000  -2.2795
          1        0      -11      -14        0        4       0

015    3.7785   0.7108   0.0000  -1.0000  -1.4893  -2.0000
          1        0      -11      -14        4        8       0
```

| | | | | | | | |
|---|---|---|---|---|---|---|---|
| 016 | 3.7664 | 0.0000 | 0.0000 | 0.0000 | -1.2828 | -2.4836 | |
| | 1 | 0 | -11 | -12 | 0 | 0 | 0 |
| 017 | 3.7321 | 0.4142 | 0.2679 | -1.0000 | -1.0000 | -2.4142 | |
| | 1 | 0 | -11 | -12 | 3 | 4 | -1 |
| 018 | 3.7136 | 0.6180 | 0.0000 | -0.4829 | -1.6180 | -2.2307 | |
| | 1 | 0 | -11 | -12 | 5 | 4 | 0 |
| 019 | 3.7105 | 0.4408 | 0.0000 | -1.0000 | -1.3842 | -1.7670 | |
| | 1 | 0 | -10 | -14 | -1 | 4 | 0 |
| 020 | 3.6903 | 0.7534 | -0.5784 | -1.0000 | -1.0000 | -1.8653 | |
| | 1 | 0 | -10 | -14 | 0 | 8 | 3 |
| 021 | 3.6262 | 0.5151 | 0.0000 | -1.0000 | -1.0000 | -2.1413 | |
| | 1 | 0 | -10 | -12 | 1 | 4 | 0 |
| 022 | 3.5926 | 0.6180 | 0.1589 | -1.0000 | -1.6180 | -1.7515 | |
| | 1 | 0 | -10 | -12 | 4 | 6 | -1 |
| 023 | 3.5616 | 1.0000 | -0.5616 | -1.0000 | -1.0000 | -2.0000 | |
| | 1 | 0 | -10 | -12 | 5 | 12 | 4 |
| 024 | 3.5344 | 1.0827 | -0.4071 | -1.0000 | -1.5111 | -1.6990 | |
| | 1 | 0 | -10 | -12 | 7 | 14 | 4 |
| 025 | 3.5141 | 0.6694 | 0.0000 | -0.5284 | -1.4782 | -2.1769 | |
| | 1 | 0 | -10 | -10 | 5 | 4 | 0 |
| 026 | 3.4979 | 0.7299 | 0.1505 | -1.0000 | -1.1876 | -2.1907 | |
| | 1 | 0 | -10 | -10 | 6 | 6 | -1 |
| 027 | 3.4679 | 0.9128 | 0.0000 | -0.7989 | -1.5818 | -2.0000 | |
| | 1 | 0 | -10 | -10 | 8 | 8 | 0 |
| 028 | 3.4495 | 0.6180 | 0.6180 | -1.4495 | -1.6180 | -1.6180 | |
| | 1 | 0 | -10 | -10 | 10 | 8 | -5 |
| 029 | 3.4609 | 0.3493 | 0.0000 | 0.0000 | -1.3387 | -2.4715 | |
| | 1 | 0 | -10 | -8 | 4 | 0 | 0 |
| 030 | 3.3885 | 0.8019 | 0.1873 | -0.5550 | -1.5758 | -2.2470 | |
| | 1 | 0 | -10 | -8 | 9 | 4 | -1 |
| 031 | 3.3723 | 1.0000 | 0.0000 | -1.0000 | -1.0000 | -2.3723 | |
| | 1 | 0 | -10 | -8 | 9 | 8 | 0 |
| 032 | 3.3923 | 0.3254 | 0.0000 | 0.0000 | -1.0000 | -2.7177 | |
| | 1 | 0 | -10 | -6 | 3 | 0 | 0 |
| 033 | 3.4037 | 0.4897 | 0.2512 | -1.0000 | -1.2827 | -1.8619 | |
| | 1 | 0 | -9 | -10 | 3 | 4 | -1 |

```
034   3.3839  0.7424   0.0000  -1.0000  -1.3279  -1.7985
        1       0       -9      -10       4        6        0

035   3.3539  1.0000  -0.4765  -1.0000  -1.0000  -1.8774
        1       0       -9      -10       5       10        3

036   3.2618  1.3399  -1.0000  -1.0000  -1.0000  -1.6017
        1       0       -9      -10       9       18        7

037   3.3723  0.0000   0.0000   0.0000  -1.0000  -2.3723
        1       0       -9       -8       0        0        0

038   3.3234  0.3579   0.0000   0.0000  -1.6813  -2.0000
        1       0       -9       -8       4        0        0

039   3.2948  0.7347   0.0000  -0.5975  -1.2927  -2.1392
        1       0       -9       -8       5        4        0

040   3.2814  0.7719   0.0000  -0.5125  -1.5408  -2.0000
        1       0       -9       -8       6        4        0

041   3.2361  1.0000   0.0000  -1.0000  -1.2361  -2.0000
        1       0       -9       -8       8        8        0

042   3.2361  0.6180   0.6180  -1.2361  -1.6180  -1.6180
        1       0       -9       -8       9        6       -4

043   3.2227  1.0000   0.1124  -1.0000  -1.5266  -1.8085
        1       0       -9       -8       9        8       -1

044   3.1819  1.2470  -0.4450  -0.5936  -1.5884  -1.8019
        1       0       -9       -8      10       12        3

045   3.1888  0.8347   0.0000  -0.6272  -1.0000  -2.3962
        1       0       -9       -6       6        4        0

046   3.1692  0.7282   0.2798  -0.4663  -1.5058  -2.2052
        1       0       -9       -6       8        2       -1

047   3.1149  0.7459   0.6180  -0.8608  -1.6180  -2.0000
        1       0       -9       -6      11        4       -4

048   3.0868  1.1558   0.1096  -1.0000  -1.1736  -2.1787
        1       0       -9       -6      11        8       -1

049   3.1413  0.4849   0.0000   0.0000  -1.0000  -2.6262
        1       0       -9       -4       4        0        0

050   3.0922  0.7020   0.0000   0.0000  -1.2855  -2.5086
        1       0       -9       -4       7        0        0
```

| 051 | 3.0000 | 1.0000 | 0.0000 | 0.0000 | -2.0000 | -2.0000 | |
| | 1 | 0 | -9 | -4 | 12 | 0 | 0 |

| 052 | 3.0000 | 0.0000 | 0.0000 | 0.0000 | 0.0000 | -3.0000 | |
| | 1 | 0 | -9 | 0 | 0 | 0 | 0 |

| 053 | 3.1774 | 0.6784 | 0.0000 | -1.0000 | -1.0000 | -1.8558 | |
| | 1 | 0 | -8 | -8 | 3 | 4 | 0 |

| 054 | 3.1642 | 0.6180 | 0.2271 | -1.0000 | -1.3914 | -1.6180 | |
| | 1 | 0 | -8 | -8 | 4 | 4 | -1 |

| 055 | 3.0965 | 1.1169 | -0.5089 | -1.0000 | -1.0000 | -1.7045 | |
| | 1 | 0 | -8 | -8 | 6 | 10 | 3 |

| 056 | 3.1020 | 0.3443 | 0.0000 | 0.0000 | -1.3228 | -2.1235 | |
| | 1 | 0 | -8 | -6 | 3 | 0 | 0 |

| 057 | 3.0478 | 0.8214 | 0.0000 | -0.7562 | -1.0000 | -2.1129 | |
| | 1 | 0 | -8 | -6 | 5 | 4 | 0 |

| 058 | 3.0437 | 0.6180 | 0.3285 | -0.5482 | -1.6180 | -1.8241 | |
| | 1 | 0 | -8 | -6 | 6 | 2 | -1 |

| 059 | 3.0143 | 0.8481 | 0.1967 | -0.7248 | -1.4780 | -1.8563 | |
| | 1 | 0 | -8 | -6 | 7 | 4 | -1 |

| 060 | 2.9809 | 1.0420 | 0.0000 | -0.7062 | -1.5371 | -1.7796 | |
| | 1 | 0 | -8 | -6 | 8 | 6 | 0 |

| 061 | 2.9474 | 1.1593 | 0.0000 | -1.0000 | -1.2859 | -1.8208 | |
| | 1 | 0 | -8 | -6 | 9 | 8 | 0 |

| 062 | 2.8422 | 1.5069 | -0.5069 | -1.0000 | -1.0000 | -1.8422 | |
| | 1 | 0 | -8 | -6 | 11 | 14 | 4 |

| 063 | 2.9439 | 0.6648 | 0.0000 | 0.0000 | -1.3684 | -2.2403 | |
| | 1 | 0 | -8 | -4 | 6 | 0 | 0 |

| 064 | 2.9327 | 0.7272 | 0.3088 | -0.6570 | -1.0000 | -2.3117 | |
| | 1 | 0 | -8 | -4 | 6 | 2 | -1 |

| 065 | 2.8951 | 1.0000 | 0.0000 | -0.6027 | -1.0000 | -2.2924 | |
| | 1 | 0 | -8 | -4 | 7 | 4 | 0 |

| 066 | 2.9032 | 0.8061 | 0.0000 | 0.0000 | -1.7093 | -2.0000 | |
| | 1 | 0 | -8 | -4 | 8 | 0 | 0 |

| 067 | 2.8529 | 1.0554 | 0.1830 | -0.6611 | -1.2718 | -2.1584 | |
| | 1 | 0 | -8 | -4 | 9 | 4 | -1 |

| | | | | | | | |
|---|---|---|---|---|---|---|---|
| 068 | 2.8136 | 1.0000 | 0.5293 | -1.0000 | -1.3429 | -2.0000 | |
| | 1 | 0 | -8 | -4 | 11 | 4 | -4 |
| 069 | 2.7913 | 1.0000 | 0.6180 | -1.0000 | -1.6180 | -1.7913 | |
| | 1 | 0 | -8 | -4 | 12 | 4 | -5 |
| 070 | 2.7321 | 1.4142 | 0.0000 | -0.7321 | -1.4142 | -2.0000 | |
| | 1 | 0 | -8 | -4 | 12 | 8 | 0 |
| 071 | 2.7964 | 0.8532 | 0.0000 | 0.0000 | -1.1955 | -2.4541 | |
| | 1 | 0 | -8 | -2 | 7 | 0 | 0 |
| 072 | 2.7411 | 0.7103 | 0.6180 | -0.2314 | -1.6180 | -2.2200 | |
| | 1 | 0 | -8 | -2 | 10 | -2 | -1 |
| 073 | 2.8284 | 0.0000 | 0.0000 | 0.0000 | 0.0000 | -2.8284 | |
| | 1 | 0 | -8 | 0 | 0 | 0 | 0 |
| 074 | 2.7321 | 0.7321 | 0.0000 | 0.0000 | -0.7321 | -2.7321 | |
| | 1 | 0 | -8 | 0 | 4 | 0 | 0 |
| 075 | 2.8136 | 0.5293 | 0.0000 | 0.0000 | -1.3429 | -2.0000 | |
| | 1 | 0 | -7 | -4 | 4 | 0 | 0 |
| 076 | 2.7913 | 0.6180 | 0.0000 | 0.0000 | -1.6180 | -1.7913 | |
| | 1 | 0 | -7 | -4 | 5 | 0 | 0 |
| 077 | 2.7537 | 0.7727 | 0.3064 | -0.6093 | -1.3293 | -1.8942 | |
| | 1 | 0 | -7 | -4 | 6 | 2 | -1 |
| 078 | 2.7321 | 1.0000 | 0.0000 | -0.7321 | -1.0000 | -2.0000 | |
| | 1 | 0 | -7 | -4 | 6 | 4 | 0 |
| 079 | 2.7093 | 1.0000 | 0.1939 | -1.0000 | -1.0000 | -1.9032 | |
| | 1 | 0 | -7 | -4 | 7 | 4 | -1 |
| 080 | 2.7093 | 1.0000 | 0.1939 | -1.0000 | -1.0000 | -1.9032 | |
| | 1 | 0 | -7 | -4 | 7 | 4 | -1 |
| 081 | 2.7056 | 1.0561 | 0.0000 | -0.5600 | -1.3504 | -1.8513 | |
| | 1 | 0 | -7 | -4 | 7 | 4 | 0 |
| 082 | 2.6554 | 1.2108 | 0.0000 | -1.0000 | -1.0000 | -1.8662 | |
| | 1 | 0 | -7 | -4 | 8 | 6 | 0 |
| 083 | 2.6287 | 1.2297 | 0.1397 | -1.0000 | -1.3198 | -1.6783 | |
| | 1 | 0 | -7 | -4 | 9 | 6 | -1 |
| 084 | 2.4142 | 1.7321 | -0.4142 | -1.0000 | -1.0000 | -1.7321 | |
| | 1 | 0 | -7 | -4 | 11 | 12 | 3 |

| 085 | 2.5991 | 0.7661 | 0.4669 | -0.3848 | -1.3053 | -2.1420 | |
| | 1 | 0 | -7 | -2 | 7 | 0 | -1 |
| 086 | 2.5616 | 1.0000 | 0.0000 | 0.0000 | -1.5616 | -2.0000 | |
| | 1 | 0 | -7 | -2 | 8 | 0 | 0 |
| 087 | 2.5395 | 1.0825 | 0.2611 | -0.5406 | -1.2061 | -2.1364 | |
| | 1 | 0 | -7 | -2 | 8 | 2 | -1 |
| 088 | 2.5035 | 1.2644 | 0.0000 | -0.5767 | -1.0000 | -2.1912 | |
| | 1 | 0 | -7 | -2 | 8 | 4 | 0 |
| 089 | 2.4383 | 1.1386 | 0.6180 | -0.8202 | -1.6180 | -1.7566 | |
| | 1 | 0 | -7 | -2 | 11 | 2 | -4 |
| 090 | 2.5576 | 0.6772 | 0.0000 | 0.0000 | -0.6772 | -2.5576 | |
| | 1 | 0 | -7 | 0 | 3 | 0 | 0 |
| 091 | 2.5243 | 0.7923 | 0.0000 | 0.0000 | -0.7923 | -2.5243 | |
| | 1 | 0 | -7 | 0 | 4 | 0 | 0 |
| 092 | 2.4142 | 1.0000 | 0.4142 | -0.4142 | -1.0000 | -2.4142 | |
| | 1 | 0 | -7 | 0 | 7 | 0 | -1 |
| 093 | 2.3914 | 0.7729 | 0.6180 | 0.0000 | -1.6180 | -2.1642 | |
| | 1 | 0 | -7 | 0 | 9 | -4 | 0 |
| 094 | 2.5141 | 0.5720 | 0.0000 | 0.0000 | -1.0000 | -2.0861 | |
| | 1 | 0 | -6 | -2 | 3 | 0 | 0 |
| 095 | 2.4458 | 0.7968 | 0.0000 | 0.0000 | -1.3703 | -1.8723 | |
| | 1 | 0 | -6 | -2 | 5 | 0 | 0 |
| 096 | 2.4142 | 0.6180 | 0.6180 | -0.4142 | -1.6180 | -1.6180 | |
| | 1 | 0 | -6 | -2 | 6 | 0 | -1 |
| 097 | 2.3799 | 1.0000 | 0.2914 | -0.7510 | -1.0000 | -1.9202 | |
| | 1 | 0 | -6 | -2 | 6 | 2 | -1 |
| 098 | 2.3342 | 1.0996 | 0.2742 | -0.5945 | -1.3738 | -1.7397 | |
| | 1 | 0 | -6 | -2 | 7 | 2 | -1 |
| 099 | 2.2784 | 1.3174 | 0.0000 | -0.7046 | -1.0000 | -1.8912 | |
| | 1 | 0 | -6 | -2 | 7 | 4 | 0 |
| 100 | 2.2283 | 1.3604 | 0.1859 | -1.0000 | -1.0000 | -1.7746 | |
| | 1 | 0 | -6 | -2 | 8 | 4 | -1 |
| 101 | 2.2882 | 0.8740 | 0.0000 | 0.0000 | -0.8740 | -2.2882 | |
| | 1 | 0 | -6 | 0 | 4 | 0 | 0 |

| 102 | 2.2470 | 0.8019 | 0.5550 | -0.5550 | -0.8019 | -2.2470 |     |
|     | 1      | 0      | -6     | 0       | 5       | 0       | -1  |
| 103 | 2.2361 | 1.0000 | 0.0000 | 0.0000  | -1.0000 | -2.2361 |     |
|     | 1      | 0      | -6     | 0       | 5       | 0       | 0   |
| 104 | 2.1753 | 1.1260 | 0.0000 | 0.0000  | -1.1260 | -2.1753 |     |
|     | 1      | 0      | -6     | 0       | 6       | 0       | 0   |
| 105 | 2.1149 | 1.0000 | 0.6180 | -0.2541 | -1.6180 | -1.8608 |     |
|     | 1      | 0      | -6     | 0       | 8       | -2      | -1  |
| 106 | 2.0000 | 1.0000 | 1.0000 | -1.0000 | -1.0000 | -2.0000 |     |
|     | 1      | 0      | -6     | 0       | 9       | 0       | -4  |
| 107 | 2.2361 | 0.0000 | 0.0000 | 0.0000  | 0.0000  | -2.2361 |     |
|     | 1      | 0      | -5     | 0       | 0       | 0       | 0   |
| 108 | 2.0743 | 0.8350 | 0.0000 | 0.0000  | -0.8350 | -2.0743 |     |
|     | 1      | 0      | -5     | 0       | 3       | 0       | 0   |
| 109 | 2.0000 | 1.0000 | 0.0000 | 0.0000  | -1.0000 | -2.0000 |     |
|     | 1      | 0      | -5     | 0       | 4       | 0       | 0   |
| 110 | 1.9319 | 1.0000 | 0.5176 | -0.5176 | -1.0000 | -1.9319 |     |
|     | 1      | 0      | -5     | 0       | 5       | 0       | -1  |
| 111 | 1.9021 | 1.1756 | 0.0000 | 0.0000  | -1.1756 | -1.9021 |     |
|     | 1      | 0      | -5     | 0       | 5       | 0       | 0   |
| 112 | 1.8019 | 1.2470 | 0.4450 | -0.4450 | -1.2470 | -1.8019 |     |
|     | 1      | 0      | -5     | 0       | 6       | 0       | -1  |

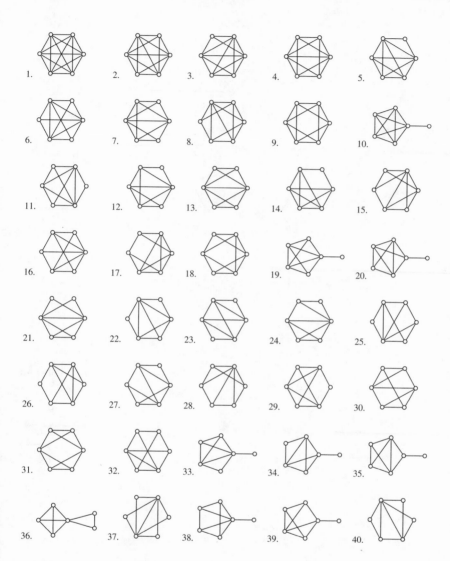

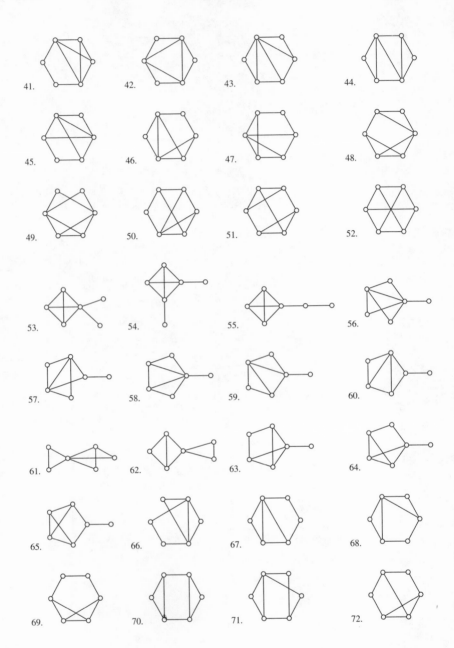

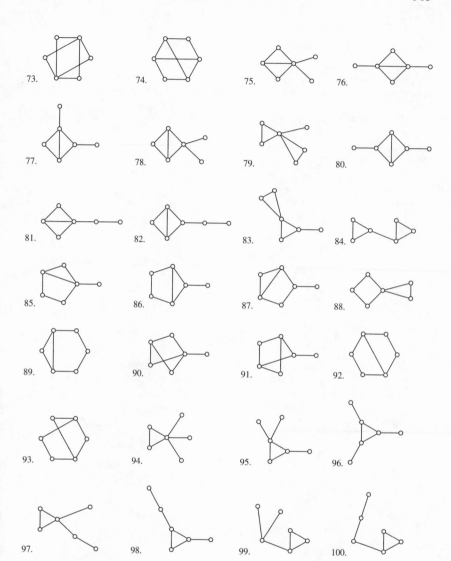

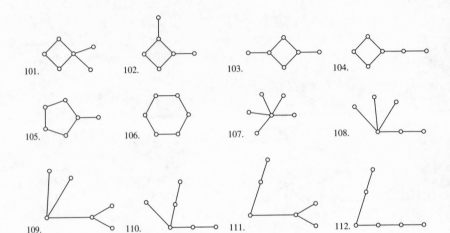

101.   102.   103.   104.

105.   106.   107.   108.

109.   110.   111.   112.

# A4.  TABLE A4

The spectra and characteristic polynomials of the adjacency matrix for trees with at most 9 vertices

```
01    1.0000 -1.0000
         1        0       -1

02    1.4142  0.0000 -1.4142
         1        0       -2        0

03    1.7321  0.0000  0.0000 -1.7321
         1        0       -3        0        0

04    1.6180  0.6180 -0.6180 -1.6180
         1        0       -3        0        1

05    2.0000  0.0000  0.0000  0.0000 -2.0000
         1        0       -4        0        0        0

06    1.8478  0.7654  0.0000 -0.7654 -1.8478
         1        0       -4        0        2        0

07    1.7321  1.0000  0.0000 -1.0000 -1.7321
         1        0       -4        0        3        0

08    2.2361  0.0000  0.0000  0.0000  0.0000 -2.2361
         1        0       -5        0        0        0        0

09    2.0743  0.8350  0.0000  0.0000 -0.8350 -2.0743
         1        0       -5        0        3        0        0

10    2.0000  1.0000  0.0000  0.0000 -1.0000 -2.0000
         1        0       -5        0        4        0        0

11    1.9319  1.0000  0.5176 -0.5176 -1.0000 -1.9319
         1        0       -5        0        5        0       -1

12    1.9021  1.1756  0.0000  0.0000 -1.1756 -1.9021
         1        0       -5        0        5        0        0

13    1.8019  1.2470  0.4450 -0.4450 -1.2470 -1.8019
         1        0       -5        0        6        0       -1

14    2.4495  0.0000  0.0000  0.0000  0.0000  0.0000 -2.4495
         1        0       -6        0        0        0        0        0

15    2.2882  0.8740  0.0000  0.0000  0.0000 -0.8740 -2.2882
         1        0       -6        0        4        0        0        0

16    2.1753  1.1260  0.0000  0.0000  0.0000 -1.1260 -2.1753
         1        0       -6        0        6        0        0        0

17    2.1358  1.0000  0.6622  0.0000 -0.6622 -1.0000 -2.1358
         1        0       -6        0        7        0       -2        0

18    2.1010  1.2593  0.0000  0.0000  0.0000 -1.2593 -2.1010
         1        0       -6        0        7        0        0        0

19    2.0529  1.2086  0.5700  0.0000 -0.5700 -1.2086 -2.0529
         1        0       -6        0        8        0       -2        0
```

```
20   2.0000  1.4142  0.0000  0.0000  0.0000 -1.4142 -2.0000
        1       0      -6       0       8       0       0       0

21   2.0000  1.0000  1.0000  0.0000 -1.0000 -1.0000 -2.0000
        1       0      -6       0       9       0      -4       0

22   1.9696  1.2856  0.6840  0.0000 -0.6840 -1.2856 -1.9696
        1       0      -6       0       9       0      -3       0

23   1.9319  1.4142  0.5176  0.0000 -0.5176 -1.4142 -1.9319
        1       0      -6       0       9       0      -2       0

24   1.8478  1.4142  0.7654  0.0000 -0.7654 -1.4142 -1.8478
        1       0      -6       0      10       0      -4       0

25   2.6458  0.0000  0.0000  0.0000  0.0000  0.0000  0.0000 -2.6458
        1       0      -7       0       0       0       0       0       0

26   2.4885  0.8986  0.0000  0.0000  0.0000  0.0000 -0.8986 -2.4885
        1       0      -7       0       5       0       0       0       0

27   2.3583  1.1994  0.0000  0.0000  0.0000  0.0000 -1.1994 -2.3583
        1       0      -7       0       8       0       0       0       0

28   2.3344  1.0000  0.7420  0.0000  0.0000 -0.7420 -1.0000 -2.3344
        1       0      -7       0       9       0      -3       0       0

29   2.3028  1.3028  0.0000  0.0000  0.0000  0.0000 -1.3028 -2.3028
        1       0      -7       0       9       0       0       0       0

30   2.3028  1.3028  0.0000  0.0000  0.0000  0.0000 -1.3028 -2.3028
        1       0      -7       0       9       0       0       0       0

31   2.2216  1.2399  0.7261  0.0000  0.0000 -0.7261 -1.2399 -2.2216
        1       0      -7       0      11       0      -4       0       0

32   2.2059  1.3376  0.5870  0.0000  0.0000 -0.5870 -1.3376 -2.2059
        1       0      -7       0      11       0      -3       0       0

33   2.1490  1.5434  0.0000  0.0000  0.0000  0.0000 -1.5434 -2.1490
        1       0      -7       0      11       0       0       0       0

34   2.1889  1.0000  1.0000  0.4569 -0.4569 -1.0000 -1.0000 -2.1889
        1       0      -7       0      12       0      -7       0       1

35   2.1566  1.3138  0.7892  0.0000  0.0000 -0.7892 -1.3138 -2.1566
        1       0      -7       0      12       0      -5       0       0

36   2.1358  1.4142  0.6622  0.0000  0.0000 -0.6622 -1.4142 -2.1358
        1       0      -7       0      12       0      -4       0       0

37   2.1120  1.4964  0.5481  0.0000  0.0000 -0.5481 -1.4964 -2.1120
        1       0      -7       0      12       0      -3       0       0

38   2.1010  1.2593  1.0000  0.0000  0.0000 -1.0000 -1.2593 -2.1010
        1       0      -7       0      13       0      -7       0       0

39   2.0953  1.3557  0.7376  0.4773 -0.4773 -0.7376 -1.3557 -2.0953
        1       0      -7       0      13       0      -7       0       1

40   2.0743  1.4142  0.8350  0.0000  0.0000 -0.8350 -1.4142 -2.0743
        1       0      -7       0      13       0      -6       0       0
```

```
41   2.0421  1.5202  0.7203  0.0000  0.0000 -0.7203 -1.5202 -2.0421
        1       0      -7       0      13       0      -5       0       0

42   2.0000  1.6180  0.6180  0.0000  0.0000 -0.6180 -1.6180 -2.0000
        1       0      -7       0      13       0      -4       0       0

43   2.0285  1.3213  1.0000  0.3731 -0.3731 -1.0000 -1.3213 -2.0285
        1       0      -7       0      14       0      -9       0       1

44   2.0000  1.4142  1.0000  0.0000  0.0000 -1.0000 -1.4142 -2.0000
        1       0      -7       0      14       0      -8       0       0

45   1.9890  1.4863  0.8135  0.4158 -0.4158 -0.8135 -1.4863 -1.9890
        1       0      -7       0      14       0      -8       0       1

46   1.9499  1.5637  0.8678  0.0000  0.0000 -0.8678 -1.5637 -1.9499
        1       0      -7       0      14       0      -7       0       0

47   1.8794  1.5321  1.0000  0.3473 -0.3473 -1.0000 -1.5321 -1.8794
        1       0      -7       0      15       0     -10       0       1

48   2.8284  0.0000  0.0000  0.0000  0.0000  0.0000  0.0000  0.0000 -2.8284
        1       0      -8       0       0       0       0       0       0       0

49   2.6762  0.9153  0.0000  0.0000  0.0000  0.0000  0.0000 -0.9153 -2.6762
        1       0      -8       0       6       0       0       0       0       0

50   2.5396  1.2452  0.0000  0.0000  0.0000  0.0000  0.0000 -1.2452 -2.5396
        1       0      -8       0      10       0       0       0       0       0

51   2.5243  1.0000  0.7923  0.0000  0.0000  0.0000 -0.7923 -1.0000 -2.5243
        1       0      -8       0      11       0      -4       0       0       0

52   2.4972  1.3281  0.0000  0.0000  0.0000  0.0000  0.0000 -1.3281 -2.4972
        1       0      -8       0      11       0       0       0       0       0

53   2.4495  1.4142  0.0000  0.0000  0.0000  0.0000  0.0000 -1.4142 -2.4495
        1       0      -8       0      12       0       0       0       0       0

54   2.3968  1.2665  0.8069  0.0000  0.0000  0.0000 -0.8069 -1.2665 -2.3968
        1       0      -8       0      14       0      -6       0       0       0

55   2.3761  1.4142  0.5952  0.0000  0.0000  0.0000 -0.5952 -1.4142 -2.3761
        1       0      -8       0      14       0      -4       0       0       0

56   2.3268  1.6080  0.0000  0.0000  0.0000  0.0000  0.0000 -1.6080 -2.3268
        1       0      -8       0      14       0       0       0       0       0

57   2.3761  1.0000  1.0000  0.5952  0.0000 -0.5952 -1.0000 -1.0000 -2.3761
        1       0      -8       0      15       0     -10       0       2       0

58   2.3467  1.3335  0.8455  0.0000  0.0000  0.0000 -0.8455 -1.3335 -2.3467
        1       0      -8       0      15       0      -7       0       0       0

59   2.3344  1.4142  0.7420  0.0000  0.0000  0.0000 -0.7420 -1.4142 -2.3344
        1       0      -8       0      15       0      -6       0       0       0

60   2.3073  1.5356  0.5645  0.0000  0.0000  0.0000 -0.5645 -1.5356 -2.3073
        1       0      -8       0      15       0      -4       0       0       0

61   2.2361  1.7321  0.0000  0.0000  0.0000  0.0000  0.0000 -1.7321 -2.2361
        1       0      -8       0      15       0       0       0       0       0
```

| # | | | | | | | | | | |
|---|---|---|---|---|---|---|---|---|---|---|
| 62 | 2.2882 | 1.4142 | 0.8740 | 0.0000 | 0.0000 | 0.0000 | -0.8740 | -1.4142 | -2.2882 | |
| | 1 | 0 | -8 | 0 | 16 | 0 | -8 | 0 | 0 | 0 |
| 63 | 2.2552 | 1.5582 | 0.6970 | 0.0000 | 0.0000 | 0.0000 | -0.6970 | -1.5582 | -2.2552 | |
| | 1 | 0 | -8 | 0 | 16 | 0 | -6 | 0 | 0 | 0 |
| 64 | 2.2638 | 1.2793 | 1.0000 | 0.4883 | 0.0000 | -0.4883 | -1.0000 | -1.2793 | -2.2638 | |
| | 1 | 0 | -8 | 0 | 17 | 0 | -12 | 0 | 2 | 0 |
| 65 | 2.2470 | 1.4142 | 0.8019 | 0.5550 | 0.0000 | -0.5550 | -0.8019 | -1.4142 | -2.2470 | |
| | 1 | 0 | -8 | 0 | 17 | 0 | -11 | 0 | 2 | 0 |
| 66 | 2.2361 | 1.4142 | 1.0000 | 0.0000 | 0.0000 | 0.0000 | -1.0000 | -1.4142 | -2.2361 | |
| | 1 | 0 | -8 | 0 | 17 | 0 | -10 | 0 | 0 | 0 |
| 67 | 2.2361 | 1.4142 | 1.0000 | 0.0000 | 0.0000 | 0.0000 | -1.0000 | -1.4142 | -2.2361 | |
| | 1 | 0 | -8 | 0 | 17 | 0 | -10 | 0 | 0 | 0 |
| 68 | 2.2164 | 1.5121 | 0.8952 | 0.0000 | 0.0000 | 0.0000 | -0.8952 | -1.5121 | -2.2164 | |
| | 1 | 0 | -8 | 0 | 17 | 0 | -9 | 0 | 0 | 0 |
| 69 | 2.1940 | 1.5904 | 0.8106 | 0.0000 | 0.0000 | 0.0000 | -0.8106 | -1.5904 | -2.1940 | |
| | 1 | 0 | -8 | 0 | 17 | 0 | -8 | 0 | 0 | 0 |
| 70 | 2.1679 | 1.6616 | 0.7345 | 0.0000 | 0.0000 | 0.0000 | -0.7345 | -1.6616 | -2.1679 | |
| | 1 | 0 | -8 | 0 | 17 | 0 | -7 | 0 | 0 | 0 |
| 71 | 2.1358 | 1.7321 | 0.6622 | 0.0000 | 0.0000 | 0.0000 | -0.6622 | -1.7321 | -2.1358 | |
| | 1 | 0 | -8 | 0 | 17 | 0 | -6 | 0 | 0 | 0 |
| 72 | 2.2361 | 1.0000 | 1.0000 | 1.0000 | 0.0000 | -1.0000 | -1.0000 | -1.0000 | -2.2361 | |
| | 1 | 0 | -8 | 0 | 18 | 0 | -16 | 0 | 5 | 0 |
| 73 | 2.2059 | 1.3376 | 1.0000 | 0.5870 | 0.0000 | -0.5870 | -1.0000 | -1.3376 | -2.2059 | |
| | 1 | 0 | -8 | 0 | 18 | 0 | -14 | 0 | 3 | 0 |
| 74 | 2.1753 | 1.4142 | 1.1260 | 0.0000 | 0.0000 | 0.0000 | -1.1260 | -1.4142 | -2.1753 | |
| | 1 | 0 | -8 | 0 | 18 | 0 | -12 | 0 | 0 | 0 |
| 75 | 2.1753 | 1.4142 | 1.1260 | 0.0000 | 0.0000 | 0.0000 | -1.1260 | -1.4142 | -2.1753 | |
| | 1 | 0 | -8 | 0 | 18 | 0 | -12 | 0 | 0 | 0 |
| 76 | 2.1646 | 1.5280 | 0.8536 | 0.5009 | 0.0000 | -0.5009 | -0.8536 | -1.5280 | -2.1646 | |
| | 1 | 0 | -8 | 0 | 18 | 0 | -12 | 0 | 2 | 0 |
| 77 | 2.1646 | 1.5280 | 0.8536 | 0.5009 | 0.0000 | -0.5009 | -0.8536 | -1.5280 | -2.1646 | |
| | 1 | 0 | -8 | 0 | 18 | 0 | -12 | 0 | 2 | 0 |
| 78 | 2.1169 | 1.6398 | 0.9110 | 0.0000 | 0.0000 | 0.0000 | -0.9110 | -1.6398 | -2.1169 | |
| | 1 | 0 | -8 | 0 | 18 | 0 | -10 | 0 | 0 | 0 |
| 79 | 2.1169 | 1.6398 | 0.9110 | 0.0000 | 0.0000 | 0.0000 | -0.9110 | -1.6398 | -2.1169 | |
| | 1 | 0 | -8 | 0 | 18 | 0 | -10 | 0 | 0 | 0 |
| 80 | 2.1358 | 1.4142 | 1.0000 | 0.6622 | 0.0000 | -0.6622 | -1.0000 | -1.4142 | -2.1358 | |
| | 1 | 0 | -8 | 0 | 19 | 0 | -16 | 0 | 4 | 0 |
| 81 | 2.1192 | 1.4142 | 1.1590 | 0.4071 | 0.0000 | -0.4071 | -1.1590 | -1.4142 | -2.1192 | |
| | 1 | 0 | -8 | 0 | 19 | 0 | -15 | 0 | 2 | 0 |
| 82 | 2.1120 | 1.4964 | 1.0000 | 0.5481 | 0.0000 | -0.5481 | -1.0000 | -1.4964 | -2.1120 | |
| | 1 | 0 | -8 | 0 | 19 | 0 | -15 | 0 | 3 | 0 |

| | | | | | | | | | | |
|---|---|---|---|---|---|---|---|---|---|---|
| 83 | 2.0840 | 1.5718 | 1.0000 | 0.4317 | 0.0000 | -0.4317 | -1.0000 | -1.5718 | -2.0840 | |
| | 1 | 0 | -8 | 0 | 19 | 0 | -14 | 0 | 2 | 0 |
| 84 | 2.0840 | 1.5718 | 1.0000 | 0.4317 | 0.0000 | -0.4317 | -1.0000 | -1.5718 | -2.0840 | |
| | 1 | 0 | -8 | 0 | 19 | 0 | -14 | 0 | 2 | 0 |
| 85 | 2.0743 | 1.6180 | 0.8350 | 0.6180 | 0.0000 | -0.6180 | -0.8350 | -1.6180 | -2.0743 | |
| | 1 | 0 | -8 | 0 | 19 | 0 | -14 | 0 | 3 | 0 |
| 86 | 2.0608 | 1.5984 | 1.0946 | 0.0000 | 0.0000 | 0.0000 | -1.0946 | -1.5984 | -2.0608 | |
| | 1 | 0 | -8 | 0 | 19 | 0 | -13 | 0 | 0 | 0 |
| 87 | 2.0356 | 1.6907 | 0.8841 | 0.4648 | 0.0000 | -0.4648 | -0.8841 | -1.6907 | -2.0356 | |
| | 1 | 0 | -8 | 0 | 19 | 0 | -13 | 0 | 2 | 0 |
| 88 | 2.0000 | 1.7321 | 1.0000 | 0.0000 | 0.0000 | 0.0000 | -1.0000 | -1.7321 | -2.0000 | |
| | 1 | 0 | -8 | 0 | 19 | 0 | -12 | 0 | 0 | 0 |
| 89 | 2.0529 | 1.4142 | 1.2086 | 0.5700 | 0.0000 | -0.5700 | -1.2086 | -1.4142 | -2.0529 | |
| | 1 | 0 | -8 | 0 | 20 | 0 | -18 | 0 | 4 | 0 |
| 90 | 2.0421 | 1.5202 | 1.0000 | 0.7203 | 0.0000 | -0.7203 | -1.0000 | -1.5202 | -2.0421 | |
| | 1 | 0 | -8 | 0 | 20 | 0 | -18 | 0 | 5 | 0 |
| 91 | 2.0153 | 1.5480 | 1.1429 | 0.4858 | 0.0000 | -0.4858 | -1.1429 | -1.5480 | -2.0153 | |
| | 1 | 0 | -8 | 0 | 20 | 0 | -17 | 0 | 3 | 0 |
| 92 | 2.0000 | 1.6180 | 1.0000 | 0.6180 | 0.0000 | -0.6180 | -1.0000 | -1.6180 | -2.0000 | |
| | 1 | 0 | -8 | 0 | 20 | 0 | -17 | 0 | 4 | 0 |
| 93 | 1.9616 | 1.6629 | 1.1111 | 0.3902 | 0.0000 | -0.3902 | -1.1111 | -1.6629 | -1.9616 | |
| | 1 | 0 | -8 | 0 | 20 | 0 | -16 | 0 | 2 | 0 |
| 94 | 1.9021 | 1.6180 | 1.1756 | 0.6180 | 0.0000 | -0.6180 | -1.1756 | -1.6180 | -1.9021 | |
| | 1 | 0 | -8 | 0 | 21 | 0 | -20 | 0 | 5 | 0 |

1.
2.
3.
4.
5.
6.
7.
8.
9.
10.
11.
12.
13.
14.
15.
16.
17.
18.
19.
20.

21.  22.  23.  24.

25.  26.  27.  28.

29.  30.  31.  32.

33.  34.  35.  36.

37.  38.  39.  40.

41.

42.

43.

44.

45.

46.

47.

48.

49.

50.

51.

52.

53.

54.

55.

56.

57.

58.

59.

60.

61.

62.

63.

64.

65.

66.

67.

68.

69.

70.

71.

72.

73.

74.

75.

76.

77.

78.

79.

80.

81.

82.

83.

84.

85.

86.

87.

88.

89.

90.

91.

92.

93.

94.

# A.5 TABLE A5

The spectra and characteristic polynomials of the adjacency matrix for cubic graphs with at most 12 vertices

```
001   3.0000 -1.0000 -1.0000 -1.0000
         1       0      -6      -8      -3

002   3.0000  1.0000  0.0000 -2.0000 -2.0000
         1       0      -9      -4      12       0

003   3.0000  0.0000  0.0000 -3.0000
         1       0      -9       0       0       0

004   3.0000  2.2361  1.0000 -1.0000 -1.0000 -2.2361
         1       0     -12      -8      38      48     -12     -40     -15

005   3.0000  1.7321  1.0000  0.4142 -1.0000 -1.7321 -2.4142
         1       0     -12      -4      38      16     -36     -12       9

006   3.0000  1.5616  0.6180  0.0000 -1.6180 -2.5616
         1       0     -12      -2      36       0     -31      12       0

007   3.0000  1.0000  1.0000 -1.0000 -1.0000 -3.0000
         1       0     -12       0      30       0     -28       0       9

008   3.0000  1.0000  0.4142  0.4142 -1.0000 -2.4142 -2.4142
         1       0     -12       0      34      16     -20      16      -3

009   3.0000  2.7785  0.0000  0.0000 -0.2892 -1.0000 -2.0000 -2.4893
         1       0     -15      -8      63      64     -37     -56     -12       0

010   3.0000  2.5616  1.0000  0.0000 -1.0000 -1.0000 -1.5616 -2.0000
         1       0     -15      -8      71      64    -101    -104      44      48       0
```

| | | | | | | | | | | | |
|---|---|---|---|---|---|---|---|---|---|---|---|
| 011 | 3.0000 1 | 2.4381 0 | 1.2470 -15 | 0.7255 -6 | -0.1485 69 | -0.4450 48 | -1.0000 -96 | -1.5350 -76 | -1.8019 30 | -2.4801 26 | 3 |
| 012 | 3.0000 1 | 2.4142 0 | 1.7321 -15 | 0.0000 -8 | 0.0000 71 | -0.4142 68 | -1.0000 -93 | -1.7321 -132 | -2.0000 -36 | -2.0000 0 | 0 |
| 013 | 3.0000 1 | 2.4142 0 | 1.3429 -15 | 0.0000 -4 | 0.0000 63 | -0.4142 36 | -0.5293 -61 | -1.0000 -56 | -2.0000 -12 | -2.8136 0 | 0 |
| 014 | 3.0000 1 | 2.1466 0 | 1.2831 -15 | 1.0000 -4 | 0.0000 71 | -0.3683 28 | -1.0000 -121 | -1.6053 -48 | -2.0000 64 | -2.4562 24 | 0 |
| 015 | 3.0000 1 | 2.1149 0 | 1.6180 -15 | 0.6180 -4 | -0.2541 69 | -0.3820 32 | -0.6180 -105 | -1.6180 -64 | -1.8608 23 | -2.6180 20 | 3 |
| 016 | 3.0000 1 | 2.0777 0 | 1.3094 -15 | 0.8019 -2 | 0.0000 67 | -0.4260 12 | -0.5550 -96 | -1.2941 -22 | -2.2470 35 | -2.6670 12 | 0 |
| 017 | 3.0000 1 | 2.0000 0 | 1.0000 -15 | 1.0000 -4 | 1.0000 75 | -1.0000 24 | -1.0000 -157 | -2.0000 -36 | -2.0000 144 | -2.0000 16 | -48 |
| 018 | 3.0000 1 | 2.0000 0 | 1.0000 -15 | 1.0000 0 | 0.0000 63 | 0.0000 0 | -1.0000 -85 | -1.0000 0 | -2.0000 36 | -3.0000 0 | 0 |
| 019 | 3.0000 1 | 1.9354 0 | 1.6180 -15 | 0.6180 -4 | 0.6180 73 | -0.6180 28 | -1.4626 -141 | -1.6180 -52 | -1.6180 99 | -2.4728 16 | -21 |
| 020 | 3.0000 1 | 1.9032 0 | 1.2470 -15 | 1.2470 -2 | -0.1939 69 | -0.4450 12 | -0.4450 -116 | -1.8019 -24 | -1.8019 54 | -2.7093 26 | 3 |
| 021 | 3.0000 1 | 1.8794 0 | 1.8794 -15 | 1.0000 -6 | -0.3473 75 | -0.3473 48 | -1.5321 -144 | -1.5321 -114 | -2.0000 75 | -2.0000 68 | 12 |
| 022 | 3.0000 1 | 1.8794 0 | 1.2631 -15 | 1.0000 -2 | 0.5157 71 | -0.3473 8 | -1.1826 -132 | -1.5321 -2 | -2.0000 91 | -2.5962 -8 | -12 |
| 023 | 3.0000 1 | 1.6180 0 | 1.6180 -15 | 1.0000 0 | -0.3820 65 | -0.3820 -4 | -0.6180 -85 | -0.6180 -20 | -2.6180 35 | -2.6180 20 | 3 |

| | | | | | | | | | | | | | |
|---|---|---|---|---|---|---|---|---|---|---|---|---|---|
| 024 | 3.0000 / 1 | 1.6180 / 0 | 1.6180 / -15 | 0.6180 / 0 | 0.6180 / 65 | -0.6180 / 0 | -0.6180 / -105 | -1.6180 / 0 | -1.6180 / 55 | -3.0000 / 0 | -9 | | |
| 025 | 3.0000 / 1 | 1.6180 / 0 | 1.3028 / -15 | 1.0000 / 0 | 0.6180 / 69 | -0.3820 / -12 | -0.6180 / -117 | -1.6180 / 36 | -2.3028 / 59 | -2.6180 / -12 | -9 | | |
| 026 | 3.0000 / 1 | 1.5616 / 0 | 1.0000 / -15 | 1.0000 / 0 | 1.0000 / 71 | 0.0000 / -16 | -1.0000 / -133 | -2.0000 / 64 | -2.0000 / 76 | -2.5616 / -48 | 0 | | |
| 027 | 3.0000 / 1 | 1.0000 / 0 | 1.0000 / -15 | 1.0000 / 0 | 1.0000 / 75 | 1.0000 / -24 | -2.0000 / -165 | -2.0000 / 120 | -2.0000 / 120 | -2.0000 / -160 | 48 | | |
| 028 | 3.0000 / 1 | 2.8323 / 0 | 1.9052 / -18 | 0.6180 / -10 | 0.5014 / 109 | 0.0000 / 112 | -1.0000 / -223 | -1.0000 / -326 | -1.0000 / 58 | -1.6180 / 196 | -1.8814 / 9 | -2.3574 / -36 | 0 |
| 029 | 3.0000 / 1 | 2.8208 / 0 | 1.4322 / -18 | 0.6180 / -6 | 0.5602 / 105 | 0.0000 / 60 | 0.0000 / -211 | -1.0000 / -122 | -1.0000 / 146 | -1.6180 / 52 | -2.1891 / -39 | -2.6240 / 0 | 0 |
| 030 | 3.0000 / 1 | 2.8192 / 0 | 1.4142 / -18 | 1.2427 / -8 | 0.0000 / 109 | 0.0000 / 84 | 0.0000 / -240 | -1.0000 / -220 | -1.4142 / 172 | -1.6719 / 168 | -2.0000 / 0 | -2.3901 / 0 | 0 |
| 031 | 3.0000 / 1 | 2.8192 / 0 | 1.2427 / -18 | 0.7321 / -4 | 0.0000 / 101 | 0.0000 / 36 | 0.0000 / -176 | 0.0000 / -40 | -1.0000 / 84 | -1.6719 / 0 | -2.3901 / 0 | -2.7321 / 0 | 0 |
| 032 | 3.0000 / 1 | 2.7093 / 0 | 1.7321 / -18 | 1.0000 / -8 | 0.4142 / 111 | 0.1939 / 88 | -1.0000 / -260 | -1.0000 / -264 | -1.0000 / 199 | -1.7321 / 232 | -1.9032 / -42 | -2.4142 / -48 | 9 |
| 033 | 3.0000 / 1 | 2.6628 / 0 | 1.3646 / -18 | 1.1935 / -6 | 0.4928 / 111 | 0.2950 / 60 | -0.4033 / -271 | -1.0000 / -152 | -1.2950 / 273 | -1.7695 / 124 | -2.1935 / -97 | -2.3474 / -18 | 9 |
| 034 | 3.0000 / 1 | 2.6554 / 0 | 1.6751 / -18 | 1.2108 / -8 | 0.5392 / 113 | 0.0000 / 88 | -1.0000 / -280 | -1.0000 / -280 | -1.0000 / 244 | -1.8662 / 296 | -2.0000 / -36 | -2.2143 / -72 | 0 |
| 035 | 3.0000 / 1 | 2.6554 / 0 | 1.2784 / -18 | 1.2108 / -4 | 0.3174 / 105 | 0.0000 / 44 | 0.0000 / -228 | -1.0000 / -104 | -1.0000 / 184 | -1.7046 / 72 | -1.8662 / -36 | -2.8912 / -36 | 0 |
| 036 | 3.0000 / 1 | 2.6458 / 0 | 1.7321 / -18 | 1.0000 / -8 | 1.0000 / 111 | -1.0000 / 96 | -1.0000 / -268 | -1.0000 / -336 | -1.0000 / 207 | -1.0000 / 416 | -1.7321 / 30 | -2.6458 / -168 | -63 |

| | | | | | | | | | | | | | |
|---|---|---|---|---|---|---|---|---|---|---|---|---|---|
| 037 | 3.0000<br>1 | 2.6180<br>0 | 2.0000<br>-18 | 1.3028<br>-10 | 0.3820<br>113 | -1.0000<br>120 | -1.0000<br>-263 | -1.0000<br>-434 | -1.0000<br>90 | -1.0000<br>468 | -2.0000<br>209 | -2.3028<br>-48 | -36 |
| 038 | 3.0000<br>1 | 2.5887<br>0 | 1.4142<br>-18 | 1.0000<br>-4 | 0.5463<br>109 | 0.0000<br>36 | 0.0000<br>-256 | -0.5463<br>-64 | -1.4142<br>228 | -2.0000<br>16 | -2.0000<br>-48 | -2.5887<br>0 | 0 |
| 039 | 3.0000<br>1 | 2.5758<br>0 | 1.8019<br>-18 | 0.8127<br>-6 | 0.4450<br>111 | 0.0000<br>62 | 0.0000<br>-265 | -1.0000<br>-166 | -1.2470<br>213 | -2.0000<br>92 | -2.0000<br>-60 | -2.3885<br>0 | 0 |
| 040 | 3.0000<br>1 | 2.5758<br>0 | 1.4909<br>-18 | 0.8127<br>-2 | 0.0000<br>103 | 0.0000<br>18 | 0.0000<br>-201 | 0.0000<br>-26 | -1.0000<br>105 | -1.6566<br>0 | -2.3885<br>0 | -2.8342<br>0 | 0 |
| 041 | 3.0000<br>1 | 2.5616<br>0 | 2.0000<br>-18 | 1.0000<br>-8 | 0.0000<br>113 | 0.0000<br>88 | 0.0000<br>-272 | -1.0000<br>-272 | -1.5616<br>176 | -2.0000<br>192 | -2.0000<br>0 | -2.0000<br>0 | 0 |
| 042 | 3.0000<br>1 | 2.5616<br>0 | 1.8422<br>-18 | 0.5069<br>-4 | 0.0000<br>105 | 0.0000<br>44 | 0.0000<br>-216 | 0.0000<br>-104 | -1.5069<br>96 | -1.5616<br>0 | -2.0000<br>0 | -2.8422<br>0 | 0 |
| 043 | 3.0000<br>1 | 2.5616<br>0 | 1.5616<br>-18 | 0.0000<br>0 | 0.0000<br>97 | 0.0000<br>0 | 0.0000<br>-144 | 0.0000<br>0 | 0.0000<br>0 | -1.5616<br>0 | -2.5616<br>0 | -3.0000<br>0 | 0 |
| 044 | 3.0000<br>1 | 2.5616<br>0 | 1.3028<br>-18 | 1.3028<br>-6 | 1.0000<br>113 | 0.0000<br>64 | -1.0000<br>-295 | -1.0000<br>-202 | -1.0000<br>334 | -1.5616<br>252 | -2.3028<br>-135 | -2.3028<br>-108 | 0 |
| 045 | 3.0000<br>1 | 2.5529<br>0 | 1.6337<br>-18 | 1.2577<br>-6 | 0.4733<br>111 | 0.1582<br>68 | -1.0000<br>-275 | -1.0000<br>-220 | -1.0000<br>257 | -1.4733<br>236 | -1.9688<br>-61 | -2.6337<br>-54 | 9 |
| 046 | 3.0000<br>1 | 2.5471<br>0 | 1.4142<br>-18 | 1.1865<br>-4 | 0.4993<br>109 | 0.0000<br>40 | -1.0000<br>-260 | -1.0000<br>-100 | -1.3331<br>248 | -1.4142<br>72 | -2.2581<br>-72 | -2.6418<br>0 | 0 |
| 047 | 3.0000<br>1 | 2.5226<br>0 | 2.0000<br>-18 | 1.1164<br>-8 | 0.3653<br>113 | 0.0000<br>92 | -1.0000<br>-276 | -1.0000<br>-312 | -1.6557<br>188 | -2.0000<br>300 | -2.0000<br>16 | -2.3485<br>-48 | 0 |
| 048 | 3.0000<br>1 | 2.5200<br>0 | 1.6408<br>-18 | 1.2220<br>-6 | 0.6180<br>113 | 0.0000<br>64 | -0.4344<br>-291 | -1.0000<br>-198 | -1.4418<br>294 | -1.6180<br>204 | -2.1084<br>-83 | -2.3982<br>-48 | 0 |
| 049 | 3.0000<br>1 | 2.5141<br>0 | 2.5141<br>-18 | 0.5720<br>-12 | 0.5720<br>111 | -1.0000<br>144 | -1.0000<br>-216 | -1.0000<br>-480 | -1.0000<br>-117 | -1.0000<br>256 | -2.0861<br>138 | -2.0861<br>-36 | -27 |

| | | | | | | | | | | | | | |
|---|---|---|---|---|---|---|---|---|---|---|---|---|---|
| 050 | 3.0000 / 1 | 2.5141 / 0 | 2.1701 / −18 | 0.5720 / −8 | 0.4142 / 111 | 0.3111 / 92 | −1.0000 / −252 | −1.0000 / −292 | −1.0000 / 119 | −1.4812 / 180 | −2.0861 / −34 | −2.4142 / −36 | 9 |
| 051 | 3.0000 / 1 | 2.5141 / 0 | 1.7321 / −18 | 1.4812 / −8 | 0.5720 / 115 | −0.3111 / 92 | −1.0000 / −300 | −1.0000 / −332 | −1.0000 / 263 | −1.7321 / 420 | −2.0861 / 30 | −2.1701 / −108 | −27 |
| 052 | 3.0000 / 1 | 2.5141 / 0 | 1.6554 / −18 | 1.0000 / −4 | 0.5720 / 107 | 0.2108 / 48 | −1.0000 / −248 | −1.0000 / −152 | −1.0000 / 219 | −1.0000 / 144 | −2.0861 / −70 | −2.8662 / −36 | 9 |
| 053 | 3.0000 / 1 | 2.5141 / 0 | 1.4812 / −18 | 1.0000 / −4 | 0.5720 / 111 | 0.4142 / 36 | −0.3111 / −276 | −1.0000 / −76 | −1.0000 / 279 | −2.0861 / 44 | −2.1701 / −106 | −2.4142 / 0 | 9 |
| 054 | 3.0000 / 1 | 2.5100 / 0 | 2.0198 / −18 | 0.6180 / −6 | 0.3750 / 109 | 0.0000 / 68 | 0.0000 / −247 | −1.0000 / −198 | −1.3929 / 146 | −1.6180 / 88 | −1.8314 / −39 | −2.6806 / 0 | 0 |
| 055 | 3.0000 / 1 | 2.5088 / 0 | 1.6751 / −18 | 0.8671 / −4 | 0.5392 / 109 | 0.0000 / 40 | 0.0000 / −256 | −1.0000 / −100 | −1.0000 / 216 | −1.7520 / 56 | −2.2143 / −60 | −2.6239 / 0 | 0 |
| 056 | 3.0000 / 1 | 2.3931 / 0 | 1.4142 / −18 | 1.2250 / −4 | 1.0000 / 113 | 0.0000 / 40 | −0.3061 / −304 | −1.0000 / −116 | −1.4142 / 360 | −1.7190 / 128 | −2.0000 / −152 | −2.5931 / −48 | 0 |
| 057 | 3.0000 / 1 | 2.3877 / 0 | 1.5321 / −18 | 1.3028 / −4 | 0.4790 / 113 | 0.3473 / 38 | −0.3071 / −298 | −1.0000 / −102 | −1.2141 / 326 | −1.8794 / 88 | −2.3028 / −119 | −2.3455 / −6 | 9 |
| 058 | 3.0000 / 1 | 2.3717 / 0 | 1.7672 / −18 | 1.1561 / −4 | 0.3728 / 111 | 0.0000 / 42 | 0.0000 / −278 | −1.0000 / −126 | −1.3121 / 261 | −1.5365 / 102 | −2.2080 / −63 | −2.6113 / 0 | 0 |
| 059 | 3.0000 / 1 | 2.3601 / 0 | 1.5037 / −18 | 1.1922 / −2 | 0.4654 / 109 | 0.0000 / 16 | 0.0000 / −263 | −0.4592 / −26 | −1.3337 / 234 | −1.7681 / 4 | −2.2438 / −39 | −2.7166 / 0 | 0 |
| 060 | 3.0000 / 1 | 2.3429 / 0 | 2.0000 / −18 | 0.7321 / −4 | 0.4707 / 109 | 0.0000 / 44 | 0.0000 / −256 | −1.0000 / −128 | −1.0000 / 188 | −1.8136 / 64 | −2.0000 / −48 | −2.7321 / 0 | 0 |
| 061 | 3.0000 / 1 | 2.3358 / 0 | 1.8174 / −18 | 0.8794 / −2 | 0.0000 / 107 | 0.0000 / 18 | 0.0000 / −237 | 0.0000 / −42 | −1.3473 / 153 | −1.5217 / 0 | −2.5321 / 0 | −2.6316 / 0 | 0 |
| 062 | 3.0000 / 1 | 2.3234 / 0 | 1.5616 / −18 | 1.0000 / 0 | 0.0000 / 105 | 0.0000 / −8 | 0.0000 / −216 | 0.0000 / 40 | −0.6421 / 96 | −2.0000 / 0 | −2.5616 / 0 | −2.6813 / 0 | 0 |

| id | | | | | | | | | | | | | |
|---|---|---|---|---|---|---|---|---|---|---|---|---|---|
| 063 | 3.0000 1 | 2.3083 0 | 1.5096 -18 | 1.1682 -4 | 1.0953 115 | 0.0000 38 | -0.2624 -322 | -1.0000 -110 | -1.4773 401 | -1.7886 122 | -2.1975 -179 | -2.3557 -48 | 0 |
| 064 | 3.0000 1 | 2.2855 0 | 1.7495 -18 | 1.2414 -4 | 0.6180 113 | 0.1939 42 | -0.4206 -302 | -1.0000 -134 | -1.3735 334 | -1.6180 140 | -2.0733 -123 | -2.6029 -30 | 9 |
| 065 | 3.0000 1 | 2.2793 0 | 1.5909 -18 | 1.3028 -2 | 0.4496 109 | 0.0000 20 | 0.0000 -267 | -1.0000 -62 | -1.0000 254 | -1.5508 60 | -2.3028 -63 | -2.7689 0 | 0 |
| 066 | 3.0000 1 | 2.2735 0 | 1.8996 -18 | 1.4378 -6 | 0.4288 115 | 0.1334 68 | -1.0000 -311 | -1.0000 -248 | -1.0000 317 | -1.6694 308 | -2.1401 -57 | -2.3636 -66 | 9 |
| 067 | 3.0000 1 | 2.2735 0 | 1.4378 -18 | 1.3226 -2 | 0.5450 111 | 0.4288 16 | -0.2707 -287 | -1.0000 -32 | -1.0000 309 | -1.9016 20 | -2.1401 -117 | -2.6952 6 | 9 |
| 068 | 3.0000 1 | 2.2724 0 | 1.2470 -18 | 1.2470 -2 | 1.1573 111 | 0.0000 18 | -0.4450 -293 | -0.4450 -42 | -1.6295 333 | -1.8019 44 | -1.8019 -120 | -2.8003 -36 | 0 |
| 069 | 3.0000 1 | 2.2706 0 | 2.0000 -18 | 1.2470 -6 | 0.5191 115 | 0.0000 66 | -0.4450 -309 | -1.0000 -226 | -1.4511 309 | -1.8019 244 | -2.0000 -68 | -2.3387 -48 | 0 |
| 070 | 3.0000 1 | 2.2671 0 | 1.6055 -18 | 1.1604 -2 | 0.5996 111 | 0.0000 14 | 0.0000 -281 | -0.5301 -18 | -1.3007 269 | -2.0000 -4 | -2.2071 -60 | -2.5947 0 | 0 |
| 071 | 3.0000 1 | 2.2643 0 | 1.9421 -18 | 0.8019 -4 | 0.6180 113 | 0.3741 38 | -0.4325 -294 | -0.5550 -98 | -1.6180 290 | -1.7818 44 | -2.2470 -95 | -2.3663 -6 | 9 |
| 072 | 3.0000 1 | 2.2361 0 | 1.7913 -18 | 1.0000 -2 | 0.6180 109 | 0.0000 20 | 0.0000 -267 | -1.0000 -58 | -1.0000 250 | -1.6180 40 | -2.2361 -75 | -2.7913 0 | 0 |
| 073 | 3.0000 1 | 2.2361 0 | 1.4142 -18 | 1.4142 -4 | 1.0000 117 | 0.0000 36 | 0.0000 -344 | -1.4142 -96 | -1.4142 468 | -2.0000 80 | -2.0000 -240 | -2.2361 0 | 0 |
| 074 | 3.0000 1 | 2.2361 0 | 1.4142 -18 | 1.4142 0 | 0.0000 105 | 0.0000 0 | 0.0000 -236 | 0.0000 0 | -1.4142 180 | -1.4142 0 | -2.2361 0 | -3.0000 0 | 0 |
| 075 | 3.0000 1 | 2.2240 0 | 1.9563 -18 | 1.2409 -4 | 0.2091 111 | 0.0000 46 | 0.0000 -282 | -1.0000 -154 | -1.3383 257 | -1.7098 142 | -1.8271 -39 | -2.7551 0 | 0 |

| | | | | | | | | | | | | | |
|---|---|---|---|---|---|---|---|---|---|---|---|---|---|
| 076 | 3.0000 1 | 2.2240 0 | 1.4413 -18 | 1.2409 0 | 0.5669 107 | 0.0000 -6 | 0.0000 -246 | -0.4851 26 | -1.0000 201 | -1.7098 -14 | -2.5231 -39 | -2.7551 0 | 0 |
| 077 | 3.0000 1 | 2.1955 0 | 1.5321 -18 | 1.3028 -4 | 1.0646 117 | 0.3473 38 | -0.6982 -346 | -1.0000 -118 | -1.4527 482 | -1.8794 148 | -2.1092 -283 | -2.3028 -66 | 45 |
| 078 | 3.0000 1 | 2.1701 0 | 1.7321 -18 | 1.4812 -4 | 0.4142 115 | 0.3111 40 | -0.3111 -320 | -1.0000 -128 | -1.4812 371 | -1.7321 136 | -2.1701 -126 | -2.4142 -12 | 9 |
| 079 | 3.0000 1 | 2.1326 0 | 1.7321 -18 | 1.3563 -4 | 1.0000 115 | 0.0681 44 | -1.0000 -328 | -1.0000 -164 | -1.0000 419 | -1.7321 244 | -1.9432 -198 | -2.6138 -120 | 9 |
| 080 | 3.0000 1 | 2.1227 0 | 1.7625 -18 | 1.3417 -2 | 0.3859 111 | 0.0000 18 | 0.0000 -285 | -0.5634 -50 | -1.4832 277 | -1.6673 40 | -2.1829 -48 | -2.7159 0 | 0 |
| 081 | 3.0000 1 | 2.1227 0 | 1.5085 -18 | 1.3417 -2 | 0.6796 115 | 0.3859 10 | 0.0000 -325 | -0.8258 10 | -1.6673 397 | -2.0000 -76 | -2.1829 -148 | -2.3623 48 | 0 |
| 082 | 3.0000 1 | 2.1202 0 | 1.7640 -18 | 1.2206 -2 | 0.6938 111 | 0.2329 20 | -0.3963 -291 | -1.0000 -64 | -1.0000 317 | -1.7858 72 | -2.0615 -121 | -2.7878 -18 | 9 |
| 083 | 3.0000 1 | 2.1149 0 | 2.0000 -18 | 1.3028 -6 | 1.0000 117 | -0.2541 68 | -1.0000 -335 | -1.0000 -262 | -1.0000 398 | -1.8608 392 | -2.0000 -127 | -2.3028 -192 | -36 |
| 084 | 3.0000 1 | 2.0907 0 | 1.5840 -18 | 1.2396 -2 | 1.0800 113 | 0.1488 18 | -0.3751 -313 | -1.0000 -56 | -1.2642 390 | -1.6543 74 | -2.1413 -184 | -2.7082 -36 | 9 |
| 085 | 3.0000 1 | 2.0821 0 | 1.9653 -18 | 1.1852 -4 | 0.7538 115 | 0.1612 40 | -0.3944 -320 | -1.0000 -128 | -1.3668 375 | -1.7957 136 | -2.2014 -154 | -2.3894 -36 | 9 |
| 086 | 3.0000 1 | 2.0814 0 | 1.4142 -18 | 1.2470 -2 | 1.1533 115 | 0.4586 14 | -0.4450 -333 | -1.0000 -26 | -1.4142 453 | -1.8019 12 | -2.1080 -256 | -2.5853 0 | 36 |
| 087 | 3.0000 1 | 2.0664 0 | 2.0000 -18 | 1.4142 -4 | 0.2222 113 | 0.0000 44 | 0.0000 -300 | -1.0000 -152 | -1.4142 300 | -1.6522 160 | -2.0000 -48 | -2.6364 -48 | 0 |
| 088 | 3.0000 1 | 2.0647 0 | 1.6058 -18 | 1.1935 -2 | 1.0000 115 | 0.2950 12 | -0.1803 -327 | -1.0000 -12 | -1.2950 413 | -2.0948 -16 | -2.1935 -193 | -2.3953 18 | 9 |

| | | | | | | | | | | | | |
|---|---|---|---|---|---|---|---|---|---|---|---|---|
| 089 | 3.0000<br>1 | 2.0545<br>0 | 1.7321<br>-18 | 1.3028<br>-2 | 0.7631<br>113 | 0.0000<br>16 | 0.0000<br>-307 | -1.0000<br>-42 | -1.2346<br>354 | -1.7321<br>36 | -2.3028<br>-135 | -2.5831<br>0 | 0 |
| 090 | 3.0000<br>1 | 2.0000<br>0 | 2.0000<br>-18 | 2.0000<br>-8 | 0.0000<br>117 | 0.0000<br>96 | -1.0000<br>-316 | -1.0000<br>-384 | -1.0000<br>240 | -2.0000<br>512 | -2.0000<br>192 | -2.0000<br>0 | 0 |
| 091 | 3.0000<br>1 | 2.0000<br>0 | 2.0000<br>-18 | 1.4142<br>-4 | 0.7321<br>113 | 0.0000<br>48 | -1.0000<br>-308 | -1.0000<br>-188 | -1.0000<br>348 | -1.4142<br>264 | -2.0000<br>-112 | -2.7321<br>-96 | 0 |
| 092 | 3.0000<br>1 | 2.0000<br>0 | 2.0000<br>-18 | 1.3028<br>-6 | 1.3028<br>117 | -1.0000<br>72 | -1.0000<br>-339 | -1.0000<br>-306 | -1.0000<br>414 | -1.0000<br>532 | -2.3028<br>-99 | -2.3028<br>-324 | -108 |
| 093 | 3.0000<br>1 | 2.0000<br>0 | 2.0000<br>-18 | 1.0000<br>0 | 0.0000<br>105 | 0.0000<br>0 | 0.0000<br>-232 | 0.0000<br>0 | -1.0000<br>144 | -2.0000<br>0 | -2.0000<br>0 | -3.0000<br>0 | 0 |
| 094 | 3.0000<br>1 | 2.0000<br>0 | 2.0000<br>-18 | 0.7321<br>0 | 0.7321<br>105 | 0.0000<br>0 | 0.0000<br>-228 | -1.0000<br>-24 | -1.0000<br>180 | -1.0000<br>16 | -2.7321<br>-48 | -2.7321<br>0 | 0 |
| 095 | 3.0000<br>1 | 2.0000<br>0 | 1.8136<br>-18 | 1.0000<br>0 | 0.7321<br>109 | 0.0000<br>-8 | 0.0000<br>-264 | -0.4707<br>40 | -1.0000<br>220 | -2.0000<br>-32 | -2.3429<br>-48 | -2.7321<br>0 | 0 |
| 096 | 3.0000<br>1 | 2.0000<br>0 | 1.6935<br>-18 | 1.3028<br>-2 | 1.0000<br>113 | 0.3297<br>20 | -1.0000<br>-315 | -1.0000<br>-78 | -1.0000<br>410 | -1.3297<br>120 | -2.3028<br>-227 | -2.6935<br>-60 | 36 |
| 097 | 3.0000<br>1 | 2.0000<br>0 | 1.5616<br>-18 | 1.5616<br>0 | 0.0000<br>109 | 0.0000<br>-8 | 0.0000<br>-260 | 0.0000<br>32 | -1.0000<br>192 | -2.0000<br>0 | -2.5616<br>0 | -2.5616<br>0 | 0 |
| 098 | 3.0000<br>1 | 2.0000<br>0 | 1.5616<br>-18 | 1.4142<br>0 | 0.7321<br>109 | 0.0000<br>-4 | 0.0000<br>-272 | -1.0000<br>4 | -1.0000<br>284 | -1.4142<br>8 | -2.5616<br>-96 | -2.7321<br>0 | 0 |
| 099 | 3.0000<br>1 | 2.0000<br>0 | 1.5616<br>-18 | 1.0000<br>0 | 1.0000<br>113 | 0.0000<br>-16 | 0.0000<br>-304 | 0.0000<br>112 | -2.0000<br>304 | -2.0000<br>-192 | -2.0000<br>0 | -2.5616<br>0 | 0 |
| 100 | 3.0000<br>1 | 2.0000<br>0 | 1.4142<br>-18 | 1.4142<br>0 | 1.0000<br>109 | 0.0000<br>0 | 0.0000<br>-288 | -1.0000<br>0 | -1.4142<br>340 | -2.0000<br>-144 | -2.0000<br>0 | -3.0000<br>0 | 0 |
| 101 | 3.0000<br>1 | 2.0000<br>0 | 1.3028<br>-18 | 1.3028<br>-2 | 1.0000<br>117 | 1.0000<br>12 | -1.0000<br>-355 | -1.0000<br>-18 | -1.0000<br>534 | -2.0000<br>8 | -2.3028<br>-387 | -2.3028<br>0 | 108 |

| | | | | | | | | | | | | |
|---|---|---|---|---|---|---|---|---|---|---|---|---|
| 102 | 3.0000 1 | 1.9673 0 | 1.5764 -18 | 1.3645 0 | 0.7475 111 | 0.0000 -10 | 0.0000 -286 | -0.4399 54 | -1.1971 277 | -2.1268 -54 | -2.2119 -63 | -2.6799 0 | 0 |
| 103 | 3.0000 1 | 1.9653 0 | 1.5772 -18 | 1.1852 0 | 1.0000 111 | 0.2920 -8 | -0.3944 -292 | -0.4781 40 | -1.3668 323 | -1.6677 -48 | -2.3894 -118 | -2.7235 0 | 9 |
| 104 | 3.0000 1 | 1.9338 0 | 1.4142 -18 | 1.3204 0 | 1.0000 113 | 0.3505 -12 | 0.0000 -312 | -0.7752 76 | -1.4142 368 | -2.0000 -128 | -2.1586 -136 | -2.6709 48 | 0 |
| 105 | 3.0000 1 | 1.9032 0 | 1.7321 -18 | 1.0000 0 | 1.0000 111 | 0.4142 -8 | -0.1939 -292 | -1.0000 40 | -1.0000 327 | -1.7321 -56 | -2.4142 -138 | -2.7093 24 | 9 |
| 106 | 3.0000 1 | 1.8164 0 | 1.5321 -18 | 1.3028 0 | 1.1355 113 | 0.3473 -10 | -0.1623 -314 | -1.0000 54 | -1.1188 386 | -1.8794 -76 | -2.3028 -179 | -2.6708 30 | 9 |
| 107 | 3.0000 1 | 1.8136 0 | 1.5616 -18 | 1.4142 0 | 1.0000 113 | 0.0000 -12 | 0.0000 -308 | -0.4707 68 | -1.4142 340 | -2.0000 -88 | -2.3429 -96 | -2.5616 0 | 0 |
| 108 | 3.0000 1 | 1.7321 0 | 1.7321 -18 | 1.0000 0 | 1.0000 111 | 1.0000 0 | -1.0000 -316 | -1.0000 0 | -1.0000 447 | -1.7321 0 | -1.7321 -306 | -3.0000 0 | 81 |
| 109 | 3.0000 1 | 1.7321 0 | 1.4812 -18 | 1.4812 0 | 1.0000 115 | 0.4142 -16 | -0.3111 -328 | -0.3111 104 | -1.7321 387 | -2.1701 -176 | -2.1701 -102 | -2.4142 24 | 9 |
| 110 | 3.0000 1 | 1.7321 0 | 1.4812 -18 | 1.2143 0 | 1.0000 115 | 1.0000 -12 | -0.3111 -340 | -1.0000 76 | -1.5392 479 | -1.7321 -148 | -2.1701 -282 | -2.6751 84 | 45 |
| 111 | 3.0000 1 | 1.5616 0 | 1.4142 -18 | 1.4142 0 | 1.0000 117 | 1.0000 -16 | 0.0000 -360 | -1.4142 112 | -1.4142 532 | -2.0000 -256 | -2.0000 -304 | -2.5616 192 | 0 |
| 112 | 3.0000 1 | 1.5321 0 | 1.5321 -18 | 1.3028 0 | 1.3028 117 | 0.3473 -18 | 0.3473 -354 | -1.0000 126 | -1.8794 486 | -1.8794 -272 | -2.3028 -207 | -2.3028 162 | -27 |

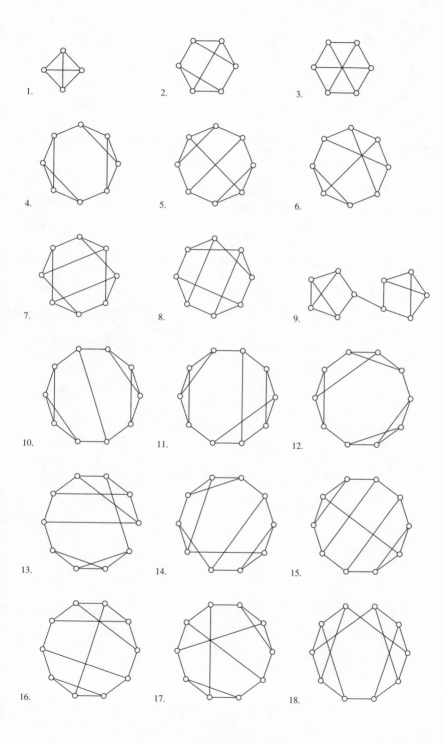

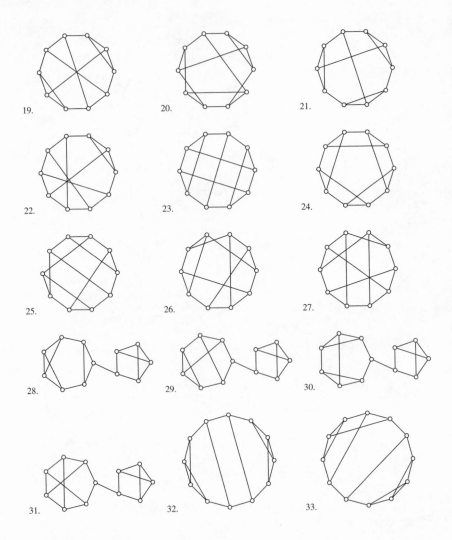

19.

20.

21.

22.

23.

24.

25.

26.

27.

28.

29.

30.

31.

32.

33.

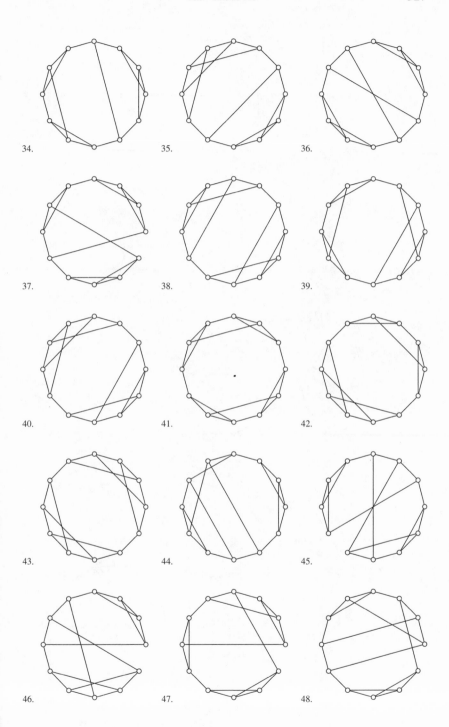

34.

35.

36.

37.

38.

39.

40.

41.

42.

43.

44.

45.

46.

47.

48.

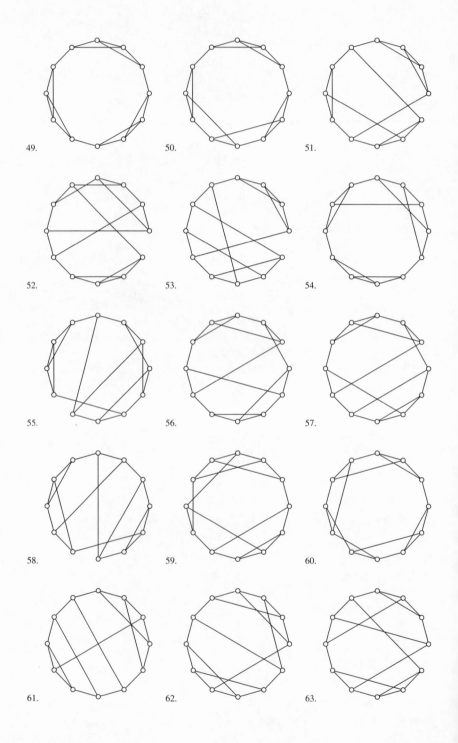

49.  50.  51.

52.  53.  54.

55.  56.  57.

58.  59.  60.

61.  62.  63.

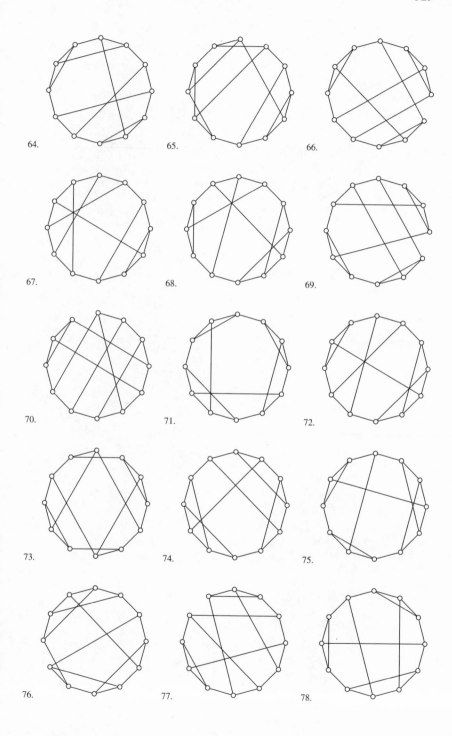

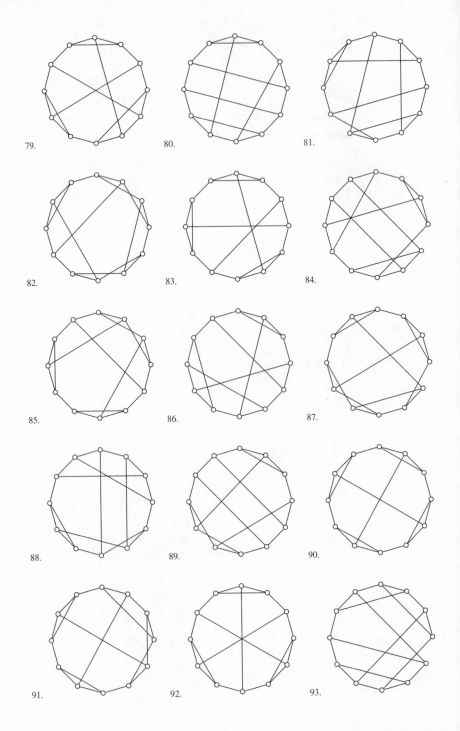

79.

80.

81.

82.

83.

84.

85.

86.

87.

88.

89.

90.

91.

92.

93.

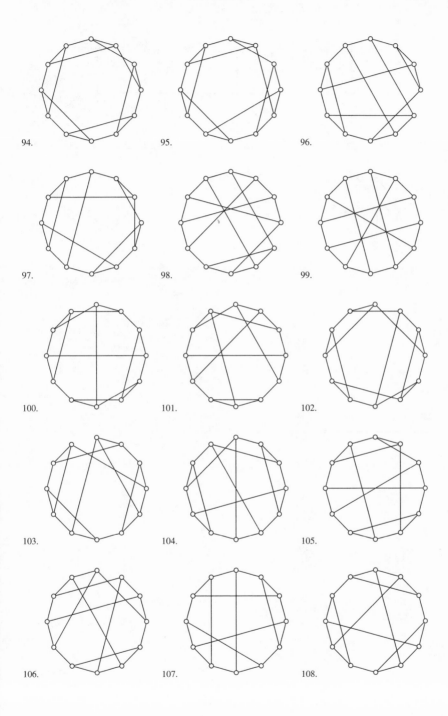

94.

95.

96.

97.

98.

99.

100.

101.

102.

103.

104.

105.

106.

107.

108.

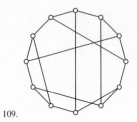

109.

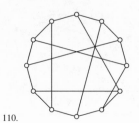

110.

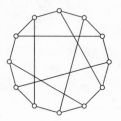

111.

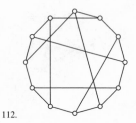

112.

# References

[Abr]  de Abreu N. M. M., Old and new results on algebraic connectivity of graphs, *Linear Algebra Appl.* **423** (2007), 53–73.

[Ach]  Acharya B. D., Spectral criterion for cycle balance in networks, *J. Graph Theory* **4** (1980), 1–11.

[Alo1]  Alon N., Eigenvalues and expanders, *Combinatorica* **6** (1986), 83–96.

[Alo2]  Alon N., Tools from higher algebra, in *Handbook of Combinatorics* (eds. Graham R., Grötchel M., Lovász L.), North-Holland Elsevier (Amsterdam) 1995, pp. 1749–1783.

[AloMi1]  Alon N., Milman V. D., Eigenvalues, expanders and superconcentrators, *Proc. 25th Annual Symp. on Foundations of Computer Science, Singer Island, Florida,* IEEE (1984), pp. 320–322.

[AloMi2]  Alon N., Milman V. D., $\lambda_1$, isoperimetric inequalities for graphs, and superconcentrators, *J. Combin. Theory Ser. B* **38** (1985), 73–88.

[AnMo]  Anderson W. N., Morley T.D., Eigenvalues of the Laplacian of a graph, *Linear Multilin. Algebra* **18** (1985), 141–145.

[ABCHRSS]  Aouchiche M., Bell F. K., Cvetković D., Hansen P., Rowlinson P., Simić S. K., Stevanović D., Variable neighborhood search for extremal graphs, 16: some conjectures related to the largest eigenvalue of a graph, *European J. Operational Research* **191** (2008), 661–676.

[Asch]  Aschbacher M., The non-existence of rank three permutation groups of degree 3250 and subdegree 57, *J. Algebra* **19** (1971), 538–540.

[Bab1]  Babai L., Automorphism group and category of cospectral graphs, *Acta Math. Acad. Sci. Hung.* **31** (1978), 295–306.

[Bab2]  Babai L., On the complexity of canonical labelling of strongly regular graphs, *SIAM J. Computing* **9** (1980), 212–216.

[Bab3]  Babai L., Kospektrale Graphen mit vorgegebenen Automorphismengruppen, *Wiss. Z.,* T.H. Ilmenau, **27** (1981) No.5, 31–37.

[Bak1]  Baker G. A., Drum shapes and isospectral graphs, BNL 10088, Brookhaven National Laboratory, Long Island, New York, 1966.

[Bak2]  Baker G. A., Drum shapes and isospectral graphs, *J. Math. Phys.* **7** (1966), 2238–2242.

[Bal]  Balaban A. T. (ed.), *Chemical Application of Graph Theory*, Academic Press (London), 1976.

[BalHa] Balaban A. T., Harary F., The characteristic polynomial does not uniquely determine the topology of a molecule, *J. Chem. Doc.* **11** (1971), 258–259.

[BalaPa] Balasubramanian K., Parthasarathy K.R., In search of a complete invariant for graphs, in *Combinatorics and Graph Theory*, Proc. 2nd Symp. held at the Indian Statistical Institute, Calcutta, February 25-29, 1980, Lecture Notes in Math. 885 (ed. Rao S.B.), Springer-Verlag (Berlin) 1981, pp. 42–59.

[BaCLS] Balińska K., Cvetković D., Lepović M., Simić S., There are exactly 150 connected integral graphs up to 10 vertices, *Univ. Beograd, Publ. Elektrotehn. Fak., Ser. Mat.* **10** (1999), 95–105.

[BaCRSS] Balińska K., Cvetković D., Radosavljević Z., Simić S., Stevanović D., A survey on integral graphs, *Univ. Beograd, Publ. Elektrotehn. Fak., Ser. Mat.* **13** (2002) 42-65 (Erratum: *loc. cit.* **15** (2004), 112).

[BaKSZ1] Balińska K. T., Kupczyk M., Simić S. K., Zwierzyński K. T., On generating all integral graphs on 11 vertices, Computer Science Center Report No. 469, Technical University of Poznań (1999/2000).

[BaKSZ2] Balińska K. T., Kupczyk M., Simić S. K., Zwierzyński K. T., On generating all integral graphs on 12 vertices, Computer Science Center Report No. 482, Technical University of Poznań (2001).

[BaKSZ3] Balińska K.T., Kupczyk M., Simić S.K., Zwierzyński K.T., On generating all integral graphs on 13 vertices, Computer Science Center Report No. 483, Technical University of Poznań (2002).

[BaSi1] Balińska K. T., Simić S. K., Some remarks on integral graphs with maximum degree four, *Novi Sad J. Math.* **31** (2001), 19–25.

[BaSi2] Balińska K. T., Simić S. K., The nonregular, bipartite, integral graphs with maximum degree four, *Discrete Math.* **236** (2001), 13–24.

[BaSZ1] Balińska K. T., Simić S. K., Zwierzińsky K. T., Which nonregular, bipartite, integral graphs with maximum degree four do not have ±1 as eigenvalues?, *Discrete Math.* **286** (2004), 15–25.

[BaSZ2] Balińska K. T., Simić S. K., Zwierzyński K. T., On generating 4-regular integral graphs, *Studia z Automatyki i Informatyki* **31** (2006), 7–16.

[BanIt] Bannai E., Ito T., *Algebraic Combinatorics I: Association Schemes*, Benjamin-Cummings (London), 1994.

[Bei] Beineke L. W., Derived graphs and digraphs, in *Beiträge zur Graphentheorie* (eds. Sachs H., Voss H., Walther H.), Teubner (Leipzig) 1968, pp. 17–33.

[BeLS] Belardo F., Li Marzi E. M, Simić S. K., Some results on the index of unicyclic graphs, *Linear Algebra Appl.* **416** (2006), 1048–1059.

[Bele] Belevitch V., Theory of 2n-terminal networks with application to conference telephony, *Electron. Comm.* **27** (1950), 231–244.

[Bel1] Bell F. K., On the maximal index of connected graphs, *Linear Algebra Appl.* **144** (1991), 135–151.

[Bel2] Bell F. K., A note on the irregularity of graphs, *Linear Algebra Appl.* **161** (1992), 45–54.

[Bel3] Bell F. K., Characterizing line graphs by star complements, *Linear Algebra Appl.* **296** (1999), 15–25.

[Bel4] Bell F. K., Line graphs of bipartite graphs with Hamiltonian paths, *J. Graph Theory* **43** (2003), 137–149.

[BelCRS1]  Bell F. K., Cvetković D., Rowlinson P., Simić S. K., Some additions to the theory of star partitions of graphs, *Discussiones Math. – Graph Theory* **19** (1999), 119–134.

[BelCRS2]  Bell F. K., Cvetković D., Rowlinson P., Simić S. K., Graphs for which the least eigenvalue is minimal I, *Linear Algebra Appl.* **429** (2008), 234–241.

[BelCRS3]  Bell F. K., Cvetković D., Rowlinson P., Simić S. K., Graphs for which the least eigenvalue is minimal II, *Linear Algebra Appl.* **429** (2008), 2168–2179. (Erratum: *loc. cit.*, to appear.)

[BelLMS]  Bell F. K., Li Marzi E. M., Simić S. K., Some new results on graphs with least eigenvalue not less than $-2$, *Rend. Sem. Mat. Messina Ser. II* **25** (2003), 11–30.

[BelRo]  Bell F. K., Rowlinson P., On the multiplicities of graph eigenvalues, *Bull. London Math. Soc.* **35** (2003), 401–408.

[BelSi]  Bell F. K., Simić S. K., On graphs whose star complement for $-2$ is a path or cycle, *Linear Algebra Appl.* **377** (2004), 249–265.

[Bens]  Benson C. T., Jacobs J. B., On hearing the shape of combinatorial drums, *J. Combin. Theory Ser. B* **13** (1972), 170–178.

[BerZh]  Berman A., Zhang X.-D., On the spectral radius of graphs with cut vertices, *J. Combin. Theory Ser. B* **83** (2001), 233–240.

[Big1]  Biggs N. L., Intersection matrices for linear graphs, in: *Combinatorial Mahematics and its Applications* (ed. Welsh, D. J. A.), Academic Press (London) 1971, pp. 15–23.

[Big2]  Biggs N. L., *Algebraic Graph Theory* (second edition), Cambridge University Press (Cambridge), 1993.

[BigBo]  Biggs N. L., Boshier A. G., Note on the girth of Ramanujan graphs, *J. Combin. Theory Ser. B* **49** (1990), 190–194.

[BiySS]  Biyikoğlu T., Simić S. K., Z. Stanić, Some notes on cographs, *Ars Combinatoria*, to appear.

[BoLS]  Boesch F. T., Li C., Suffel C., On the existence of uniformly most reliable networks, *Networks* **21** (1991), 181–194.

[BolNi]  Bollobás B., Nikiforov V., Cliques and the spectral radius, *J. Combin. Theory Ser. B* **97** (2007), 859–865.

[Bop]  Boppana R.B., Eigenvalues and graph bisection: an average case analysis, *Proc. 28th Annual Symp. Found. Comp. Sci., IEEE* 1987, pp. 280–285.

[Bos]  Bose R. C., Strongly regular graphs, partial geometries and partially balanced designs, *Pacific J. Math.* **13** (1963), 389–419.

[BosMe]  Bose R. C., Mesner D. M., On linear associative algebras corresponding to association schemes of partially balanced designs, *Ann. Math. Statist.* **30** (1959), 21–36.

[BosSh]  Bose R. C., Shrikhande S. S., Graphs in which each pair of vertices is adjacent to the same number $d$ of other vertices, *Studia Sci. Math Hung.* **5** (1970), 181–195.

[BouJo]  Boulet R., Jouve B., The lollipop graph is determined by its spectrum, *Electronic J. Combin.* **15** (2008), Paper R74.

[BGI]  Brand C., Guidili B., Imrich W., Characterization of trivalent graphs with minimal eigenvalue gap, *Croatica Chemica Acta* **80** (2007), 193–201.

[BrHS] Brankov V., Hansen P., Stevanović D., Automated conjectures on upper bounds for the largest Laplacian eigenvalue of graphs, *Linear Algebra Appl.* **414** (2006), 407–424.

[Bra] Branković Lj., *Usability of secure statistical databases*, PhD thesis, University of Newcastle (Australia), 1998.

[BraMS] Branković Lj., Miller M., Siráň J., Graphs, (0, 1)-matrices and usability of statistical databases, *Congr. Num.* **120** (1996), 169–82.

[BraCv] Branković Lj., Cvetković D., The eigenspace of the eigenvalue −2 in generalized line graphs and a problem in the security of statistical databases, *Univ. Beograd, Publ. Elektrotehn. Fak., Ser. Mat. Fiz.* **14** (2003), 37–48.

[BriMe] Bridges W. G., Mena R. A., Multiplicative cones – a family of three-eigenvalue graphs, *Aequationes Math.* **22** (1981), 208–214.

[Bro] Brouwer, A. E., Small integral trees, *Electronic J. Combin.* **15** (2008), Paper N1.

[BroCN] Brouwer A. E., Cohen A. M., Neumaier A., *Distance-Regular Graphs*, Springer-Verlag (Berlin), 1989.

[BroHae] Brouwer A. E., Haemers W. H., The integral trees with spectral radius 3, *Linear Algebra Appl.* **429** (2008), 2710–2718.

[BroLi] Brouwer A. E., van Lint J. H., Strongly regular graphs and partial geometries, in *Enumeration and Design* (eds. Jackson, D. M., Vanstone S. A.), Academic Press (Toronto) 1984, pp. 85–122.

[BroNe] Brouwer A. E., Neumaier A., The graphs with spectral radius between 2 and $\sqrt{2 + \sqrt{5}}$. *Linear Algebra Appl.* **114/115** (1989), 273–276.

[Brow] Brown W., On the non-existence of a type of regular graph of girth 5, *Canadian J. Math.* **19** (1967), 644–648.

[BruCv] Brualdi R. A., Cvetković D., *A Combinatorial Approach to Matrix Theory and its Application*, CRC Press (Boca Raton), 2008.

[BruRy] Brualdi R. A., Ryser H. J., *Combinatorial Matrix Theory*, Cambridge University Press (Cambridge), 1991.

[BruSo] Brualdi R. A., Solheid E. S., On the spectral radius of connected graphs, *Publ. Inst. Math. Beograd.* **39**(53) (1986), 45–54.

[Bruc] Bruck R. H., Finite nets, II: uniqueness and imbedding, *Pacific J. Math.* **13** (1963), 421–457.

[BuČCS] Bussemaker F. C., Čobeljić S., Cvetković D., Seidel J. J., *Computer Investigation of Cubic Graphs*, Technological University Eindhoven, T. H Report 76-WSK-01, 1976.

[BuCv] Bussemaker F. C., Cvetković D., There are exactly 13 connected, cubic, integral graphs, *Univ. Beograd, Publ. Elektrotehn. Fak., Ser. Mat. Fiz.*, **544–576** (1976), 43–48.

[BuCS1] Bussemaker F. C., Cvetković D., Seidel J. J., Graphs Related to Exceptional Root Systems, Technological University of Eindhoven, T. H. Report 76-WSK-05, 1976.

[BuCS2] Bussemaker F. C., Cvetković D., Seidel J. J., Graphs related to exceptional root systems, in: *Combinatorics I, II*, Proc. 5th Hungarian Coll. on Combinatorics, Keszthely 1976 (eds. Hajnal A., Sos V. T.), North-Holland (Amsterdam) 1978, pp. 185–191.

[Cam1] Cameron P. J., *Permutation Groups*, Cambridge University Press (Cambridge), 1991.

[Cam2] Cameron P. J., Strongly regular graphs, in *Selected Topics in Algebraic Graph Theory* (eds. Beineke L. W, Wilson R. J.), Cambridge University Press (Cambridge) 2004, pp. 203–221.

[Cam3] Cameron P. J., Automorphisms of graphs, in *Selected Topics in Algebraic Graph Theory* (eds. Beineke L. W, Wilson R. J.), Cambridge University Press (Cambridge) 2004, pp. 137–155.

[CamGS] Cameron P. J., Goethals J. M., Seidel J. J., Strongly regular graphs having strongly regular subconstituents, *J. Algebra* **55** (1978), 257–280.

[CamGSS] Cameron P. J., Goethals J. M., Seidel J. J., Shult E. E., Line graphs, root systems, and elliptic geometry, *J. Algebra* **4** (1976), 305–327.

[CamLi] Cameron P. J., van Lint J. H., *Designs, Graphs, Codes and their Links*, Cambridge University Press (Cambridge), 1991.

[CaoHo] Cao D., Hong Y., Graphs characterized by the second eigenvalue, *J. Graph Theory* **17** (1993), 325–331.

[CapCGH] Caporossi G., Cvetković D., Gutman I., Hansen P., Variable neighborhood search for extremal graphs 2: finding graphs with extremal energy, *J. Chem. Inform. Comp. Sci.* **39** (1999), 984–996.

[CarCRS] Cardoso D. M., Cvetković D., Rowlinson P., Simić S., A sharp lower bound for the least eigenvalue of the signless Laplacian of a finite graph, *Linear Algebra Appl.* **429** (2008), 2770–2780.

[Cay] Cayley A., A theorem on trees, *Quart. J. Math.* **23** (1889), 376–378.

[Cha1] Chang L. C., The uniqueness and non-uniqueness of the triangular association scheme, *Sci. Record* **3** (1959), 604–613.

[Cha2] Chang L. C, Association schemes of partially balanced block designs with parameters $v = 28, n_1 = 12, n_2 = 15, p_{11}^2 = 4$, *Sci. Record* **4** (1960), 12–18.

[Chao] Chao C.-Y., A note on the eigenvalues of a graph, *J. Combin. Theory Ser. B* **10** (1971), 301–302.

[Che] Chen Y., Properties of spectra of graphs and line graphs, *Appl. Math. J. Chinese Univ. Ser. B* **17**(3) (2002), 371–376.

[Cheng] Cheng C.S. Maximizing the total number of spanning trees in a graph: two related problems in graph theory and optimum design theory, *J. Combin. Theory Ser. B* **31** (1981), 240–248.

[ChOz] Chin, F. Y., Ozsoyoglu, G., Auditing and inference control in statistical databases, *IEEE T. Software Eng.*, SE8(6) (1982), 574–582.

[Chi] Chiu P., Cubic Ramanujan graphs, *Combinatorica* **12** (1992), 275–285.

[CDDEKL] Christandl M., Datta N., Dorlas T. C., Ekrt A., Kay A., Landahl A .J., Perfect transfer of arbitrary states in quantum spin networks, v.2, 22 Nov. 2004 <arXiv:quant-ph/0411020>.

[Chr] Christofides N., The shortest Hamiltonian chain of a graph, *SIAM J. Appl. Math.* **19** (1970), 689–696.

[Chu1] Chung F. R. K., Diameters and eigenvalues, *J. Amer. Math. Soc.* **2** (1989), 187–196.

[Chu2] Chung F. R. K., *Spectral Graph Theory*, Amer. Math. Soc. (Providence), 1997.

[ChuFM] Chung F. R. K., Faber V., Manteuffel T. A., An upper bound on the diameter of a graph from eigenvalues associated with its Laplacian, *SIAM J. Discrete Math.* **7** (1994), 443–457.

[CiGr] Cioabă S. M., Gregory D. A., Large matchings from eigenvalues, *Linear Algebra Appl.* **422** (2007), 308–317.

[CiGN] Cioabă S. M., Gregory D. A, Nikiforov V., Extreme eigenvalues of nonregular graphs, *J. Combin. Theory Ser. B* **97** (2007), 483–486.

[Clar] Clarke F. H., A graph polynomial and its application. *Discrete Math.* **3** (1972), 305–315.

[Coh] Cohen A. M., Distance-transitive graphs, in: *Selected Topics in Algebraic Graph Theory* (eds. Beineke L. W, Wilson R. J.), Cambridge University Press (Cambridge) 2004, pp. 222–249.

[Col] Collatz L., *Eigenwertaufgaben mit Technischen Anwendungen* (2nd edn.), Akademische Verlagsgesellschaft Geest & Portig K.-G. (Leipzig), 1963.

[ColSi] Collatz L., Sinogowitz U., Spektren endlicher Grafen, *Abh. Math. Sem. Univ. Hamburg* **21** (1957), 63–77.

[Con1] Constantine G. M., Schur convex functions on the spectra of graphs, *Discrete Math.* **45** (1983) 181–188.

[Con2] Constantine G., Lower bound on the spectra of symmetric matrices with non-negative entries, *Linear Algebra Appl.* **65** (1985), 171–178.

[CoCPS] Cook, W. J., Cunningham W. H., Pulleyblank W. R, Schrijver A., *Combinatorial Optimization*, Wiley (New York), 1998.

[Cou] Coulson C. A., The calculation of resonance and localization energies in aromatic molecules, *J. Chem. Soc.* 1954, 3111–3115.

[CouLM] Coulson C. A., O'Leary B., Mallion R. B., *Hückel Theory for Organic Chemists*, Academic Press (London), 1978.

[CouRu] Coulson C. A., Rushbrooke G. S., Note on the method of molecular orbitals, *Proc. Cambridge Phil. Soc.* **36** (1940), 193–200.

[Cve1] Cvetković D., Graphs and their spectra (Thesis), *Univ. Beograd, Publ. Elektrotehn. Fak. Ser. Mat. Fiz.* **354–356** (1971), 1–50.

[Cve2] Cvetković D. The spectral method for determining the number of trees, *Publ. Inst. Math. (Beograd)* **11** (1971), 135–141.

[Cve3] Cvetković D., Spectrum of the total graph of a graph, *Publ. Inst. Math. (Beograd)* **16**(30) (1973), 49–52.

[Cve4] Cvetković D., Spectra of graphs formed by some unary operations, *Publ. Inst. Math. (Beograd)* **19**(33) (1975), 37–41.

[Cve5] Cvetković D., Cubic integral graphs, *Univ. Beograd, Publ. Elektrotehn. Fak. Ser. Mat. Fiz.* **498–541** (1975), 107–113.

[Cve6] Cvetković D., The main part of the spectrum, divisors and switching of graphs, *Publ. Inst. Math. (Beograd)* **23**(37) (1978), 31–38.

[Cve7] Cvetković D., Some possible directions in further investigation of graph spectra, in *Algebraic Methods in Graph Theory*, Vols. I, II (Szeged, 1978; eds. Lovász L., Sós V. T.), *Colloq. Math. Soc. János Bolyai* 25, North Holland (Amsterdam) 1981, pp. 47–67.

[Cve8] Cvetković D., On graphs whose second largest eigenvalue does not exceed 1, *Publ. Inst. Math (Beograd)* **31**(45) (1982), 15–20.

[Cve9] Cvetković D., Constructing trees with given eigenvalues and angles, *Linear Algebra Appl.* **105** (1988), 1–8.

[Cve10] Cvetković D., Star partitions and the graph isomorphism problem, *Linear Multilin. Algebra* **39** (1995), 109–132.

[Cve11] Cvetković D., Characterizing properties of some graph invariants related to electron charges in the Hückel molecular orbital theory, in *Discrete Mathematical Chemistry (DIMACS Workshop, March 1998)*, Amer. Math. Soc. (Providence), *DIMACS Series in Discrete Mathematics and Theoretical Computer Science* **51** (2000), pp. 79–84.

[Cve12] Cvetković D., On the reconstruction of the characteristic polynomial of a graph, *Discrete Math.* **212** (2000), 45–52.

[Cve13] Cvetković D., Graphs with least eigenvalue −2; a historical survey and recent developments in maximal exceptional graphs, *Linear Algebra Appl.* **356** (2002), 189–210.

[Cve14] Cvetković D., Signless Laplacians and line graphs, *Bull. Acad. Serbe Sci. Arts, Cl. Sci. Math. Natur., Sci. Math.* **131**(2005) No. 30, 85–92.

[CvCK1] Cvetković D., Čangalović, Kovačević-Vujčić V., Semidefinite programming methods for the symmetric traveling salesman problem, in *Integer Programming and Combinatorial Optimization*, Proc. 7th Internat. IPCO Conf., Graz, Austria, June 1999 (eds. Cornuejols G., Burkard R. E., Woeginger G. J.), *Lecture Notes in Comp. Sci.* 1610, Springer (Berlin) 1999, pp. 126–136.

[CvCK2] Cvetković D., Čangalović M., Kovačević-Vujčić V., Semidefinite relaxations of the travelling salesperson problem, *YUJOR* **9** (1999), 157–168.

[CvCK3] Cvetković D., Čangalović M., Kovačević-Vujčić V., Optimization and highly informative graph invariants, in: *Two Topics in Mathematics* (ed. Stanković B.) Zbornik radova **10**(18), Matematicki Institut SANU (Beograd) 2004, pp. 5–39.

[CvDM] Cvetković D., Dimitrijević V., Milosavljević M., *Variations on the Travelling Salesman Theme*, Libra Produkt (Belgrade), 1996.

[CvDo1] Cvetković D., Doob M., On spectral characterizations and embeddings of graphs, *Linear Algebra Appl.* **27** (1979), 17–26.

[CvDo2] Cvetković D., Doob M., Root systems, forbidden subgraphs and spectral characterizations of line graphs, in *Graph Theory*, Proc. Fourth Yugoslav Seminar on Graph Theory, Novi Sad, April 15-16. 1983 (eds. Cvetković D., Gutman I., Pisanski T., Tošić R.), Univ. Novi Sad Inst. Math. (Novi Sad) 1984, pp. 69–99.

[CvDG] Cvetković D., Doob M., Gutman I., On graphs whose eigenvalues do not exceed $\sqrt{2 + \sqrt{5}}$, *Ars Combinatoria* **14** (1982), 225–239.

[CvDGT] Cvetković D., Doob M., Gutman I., Torgašev A., *Recent Results in the Theory of Graph Spectra*, North-Holland (Amsterdam), 1988.

[CvDSa] Cvetković D., Doob M., Sachs H., *Spectra of Graphs*, 3rd edition, Johann Ambrosius Barth Verlag (Heidelberg), 1995.

[CvDS1] Cvetković D., Doob M., Simić S., Some results on generalized line graphs, *Comptes Rendus Math. Rep. Acad. Sci. Canada* **2** (1980), 147–150.

[CvDS2] Cvetković D., Doob M., Simić S., Generalized line graphs, *J. Graph Theory* **5** (1981), 385–399.

[CvFRS] Cvetković D., Fowler P.W., Rowlinson P., Stevanović D., Constructing fullerene graphs from eigenvalues and angles, *Linear Algebra Appl.* **356** (2002), 37–56.

[CvGr] Cvetković D., Grout J., Maximal energy graphs should have a small number of distinct eigenvalues, *Bull. Acad. Serbe Sci. Arts, Cl. Sci. Math. Natur., Sci. Math.* **134** (2007), No. 32, 43–57.

[CvGu1]  Cvetković D., Gutman I., On the spectral structure of graphs having the maximal eigenvalue not greater than two, *Publ. Inst. Math. (Beograd)* **18**(32) (1975), 39–45.

[CvGu2]  Cvetković D., Gutman I., A new spectral method for determining the number of spanning trees, *Publ. Inst. Math. (Beograd)* **29** (1981), 49–52.

[CvGT]  Cvetković D., Gutman I., Trinajstić N., Conjugated molecules having integral graph spectra, *Chem. Phys. Letters* **29** (1974), 65–68.

[CvLe1]  Cvetković D., Lepović M., Cospectral graphs with the same angles and with a minimal number of vertices, *Univ. Beograd, Publ. Elektrotehn. Fak., Ser. Mat.* **8** (1997), 88–102.

[CvLe2]  Cvetković D., Lepović M., Seeking counterexamples to the reconstruction conjecture for the characteristic polynomial of a graph and a positive result, *Bull. Acad. Serbe Sci. Arts, Cl. Sci. Math. Natur., Sci. Math.* **116** (1998), No. 23, 91–100.

[CvLe3]  Cvetković D., Lepović M., Sets of cospectral graphs with least eigenvalue at least −2 and some related results, *Bull. Acad. Serbe Sci. Arts, Cl. Sci. Natur., Sci. Math.* **129** (2004), No. 29, 85–102.

[CvLe4]  Cvetković D., Lepović M., Towards an algebra of SINGs, *Univ. Beograd, Publ. Elektrotehn. Fak., Ser. Mat.* **16** (2005), 110–118.

[CvLe5]  Cvetković D., Lepović M., Cospectral graphs with least eigenvalue at least −2, *Publ. Inst. Math. (Beograd)*, **78**(92) (2005), 51–63.

[CvLi]  Cvetković D., van Lint J. H., An elementary proof of Lloyd's theorem, *Proc Kon. Ned. Akad. V. Wet. A*, **80** (1977), 6–10.

[CvLRS1]  Cvetković D., Lepović M., Rowlinson P., Simić S., A database of star complements of graphs, *Univ. Beograd, Publ. Elektrotehn. Fak., Ser. Mat.* **9** (1998), 103–112.

[CvLRS2]  Cvetković D., Lepović M., Rowlinson P., Simić S., The maximal exceptional graphs, *J. Combin. Theory Ser. B* **86** (2002), 347–363.

[CvPe1]  Cvetković D., Petrić M., A table of connected graphs on six vertices, *Discrete Math.* **50** (1984), 37–49.

[CvPe2]  Cvetković D., Petrić M., Tables of graph spectra, *Univ. Beograd, Publ. Elektrotehn. Fak., Ser. Mat.* **4** (1993), 49–67.

[CvRa]  Cvetković D., Radosavljević Z., A construction of the 68 connected regular graphs, non-isomorphic but cospectral to line graphs, in *Graph Theory*, Proc. Fourth Yugoslav Seminar on Graph Theory, Novi Sad, 15-16 April 1983 (eds. Cvetković D., Gutman I., Pisanski T., Tošić R.), Univ. Novi Sad Inst. Math. (Novi Sad) 1984, pp. 101–123.

[CvRo1]  Cvetković D., Rowlinson P., Some properties of graph angles, *Scientia (Valparaiso) Ser. A* **1** (1988), 41–51.

[CvRo2]  Cvetković D., Rowlinson P., On connected graphs with maximal index, *Publ. Inst. Math. Beograd.* **44**(58) (1988), 29–34.

[CvRo3]  Cvetković D., Rowlinson P., The largest eigenvalue of a graph: a survey, *Linear Multilin. Algebra* **28** (1990), 3–33.

[CvRo4]  Cvetković D., Rowlinson P., Some results in the theory of graph spectra, in: *Topics in Algebraic Graph Theory* (eds. Beineke L. W, Wilson R. J.), Cambridge University Press (Cambridge) 2004, pp. 88–112.

[CvRS1] Cvetković D., Rowlinson P., Simić S. K., On some algorithmic investigations of star partitions, *Discrete Applied Math.* **62** (1995), 119–130.

[CvRS2] Cvetković D., Rowlinson P., Simić S. K., *Eigenspaces of Graphs*, Cambridge University Press (Cambridge), 1997.

[CvRS3] Cvetković D., Rowlinson P., Simić S. K., Some characterizations of graphs by star complements, *Linear Algebra Appl.* **301** (1999), 81–97.

[CvRS4] Cvetković D., Rowlinson P., Simić S. K., Constructions of the maximal exceptional graphs with largest degree less than 28, Technical Report CSM-156, Department of Computing Science and Mathematics, University of Stirling, 2000.

[CvRS5] Cvetković D., Rowlinson P., Simić S. K., Graphs with least eigenvalue $-2$: the star complement technique, *J. Algebraic Combinatorics* **14** (2001), 5–16.

[CvRS6] Cvetković D., Rowlinson P., Simić S. K., The maximal exceptional graphs with largest degree less than 28, *Bull. Acad. Serbe Sci. Arts, Cl. Sci. Math. Natur., Sci. Math.* **22** (2001), No. 26, 115–131.

[CvRS7] Cvetković D., Rowlinson P., Simić S., *Spectral Generalizations of Line Graphs*, Cambridge University Press (Cambridge), 2004.

[CvRS8] Cvetković D., Rowlinson P., Simić S., Graphs with least eigenvalue $-2$; a new proof of the 31 forbidden subgraphs theorem, *Designs Code. Cryptogr.* **34** (2005), 229–240.

[CvRS9] Cvetković D., Rowlinson P., Simić S., Star complements and exceptional graphs, *Linear Algebra Appl.* **423** (2007), 146–154.

[CvRS10] Cvetković D., Rowlinson P., Simić S., Signless Laplacians of finite graphs. *Linear Algebra Appl.* **423** (2007), 155–171.

[CvRS11] Cvetković D., Rowlinson P., Simić S., Eigenvalue bounds for the signless Laplacian, *Publ. Inst. Math. (Beograd)* **81**(95) (2007), 11–27.

[CvSi1] Cvetković D., Simić S., Non-complete extended $p$-sum of graphs, graph angles and star partitions, *Publ. Inst. Math. (Beograd)* **53**(67) (1993), 4–16.

[CvSi2] Cvetković D., Simić S., On the graphs whose second largest eigenvalue does not exceed $(\sqrt{5} - 1)/2$, *Discrete Math.* **138** (1995), 213–227.

[CvSi3] Cvetković D., Simić S., The second largest eigenvalue of a graph – a survey, Int. Conf. on Algebra, Logic & Discrete Math., Niš, 14-16 April 1995 (eds. Bogdanović S., Ćirić M., Perović Ž.), *FILOMAT* **9** (1995) No. 3, 449–472.

[CvSi4] Cvetković D., Simić S., Minimal graphs whose second largest eigenvalue is not less than $(\sqrt{5} - 1)/2$, *Bull. Acad. Serbe Sci. Arts, Cl. Sci. Math. Natur., Sci. Math.* **121** (2000), No. 25, 47–70.

[CvSi5] Cvetković D., Simić S., Towards a spectral theory of graphs based on the signless Laplacian I, *Publ. Inst. Math. (Beograd)* **85**(99) (2009), 19–33.

[CvSi6] Cvetković D., Simić S., Towards a spectral theory of graphs based on the signless Laplacian II, *Linear Algebra Appl.*, to appear

[CvSiS] Cvetković D., Simić S., Stevanović D., 4-regular integral graphs, *Univ Beograd, Publ. Elektrotehn. Fak., Ser. Mat.* **9** (1998), 89–102.

[CvSt] Cvetković D., Stevanović D., Graphs with least eigenvalue at least $-\sqrt{3}$, *Publ. Inst. Math. (Beograd)* **73**(87) (2003), 39–51.

[Dam1] van Dam E. R., Regular graphs with four eigenvalues, *Linear Algebra Appl.* **228** (1995), 139–162.

[Dam2] van Dam E. R., Nonregular graphs with three eigenvalues, *J. Combin. Theory Ser. B*, **73** (1998), 101–118.

[Dam3] van Dam E. R., Strongly regular decompositions of the complete graph, *J. Algebraic Combinatorics* **17** (2003), 181–201.

[DamHa1] van Dam E. R., Haemers W. H., Eigenvalues and the diameter of graphs, *Center for Economic Research Discussion Paper No. 9343*, Tilburg University, The Netherlands, 1993.

[DamHa2] van Dam E. R., Haemers W. H., Eigenvalues and the diameter of graphs, *Linear Multilin. Algebra* **39** (1995), 33–44.

[DamHa3] van Dam E. R., Haemers W. H., Spectral characterizations of some distance-regular graphs, *J. Algebraic Combinatorics* **15** (2002), 189–202.

[DamHa4] van Dam E. R., Haemers W. H., Which graphs are determined by their spectrum?, *Linear Algebra Appl.* **373** (2003), 241–272.

[DamHK] van Dam E. R., Haemers W. H., Koolen J. H., Cospectral graphs and the generalized adjacency matrix, *Linear Algebra Appl.* **423** (2007), 33–41.

[DamSp] van Dam E. R., Spence E., Small regular graphs with four eigenvalues, *Discrete Math.* **189** (1998), 233–257.

[DAGT] D'Amato S. S., Gimarc B. M., Trinajstić N., Isospectral and subspectral molecules, *Croat. Chem. Acta* **54** (1981), 1–52.

[DanHa1] Daneshgar A., Hajiabolhassan H., Graph homomorphisms through random walks, *J. Graph Theory* **44** (2003), 15–38.

[DanHa2] Daneshgar A., Hajiabolhassan H., Graph homomorphisms and nodal domains, *Linear Algebra Appl.* **418** (2006), 44–52.

[Das1] Das K. Ch., An improved upper bound for Laplacian graph eigenvalues, *Linear Algebra Appl.* **368** (2003), 269–278.

[Das2] Das K. Ch., A characterization of graphs which achieve the upper bound for the largest Laplacian eigenvalue of graphs, *Linear Algebra Appl.* **376** (2004), 173–186.

[Das3] Das, K. Ch., A sharp upper bound for the number of spanning trees of a graph, *Graph. Combinator.* **23** (2007), 625–632.

[DasKu] Das K. Ch., Kumar P., Some new bounds on the spectral radius of graphs, *Discrete Math.* **281** (2004), 149–161.

[DavSV] Davidoff G., Sarnak P., Valette A., *Elementary Number Theory, Group Theory and Ramanujan Graphs*, Cambridge University Press (Cambridge), 2003.

[Ded] Dedo E., La reconstruibilita del polinomio carrateristico del comutato di un grafo, *Boll. Unione Mat. Ital.* **18A** (1981), 423–429.

[DedPo] Dedo E., Porcu L., Algebraic properties of line multidigraphs of a multidi-graph, *J. Combin. Inform. System Sci.* **12** (1987), 113–118.

[DelSo] Delorme C., Poljak S., Diameter, covering index, covering radius and eigen-values, *European J. Combin.* **12** (1991), 95–108.

[DemKM] Demetrovics J., Katona G. O. H., Miklós D., On the security of individual data, *Ann. Math. and Artificial Intelligence* **46** (2006), 98–113.

[DesRa] Desai M., Rao V., A characterization of the smallest eigenvalue of a graph, *J. Graph Theory* **18** (1994), 181–194.

[Dia] Dias J. R., *Molecular Orbital Calculations Using Chemical Graph Theory*, Springer-Verlag (Berlin), 1983.

[DieFM] Diekmann R., Frommer A., Monien B., Efficient schemes for nearest neigh-bor load balancing, *Parallel Comput.* **25** (1999), 789–812.

[DinKZ] Dinic E. A., Kel'mans A. K., Zaitsev M. A., Non-isomorphic trees with the same $T$-polynomial, *Inform. Process. Lett.* **6** (1977), 73–76.

[Doo1] Doob M., A geometrical interpretation of the least eigenvalue of a line graph., in: *Proc. Second Conference on Comb. Math. and Appl.*, Univ. North Carolina (Chapel Hill), 1970, pp. 126–135.

[Doo2] Doob M., Graphs with a small number of distinct eigenvalues, *Ann. New York Acad. Sciences* **175** (1970), 104–110.

[Doo3] Doob M., On the spectral characterization of the line graph of a BIBD., in *Proc. Second Louisiana Conf. on Comb., Graph Theory and Computing, 8-11 March 1971* (eds. Mullin R. C., Reid K. B., Roselle D. P., Thomas R. D.), Louisiana State University (Baton Rouge), 1971, pp. 225–234.

[Doo4] Doob M., On the spectral characterization of the line graph of a BIBD, II., in Proc. Manitoba Conference on Numerical Mathematics, University of Manitoba, 1971, pp. 117–126.

[Doo5] Doob M., On embedding a graph in an isospectral family, in *Proc. 2nd Manitoba Conference on Numerical Mathematics*, Utilitas Mathematica (Winnipeg, Manitoba) 1973, pp. 137–142.

[Doo6] Doob M., A spectral characterization of the line graph of a BIBD with $\lambda = 1$, *Linear Alg. Appl.* **12** (1975), 11–20.

[Doo7] Doob M., A note on eigenvalues of a line graph., in *Proc. Conf. on Algebraic Aspects of Combinatorics, University of Toronto, January 1975* (eds. Corneil D., Mendelsohn E.), *Congr. Num.* XIII, Utilitas Mathematica. (Winnipeg, Manitoba), 1975, pp. 209–211.

[Doo8] Doob M., Seidel switching and cospectral graphs with four distinct eigenvalues, *Ann. New York Acad. Sci.* **319** (1979), 164–168.

[DooCv] Doob M., Cvetković D., On spectral characterizations and embedding of graphs, *Linear Algebra Appl.* **27** (1979), 17–26.

[DooHa] Doob M., Haemers W. H., The complement of the path is determined by its spectrum, *Linear Algebra Appl.* **356** (2002), 57–65.

[Ell] Ellingham M. N., Basic subgraphs and graph spectra, *Australasian J. Combinatorics* **8** (1993), 247–265.

[ElKM] Elsässer R., Královič R., Monien B., Sparse topologies with small spectrum size, *Theor. Comput. Sci.* **307** (2003), 549–565.

[ErRS] Erdös P., Rényi A., Sós V., On a problem in graph theory, *Studies Math. Hungar.* **1** (1966), 215–235.

[Far] Faria I., Permanental roots and the star degree of a graph, *Linear Algebra Appl.* **64** (1985), 255–265.

[FaGr] Farrell E. J., Grell J. C., Some further constructions of cocircuit and cospectral graphs, *Carib. Math.* **4** (1985), 17–28.

[FavMS1] Favaron O., Mahéo M., Saclé J.-F., Some eigenvalue properties in graphs (conjectures of Graffiti, II), *Discrete Math.* **111** (1993), 197–220.

[FeHi] Feit W., Higman G., The nonexistence of certain generalized polygons, *J. Algebra* **1** (1964), 114–131.

[Fie1] Fiedler M., Algebraic connectivity of graphs, *Czech. Math. J.* **23**(98) (1973), 298–305.

[Fie2] Fiedler M., A property of eigenvectors of nonnegative symmetric matrices and its application to graph theory, *Czech. Math. J.* **25**(100) (1975), 619–633.

[Fie3] Fiedler M., An algebraic approach to connectivity of graphs, in *Recent Advances in Graph Theory* (Proc. Second Czechoslovak Sympos., Prague, 1974), Academia (Prague) 1975, pp. 193–196.

[Fie4] Fiedler M., Absolute algebraic connectivity of trees, *Linear and Multilinear Algebra* **26** (1990), 85–106.

[Fie5] Fiedler M., A geometric approach to the Laplacian matrix of a graph, in *Combinatorial and Graph-Thoeretical Problems in Linear Algebra* (eds. Brualdi R. A., Friedland S. and Klee V.), Springer-Verlag (New York) 1993, pp. 73–98.

[FieSe] Fiedler M., Sedláček J., O w-basich orientovaných grafu (Czech.), *Časopis. Pěst. Mat.* **83** (1958), 214–225.

[Fin] Finck H.-J., Vollstandiges Produkt, chromatische Zahl und characteristisches Polynom regulärer Graphen II, *Wiss. Z., T. H. Ilmenau* **11** (1965), 81–87.

[FiGr] Finck H.-J., Grohmann, G., Vollstandiges Produkt, chromatische Zahl und characteristisches Polynom regulärer Graphen I, *Wiss. Z., T. H. Ilmenau* **11** (1965), 1–3.

[Fis] Fisher M., On hearing the shape of a drum, *J. Combin. Theory* **1** (1966), 105–125.

[GabGa] Gabber O, Galil Z., Explicit constructions of linear-sized superconcentrators, *J. Comput. Syst. Sci.* **22** (1981), 307–420.

[GarJo] Garey M. R., Johnson D. S., *Computers and Intractability: A Guide to the Theory of N P-Completness*, Freeman (San Francisco), 1979.

[Gan] Gantmacher F. R., *The Theory of Matrices*, Chelsea (New York), 1959.

[Gib] Gibbons A., *Algorithmic Graph Theory*, Cambridge University Press (Cambridge), 1985.

[God] Godsil C. D. Equiarboreal graphs, *Combinatorica* **1** (1981), 163–167.

[GoHMK] Godsil C. D., Holton D. A., McKay B. D., The spectrum of a graph, in *Combinatorial Mathematics V* (ed. Little C. H. C.), Lecture Notes in Math. 622, Springer-Verlag (Berlin) 1977, pp. 91–117.

[GoMK1] Godsil C. D., McKay B. D., Some computational results on the spectra of graphs, in *Combinatorial Mathematics IV*, Proc. 4th Australian Conf. held at the Univ. of Adelaide, Aug. 27-29, 1975 (eds. Casse L. R. A., Wallis W. D.), Springer-Verlag (Berlin) 1976, pp. 73–92.

[GoMK2] Godsil C. D., McKay B. D., Constructing cospectral graphs, *Aequationes Math.* **25** (1982), 257–268.

[GoRo] Godsil C., Royle G., *Algebraic Graph Theory*, Springer (New York), 2001.

[Goe] Goemans M., Semidefinite programming in combinatorial optimization, *Math. Program.* **79** (1997), 143–161.

[GoeWi] Goemans M., Williamson D. P., Improved approximation algorithms for maximum cut and satisfiability problems using semidefinite programming, *J. ACM* **42** (1995), 1115–1145.

[GrGT] Graovac A., Gutman I., Trinajstić N., *A Topological Approach to the Chemistry of Conjugated Molecules*, Springer-Verlag (Berlin), 1977.

[Gri] Grimmett G. R., An upper bound for the number of spanning trees of a graph, *Discrete Math.* **16** (1976), 323–324.

[Gro] Grone R., Eigenvalues and the degree sequences of graphs, *Linear Multilin. Algebra* **39** (1995), 133–136.

[GroMe1] Grone R., Merris, R., A bound for the complexity of a simple graph, *Discrete Math.* **69** (1988), 97–99.

[GroMe2] Grone R., Merris, R., The Laplacian spectrum of a graph, II, *SIAM J. Discrete Math.* **7** (1994), 221–229.

[Guo1] Guo J.-M., A new upper bound for the Laplacian spectral radius of graphs, *Linear Algebra Appl.* **400** (2005), 61–66.

[Guo2] Guo J.-M., The effect on the Laplacian spectral radius of a graph by adding or grafting edges, *Linear Algebra Appl.* **413** (2006), 59–71.

[Gut1] Gutman I., Bounds for total $\pi$-election energy, *Chem. Phys. Letters* **24** (1974), 283–285.

[Gut2] Gutman I., Acyclic systems with extremal Hückel $\pi$-electron energy, *Theoretica Chimica Acta* **45** (1977), 79–87.

[Gut3] Gutman I., The energy of a graph, *Berichte Math. Stat. Sekt. Forschungszentrum Graz* **103** (1978), 1–22.

[Gut4] Gutman I., New approach to the McClelland approximation, *MATCH Commun. Math. Chem.* **14** (1983), 71–81.

[Gut5] Gutman I., The energy of a graph: old and new results, in *Algebraic Combinatorics and Applications* (eds. Betten A., Kohnert A. , Laue A. R, Wassermann A.), Springer-Verlag (Berlin) 2001, pp. 196–211.

[Gut6] Gutman I., Topology and stability of conjugated hydrocarbons: the dependence of total $\pi$-electron energy on molecular topology, *J. Serbian Chem. Soc.* **70** (2005), 441–456.

[Gut7] Gutman I., Chemical graph theory – the mathematical connection, in *Advances in Quantum Chemistry* 51 (eds. Sabin S. R., Brändas E. J.), Elsevier (Amsterdam) 2006, pp. 125–138.

[GuCv] Gutman I., Cvetković D., The reconstruction problem for characteristic polynomials of graphs, *Univ. Beograd, Publ. Elektrotehn. Fak., Ser. Mat. Fiz.*, **498–541** (1975), 45–48.

[GuPo] Gutman I., Polansky O. E. , *Mathematical Concepts in Organic Chemistry*, Springer-Verlag (Berlin), 1986.

[GuTr] Gutman I., Trinajstić N., Graph theory and molecular orbitals, *Topics Curr. Chem.* **42** (1973), 49–93.

[Hae1] Haemers W. H., A generalization of the Higman-Sims technique, *Proc. Kon. Ned. Akad. Wet. A* **81**(4) (1978), 445–447.

[Hae2] Haemers W. H., Interlacing eigenvalues and graphs, *Linear Algebra Appl.* **227–228** (1995), 593–616.

[HaeLZ] Haemers W. H., Liu X., Zhang Y., Spectral characterizations of lollipop graphs, *Linear Algebra Appl.* **428** (2008), 2415–2423.

[HaeSp] Haemers W. H., Spence E., Enumeration of cospectral graphs, *Europ. J. Comb.* **25** (2004), 199–211.

[Hag] Hagos E. M., The characteristic polynomial of a graph is reconstructible from the characteristic polynomials of its vertex-deleted subgraphs and their complements, *Electr. J. Comb.* **7** (2000), R12.

[Hak] Hakimi S., On the realizability of a set of integers as degrees of the vertices of a graph, *J. SIAM Appl. Math.* **10** (1962), 492–506.

[Hal] Halmos P. R., *Finite-dimensional Vector Spaces*, 2nd edn., van Nostrand (Princeton), 1958.

[HaKe] Hammer P. L. Kelmans A. K., Laplacian spectra and spanning trees of threshold graphs, *Discrete Appl. Math.* **65** (1996), 255–273.

[Ham]   Hammersley J. M., The friendship theorem and the love problem, in *Surveys in Combinatorics* (ed. Lloyd E. K.), Cambridge University Press (Cambridge) 1983, pp. 31–54.

[HanMS]  Hansen P., Melot H., Stevanović D., Integral complete split graphs, *Univ. Beograd, Publ. Elektrotehn. Fak., Ser. Mat.* **13** (2002), 89–95.

[Har1]   Harary F., The determinant of the adjacency matrix of a graph, *SIAM Review* **4** (1962), 202–210.

[Har2]   Harary F., *Graph Theory*, Addison-Wesley (Reading), 1969.

[HarKMR]  Harary F., King C., Mowshowitz A., Read R. C., Cospectral graphs and digraphs, *Bull. London Math. Soc.* **3** (1971), 321–328.

[HarSc]  Harary F., Schwenk A. J, Which graphs have integral spectra?, in *Graphs and Combinatorics*, Proc. Capit. Conf. Graph Theory and Combinatorics, George Washington Univ., June 1973 (eds. Bari R., Harary F.), Springer-Verlag (New York) 1974, pp. 45–51.

[Hat]   Hattori Y., Nonisomorphic graphs with the same $T$–polynomial, *Inform. Process. Lett.* **22** (1986), 133–134.

[Hay]   Haynsworth E. V., Applications of a theorem on partitioned matrices, *J. Res. Nat. Bureau Stand.* **62** (1959), 73–78.

[Hei]   Heilbronner E., Some comments on cospectral graphs, *MATCH* **5** (1979), 105–133.

[Herm]  Hermann E. C., On the relevance of isospectral nonisomorphic graphs for chemistry, *MATCH* **19** (1986), 43–52.

[Hern1]  Herndon W. C., The characteristic polynomial does not uniquely determine molecular topology, *J. Chem. Doc.* **14** (1974), 150–151.

[Hern2]  Herndon W. C., Isospectral molecules, *Tetrahedron Letters* **8** (1974), 671–674.

[HeEl1]  Herndon W. C., Ellzey M. L. Jr, Isospectral graphs and molecules, *Tetrahedron* **31** (1975), 99–107.

[HeEl2]  Herndon W. C., Ellzey M. L. Jr, The construction of isospectral graphs, *MATCH* **20** (1986), 53–79.

[HiSi]   Higman D. G., Sims, C. C., A simple group of order 44,352,000, *Math. Z.* **105** (1968), 110–113.

[Hof1]   Hoffman A. J., On the exceptional case in a characterization of the arcs of a complete graph, *IBM J. Res. Develop.* **4** (1960), 487–496.

[Hof2]   Hoffman A. J., On the uniqueness of the triangular association scheme, *Ann. Math. Stat.* **31** (1960), 492–497.

[Hof3]   Hoffman A. J., On the polynomial of a graph, *Amer. Math. Monthly* **70** (1963), 30–36.

[Hof4]   Hoffman A. J., On the line graph of a projective plane, *Proc. Amer. Math. Soc.* **16** (1965), 297–302.

[Hof5]   Hoffman A. J., Some recent results on spectral properties of graphs, in: *Beiträge zur Graphentheorie* (eds. Sachs H., Voss H-J., Walther H.), Teubner (Leipzig) 1968, pp. 75–80.

[Hof6]   Hoffman A. J., On eigenvalues and colorings of graphs, in *Graph Theory and its Applications* (ed. Harris B.) Academic Press (New York) 1970, pp. 79–91.

[Hof7]   Hoffman A. J., On graphs whose least eigenvalues exceeds $-1 - \sqrt{2}$, *Linear Algebra Appl.* **16** (1977), 153–165.

[Hof8] Hoffman A. J., On limit points of the least eigenvalue of a graph, *Ars Combinatoria* **3** (1977), 3–14.

[HofRa1] Hoffman A. J., Ray-Chaudhuri D. K., On the line graph of a finite affine plane, *Canad. J. Math.* **17** (1965), 687–694.

[HofRa2] Hoffman A. J., Ray-Chaudhuri D. K., On the line graph of a symmetric balanced incomplete block design, *Trans. Amer. Math. Soc.* **116** (1965), 238–252.

[HofRa3] Hoffman A. J., Ray-Chaudhuri D. K., On a spectral characterization of regular line graphs, unpublished manuscript.

[HofSi] Hoffman A. J., Singleton R. R., On Moore graphs with diameters 2 and 3, *IBM J. Res. Develop.* **4** (1960), 497–504.

[HofSm] Hoffman A. J., Smith J. H., On the spectral radii of topologically equivalent graphs, in *Recent Advances in Graph Theory* (ed. Fiedler M.), Academia Praha 1975, pp. 273–281.

[HofWH] Hoffman A. J., Wolfe P., Hofmeister M., A note on almost regular matrices, *Linear Algebra Appl.* **226–228** (1995), 105–108.

[Hofm] Hofmeister M., A note on almost regular graphs, *Math. Nachr.* **166** (1994), 259–262.

[Hon1] Hong Y., The $k$-th largest eigenvalue of a tree, *Linear Algebra Appl.* **73** (1986), 151–155.

[Hon2] Hong Y., Bounds of eigenvalues of a graph, *Acta Math. Appl. Sinica* **2** (1988), 165–168.

[Hon3] Hong Y., On the least eigenvalue of a graph, *System Sci. Math. Sci.* **6** (1993), 269–275.

[HonZh] Hong Y., Zhang X.-D., Sharp upper and lower bounds for largest eigenvalue of the Laplacian matrices of trees, *Discrete Math.* **296** (2005), 187–197.

[How] Howes L., On subdominantly bounded graphs – summary of results, in *Recent Trends in Graph Theory*, Proc. First New York City Graph Theory Conf., 11–13 June 1970 (eds. Capobianco M., Frechen J. B., Krolik M.), Springer-Verlag (Berlin) 1971, pp. 181–183.

[Hua] Hua H., On minimal energy of unicyclic graphs with prescribed girth and pendant vertices, *MATCH Commun. Math. Comput. Chem.* **57** (2007), 351–361.

[Hub] Hubaut X. L, Strongly regular graphs, *Discrete Math.* **13** (1975), 357–381.

[Huc] Hückel E., Quantentheoretische Beiträge zum Benzolproblem, *Z. Phys.* **70** (1931), 204–286.

[Hut] Hutschenreuter H., Einfacher Beweis des Matrix-Gerüst-Satzesder Netzwerk-theorie, *Wiss. Z. TH Ilmenau* **13** (1967), 403–404.

[JaRo] Jackson P. S., Rowlinson P., On graphs with complete bipartite star complements, *Linear Algebra Appl.* **298** (2000), 9–20.

[JeSi] Jerrum M., Sinclair A., Approximating the permanent, *SIAM J. Computing* **18** (1989), 1149–1178.

[Jia] Jiang Yuan-sheng, Problem on isospectral molecules, *Sci. Sinica (B)* **27** (1984), 236–248.

[JoMa] John P., Mallion R. B., An algorithmic approach to the number of spanning trees in Buckminsterfullerene, *J. Math. Chem.* **15** (1994), 261–271.

[JoSa1] John P., Sachs H., Calculating the characteristic polynomial, eigenvectors and number of spanning trees of hexagonal systems, *J. Chem. Soc. Faraday Trans.* **86** (1990), 1033–1039.

[JoSa2] John P., Sachs H., Calculating the numbers of perfect matchings and spanning trees, Pauling's orders, the characteristic polynomial, and the eigenvalues of a bensenoid system, in *Topics in Current Chemistry*, vol. 153, Springer-Verlag (Berlin-Heidelberg) 1990, pp. 145–179.

[Kac] Kac M., Can one hear a shape of a drum?, *Amer. Math. Monthly* **73** (April 1966), Part II, 1–23.

[KaMS] Karger D., Motwani R., Sudan M. , Approximate graph coloring by semidefinite programming, *J. ACM* **45** (1998), 246–265.

[KasÖs] Kaski P., Östergård, There are exactly five biplanes with $k = 11$, *J. Combinatorial Designs* **16** (2008), 117–127.

[Kast] Kasteleyn P. W., Graph theory and crystal physics, in *Graph Theory and Theoretical Physics* (ed. Harary F.), Academic Press (London) 1967, pp. 43–110.

[Kel1] Kel'mans A. K., The number of trees in a graph I, *Automat. i Telemeh.* 26(1965), 2194–204 (Russian); transl. *Automat. Remote Control* **26** (1965), 2118–2129.

[Kel2] Kel'mans A. K., The number of trees in a graph II, *Automat. i Telemeh.* 27(1966), 56–65 (Russian); transl. *Automat. Remote Control* **27**(1966), 233–241.

[Kel3] Kel'mans A. K., Properties of the characteristic polynomial of a graph, *Kibernetiky na sluzbu kommunizmu* (Russian) *Energija* (Moskva-Leningrad) **4** (1967), 27–41 .

[Kel4] Kel'mans A. K., Comparison of graphs by their number of spanning trees, *Discrete Math.* **16** (1976), 241–61. (Erratum: *J. Combin. Theory Ser. B* 24 (1978), 375.)

[KelCh] Kel'mans A. K., Chelnokov V. M., A certain polynomial of a graph and graphs with an extremal number of trees, *J. Combin. Theory Ser. B* **16** (1974), 197–214.

[Kirc] Kirchhoff G., Über die Auflösung der Gleichungen, auf welche man bei der Untersuchung der linearen Verteilung galvanischer Ströme geführt wird, *Ann. Phys. Chem.* **72** (1847), 497–508.

[Kir1] Kirkland S., Completion of Laplacian integral graphs via edge addition, *Discrete Math.* **295** (2005), 75–90.

[Kir2] Kirkland S., A note on a distance bound using eigenvalues of the normalized Laplacian matrix, *Electronic J. Linear Algebra* **16** (2007), 204–207.

[Kir3] Kirkland S., Laplacian integral graphs with maximum degree 3, *Electronic J. Comb.* **15** (2008), Paper R120.

[KMSTKR] Knop J. V., Muller W. R., Szymanski K., Trinajstić N., Kleiner A. F., Randić M., On irreducible endospectral graphs, *J. Math. Phys.* **27** (1986), 2601–2612.

[KoSu] Kolmykov V. A., Subotin V. F., Spectra of graphs and cospectrality, Voronezh University, Voronezh. (manuscript no. 6708-83 DEP, VINITI 12 Dec. 1983), 1983.

[KooMo1] Koolen J., Moulton V., Maximal energy graphs, *Advances in Appl. Math.* **26** (2001), 47–52.

[KooMo2] Koolen J., Moulton V., On a conjecture of Bannai and Ito: there are only finitely many distance-regular graphs with degree 5, 6 or 7, *European J. Combin.* **23** (2002), 987–1006.

[KotLo] Kotlov A., Lovász L., The rank and size of graphs, *J. Graph Theory* **23** (1996), 185–189.

[KrPa1] Krishnamoorthy V., Parthasarathy K. R., A note on non-isomorphic cospectral digraphs, *J. Combin. Theory Ser. B* **17** (1974), 39–40.

[KrPa2] Krishnamoorthy V., Parthasarathy K. R., Cospectral graphs and digraphs with given automorphism group, *J. Combin. Theory Ser. B* **19** (1975), 204–213.

[Kri] Krishnamurthy E. V., A form invariant multivariable polynomial representation of graphs, in *Combinatorics and Graph Theory*, Proc. 2nd Symp. held at the Indian Statistical Institute, Calcutta, February 25-29, 1980, Lecture Notes in Math., 885 (ed. Rao S. B.), Springer-Verlag (Berlin) 1981, pp. 18–32.

[Lap] Laporte G., The traveling salesman problem: an overview of exact and approximate algorithms, *European J. Operational Research* **59** (1992), 231–247.

[LauSc] Lauri J., Scapellato R., *Topics in Graph Automorphisms and Reconstruction*, Cambridge University Press (Cambridge), 2003.

[LawLRS] Lawler E. L., Lenstra J. K., Rinnooy Kan A. H. G., Shmoys D. B., *The Traveling Salesman Problem*, Wiley (New York), 1985.

[LemSe] Lemmens P. W. H., Seidel J. J., Equiangular lines, *J. Algebra* **24** (1973), 494–512.

[Lep1] Lepović M., Some statistical data on graphs with 10 vertices, *Univ. Beograd, Publ. Elektrotehn Fak., Ser. Mat.* **9** (1998), 79–88.

[Lep2] Lepović M., private communication.

[Lep3] Lepović M., On conjugate adjacency matrices of graphs, *Discrete Math.* **307** (2007), 730–738.

[LepGu] Lepović M., Gutman I., No starlike trees are cospectral, *Discrete Math.* **242** (2002), 291–295.

[LepSBZ] Lepović M., Simić S. K., Balińska K. T., Zwierzyński K. T., There are 93 non-regular, bipartite integral graphs with maximum degree four, Technical University of Poznań, CSC Report No. 511, Poznań, 2005.

[Li] Li J., Ph.D. thesis, The University of Manitoba, Winnipeg, 1994.

[LiPa] Li J.-S., Pan Y.L., De Caen's inequality and bounds on the largest Laplacian eigenvalue of a graph, *Linear Algebra Appl.* **328** (2001), 153–160.

[LiZh1] Li J-S., Zhang X-D., A new upper bound for eigenvalues of the Laplacian matrix of a graph, *Linear Algebra Appl.* **265** (1997), 93–100.

[LiZh2] Li J-S., Zhang X-D., On the Laplacian eigenualues of a graph, *Linear Algebra Appl.* **285** (1998), 305–307.

[LiFe] Li Q., Feng K. E., On the largest eigenvalue of graphs (Chinese), *Acta Math. Appl. Sinica* **2** (1979), 167–175.

[LiWZ] Lin Liang-tang, Wang Nan-qin, Zhang Qian-er, Isospectral molecules, *Acta Univ. Amoiensis Scientiarum Naturalium* **2** (1979), 65–75.

[LinSe] van Lint, J. H., Seidel J. J., Equilateral point sets in elliptic geometry, *Proc. Nederl. Acad. Wetensch, Ser. A* **69** (1966), 335–348.

[LiuRo] Liu B., Rowlinson P., Dominating properties of star complements, *Publ. Inst. Math. Beograd* **68**(82) (2000), 46–52.

[LuPS] Lubotzky A., Phillips R., Sarnak P., Ramanujan graphs, *Combinatorica* **8** (1988), 261–277.

[MaMi] Marcus M., Minc H., *A Survey of Matrix Theory and Matrix Inequalities*, Allyn and Bacon (Boston), 1964.

[Mar] Margulis G. A., Explicit construction of concentrators, *Problemy Peredaci Informacii* **9**(4) (1973), 71–80 (trans. Problems Inform. Transmission (1975)).

[Mas] Masuyama M., A test for graph isomorphism, *Repts. Statist. Appl. Res. Union Japan Sci. Eng.* **20** (1973), 41–64.

[McC] McClelland B. J, Properties of the latent roots of a matrix: the estimation of $\pi$-electron energies, *J. Chem. Phys.* **54** (1971), 640–643.

[McK] McKay B.D., Spanning trees in random regular graphs, Vanderbilt Univ., Computer Sci. Technical Report CS-81-05 (1981).

[McL] McLaughlin J., A simple group of order 898,128,000, in *Theory of Finite Groups* (eds. Brauer R., Sah C-H.), Benjamin (New York) 1969, pp. 109–111.

[Mer1] Merris R., Laplacian matrices of graphs: a survey, *Linear Algebra Appl.* **197–198** (1994), 143–176.

[Mer2] Merris R., Degree maximal graphs are Laplacian integral, *Linear Algebra Appl.* **199** (1994) 381–389.

[Mer3] Merris R., A survey of graph Laplacians, *Linear Multilin. Algebra* **39** (1995), 19–31.

[Mer4] Merris R., A note on Laplacian graph eigenvalues, *Linear Algebra Appl.*, **285** (1998), 33–35.

[Mer5] Merris R., *Graph Theory*, Wiley (New York), 2001.

[Mes] Mesner D. M., A new family of partially balanced incomplete block designs, *Ann. Math. Statist.* **38** (1967), 571–581.

[Mey] Meyer J.F., Algebraic isomorphism invariants for graphs of automata, in *Graph Theory and Computing* (including part of the Proc. of a Conf. held at the University of the West Indies, Kingston, Jamaica, January 1969, ed. Read R.C.), Academic Press (New York) 1972, pp. 123–152.

[Mil] Milić M., Flow-graph evaluation of the characteristic polynomials of a matrix, *IEEE Trans. Circuit Theory* **CT-11** (1964), 423–424.

[Mir] Mirsky L., Inequalities for normal and Hermitian matrices, *Duke Math. J.* **24** (1957), 591–599.

[Mnu] Mnuhin V. B., Spectra of graphs under certain unary operations (Russian), in *Nekotorye Topolologicheskie i Kombinator: Svoistva Grafov*, Akad. Nauk. Ukrain. SSR Inst. Mat. Preprint No.8 (1980), 38–44.

[Moh1] Mohar B., Isoperimetric number of graphs, *J. Combin. Theory Ser. B* **47** (1989), 274–291.

[Moh2] Mohar B., The Laplacian spectrum of graphs, in *Graph Theory, Combinatorics and Applications* (eds. Alavi Y., Chartrand G., Oellerman O. E., Schwenk A. J.), Wiley (New York) 1991, pp. 871–898.

[Moh3] Mohar B., Eigenvalues, diameter, and mean distance in graphs, *Graph. Combinator.* **7** (1991), 53–64.

[Moh4] Mohar B., Graph Laplacians, in *Topics in Algebraic Graph Theory* (eds. Beineke L. W., Wilson R. J.), Cambridge University Press (Cambridge), 2004, pp. 113–136.

[Moh5] Mohar B., Some algebraic methods in graph theory and combinatorial optimization, *Discrete Math.*, to appear.

[MohPo1] Mohar B., Poljak S., Eigenvalues and the max-cut problem, *Czech. Math. J*, **40**(115) (1990), 343–352.

[MohPo2] Mohar B., Poljak S., Eigenvalues in combinatorial optimization, *IMA Preprint series* # 939, Inst. Math. Appl., University of Minnesota, 1992.

[MohPo3]  Mohar B., Poljak S., Eigenvalues in combinatorial optimization, in *Combinatorial and Graph-Theoretical Problems in Linear Algebra*, (eds. Brualdi R. A. et al.), *IMA Volumes in Mathematics and Its Applications*, vol. 50, Springer-Verlag (Berlin) 1993, pp. 107–151.

[Mon]  Montroll E. W., Lattice statistics, in *Applied Combinatorial Mathematics* (ed. Beckenbach E. F.), Wiley (New York), 1964, pp. 96–143.

[Mor1]  Morgenstern M., Explicit construction of natural bounded concentrators, in *32nd Annual Symposium on Foundations of Computer Science* 404, IEEE Computer Society Press, 1991, pp. 392–397.

[Mor2]  Morgenstern M., Existence and explicit construction of $q + 1$ regular Ramanujan graphs for every prime power $q$, *J. Combin. Theory Ser. B* **62** (1994), 44–62.

[MotSt]  Motzkin T. S., Straus E. G., Maxima for graphs and a new proof of a theorem of Turán, *Canadian J. Math.* **17** (1965), 533–540.

[Mow]  Mowshowitz A., The adjacency matrix and the group of a graph, in *New Directions in the Theory of Graphs*, Proc. 3rd Ann Arbor Conf. on Graph Theory held at the University of Michigan, Oct. 21-23, 1971 (ed. Harary F.), Academic Press (New York) 1973, pp. 129–148.

[MuKl]  Muzychuk M., Klin M., On graphs with three eigenvalues, *Discrete Math.* **189** (1998), 191–207.

[Neu]  Neumaier A., The second largest eigenvalue of a tree, *Linear Algebra Appl.* **46** (1982), 9–25.

[NeuSe]  Neumaier A., Seidel J. J., Discrete hyperbolic geometry, *Combinatorica* **3** (1983), 219–237.

[Nik1]  Nikiforov V., Some inequalities for the largest eigenvalue of a graph, *Comb. Probab. Comput.* **11** (2002), 179–189.

[Nik2]  Nikiforov V., Walks and the spectral radius of graphs, *Linear Algebra Appl.* **418** (2006), 257–268.

[Nik3]  Nikiforov V., Graphs and matrices with maximal energy, *J. Math. Anal. Appl.* **327** (2007), 735–738.

[Nik4]  Nikiforov V., Chromatic number and spectral radius, *Linear Algebra Appl.* **426** (2007), 810–814.

[Nil]  Nilli A., On the second eigenvalue of a graph, *Discrete Math.* **91** (1991), 207–210.

[Nos]  Nosal E., On the number of spanning trees of finite graphs, University of Calgary Research paper No. 95, 1970.

[Nuff]  van Nuffelen C., On the rank of the incidence matrix of a graph, *Cahiers Centre Étud. Rech. Opér. (Bruxelles)* **15** (1973), 363–365.

[OLAH]  Oliveira C. S., de Lima L. S., de Abreu N. M. M., Hansen P., Bounds on the index of the signless Laplacian of a graph, to appear.

[OmTa]  Omidi G. R., Tajbakhsh K., Starlike trees are determined by their Laplacian spectrum, *Linear Algebra Appl.* **422** (2007), 654–658.

[Osh]  G. Oshikiri, Cheeger constant and connectivity of graphs, *Interdis. Inform. Sci.* **8** (2002), 147–150.

[Per]  Percus J. K., *Combinational Methods*, Springer-Verlag (New York), 1969.

[PeSa1]  Petersdorf M., Sachs H., Über Spektrum, Automorphismengruppe und Teiler eines Graphen, *Wiss. Z., T. H. Ilmenau*, **15** (1969), 123–128.

[PeSa2] Petersdorf M., Sachs H., Spektrum und Automorphismengruppe eines Graphen, in *Combinatorial Theory and Its Applications III* (eds. Erdös P., Rényi A., Sós V. T.), Bolyai Janos Mat. Tarsulat, Budapest, North-Holland (Amsterdam) 1970, pp. 891–907.

[Pet1] Petrović M., On graphs with exactly one eigenvalue less than $-1$, *J. Combin. Theory Ser. B* **52** (1991), 102–112.

[Pet2] Petrović M., On graphs whose second largest eigenvalue does not exceed $\sqrt{2} - 1$. *Univ. Beograd, Publ. Elektrotehn. Fak., Ser. Mat.* **4** (1993), 70–75.

[PetMi1] Petrović M., Milekić B., On the second largest eigenvalue of line graphs, *J. Graph Theory* **27** (1998), 61–66.

[PetMi2] Petrović M., Milekić B., On the second largest eigenvalue of generalized line graphs, *Publ. Inst. Math. (Beograd)* **68**(82) (2000), 37–45.

[PetRa] Petrović M., Radosavljević Z., *Spectrally Constrained Graphs*, Faculty of Science, University of Kragujevac, 2001.

[Plo] Plonka J., On Γ-regular graphs, *Colloq. Math.* **46** (1982), 131–134.

[Pow1] Powers D. L., Structure of a matrix according to its second eigenvector, in *Current Trends in Matrix Theory* (eds. Uhlig F., Grone R.), Elsevier (New York) 1987, pp. 261–266.

[Pow2] Powers D. L., Graph partitioning by eigenvectors, *Linear Algebra Appl.* **101** (1988), 121–133.

[Pow3] Powers D. L., Bounds on graph eigenvalues, *Linear Algebra Appl.* **117** (1989), 1–6.

[Pra] Prasolov V. V., *Problems and Theorems in Linear Algebra*, Amer. Math. Soc. (Providence), 1994.

[Ran] Randić M., On the characteristic equations of the characteristic polynomial, *SIAM J. Alg. Disc. Math.* **6** (1985), 145–162.

[RanKl] Randić M., Kleiner A. F., On the construction of endospectral graphs, in *Combinatorial Mathematics*, Proc. 3rd Int. Conf., New York (USA) 1985, *Ann. N. Y. Acad. Sci.* **555** (1989), pp. 320–331.

[RanTŽ] Randić M., Trinajstić N., Živković T., On molecular graphs having identical spectra, *J. C. S. Faraday II* **72** (1976), 244–256.

[RaoRa] Rao S. B., Rao A. R., A characterization of the line graph of a BIBD with $\lambda = 1$, *Sankhyā A* **31** (1969), 369–370.

[RaoSV] Rao S. B., Singhi N. M., Vijayan K. S., Spectral characterization of the line graph of $K_\ell^n$, in *Combinatorics and Graph Theory*, Proc. Symp. Indian Statistical Institute, Calcutta, 25–29 February 1980 (ed. Rao S. B.), Lecture Notes in Math. 885, Springer-Verlag (Berlin), 1981, pp. 473–480.

[ReCo] Read R. C., Corneil D. G., The graph isomorphism disease, *J. Graph Theory* **1** (1977), 339–363.

[RiMW] Rigby M. J., Mallion R. B., Waller D. A., On the quest for an isomorphism invariant which characterises finite chemical graphs, *Chem. Phys. Letters* **59** (1978), 316–320.

[Row1] Rowlinson P., On the number of simple eigenvalues of a graph, *Proc. Royal Soc. Edinburgh* **94A** (1983), 247–250.

[Row2] Rowlinson P., Certain 3-decompositions of complete graphs with an applications to finite fields, *Proc. Royal Soc. Edinburgh* **99A** (1985), 277–281.

[Row3] Rowlinson P., A deletion-contraction algorithm for the characteristic polynomial of a multigraph, *Proc. Royal Soc. Edinburgh* **105A** (1987), 153–160.

[Row4] Rowlinson P., On the maximal index of graphs with a prescribed number of edges, *Linear Algebra Appl.* **110** (1988), 43–53.

[Row5] Rowlinson P., On angles and perturbations of graphs, *Bull. London Math. Soc.* **20** (1988), 193–197.

[Row6] Rowlinson P., Graph perturbations, in *Surveys in Combinatorics 1991* (ed. Keedwell, A. D.), Cambridge University Press (Cambridge), 1991.

[Row7] Rowlinson P., The spectrum of a graph modified by the addition of a vertex, *Univ. Beograd, Publ. Elektrotehn. Fak. Ser. Mat.* **3** (1992), 67–70.

[Row8] Rowlinson P., Eutactic stars and graph spectra, in *Combinatorial and Graph-Thoeretical Problems in Linear Algebra* (eds. Brualdi R. A., Friedland S. and Klee V.), Springer-Verlag (New York) 1993, pp. 153–164.

[Row9] Rowlinson P., Dominating sets and eigenvalues of graphs, *Bull. London Math. Soc.* **26** (1994), 248–254.

[Row10] Rowlinson P., Star partitions and regularity in graphs, *Linear Algebra Appl.* **226–228** (1995), 247–265.

[Row11] Rowlinson P., The characteristic polynomials of modified graphs, *Discrete Applied Math.* **67** (1996), 209–219.

[Row12] Rowlinson P., Star sets in regular graphs, *J. Comb. Math. Comb. Computing* **34** (2000), 3–22.

[Row13] Rowlinson P., Star complements in finite graphs: a survey, *Rendiconti Sem. Mat. Messina* **8** (2002), 145–162.

[Row14] Rowlinson P., Star complements and the maximal exceptional graphs, *Publ. Inst. Math. (Beograd)* **76**(90) (2004), 25–30.

[Row15] Rowlinson P., Co-cliques and star complements in extremal strongly regular graphs, *Linear Algebra Appl.* **421** (2007), 157–162.

[Row16] Rowlinson P., The main eigenvalues of a graph: a survey, *Appl. Anal. Discrete Math.* **1** (2007), 455–471.

[RowJa] Rowlinson P., Jackson P. S., Star complements and switching in graphs, *Linear Algebra Appl.* **356** (2002), 145–156.

[RowSc] Rowlinson P., Sciriha I., Some properties of the Hoffman-Singleton graph, *Appl. Anal. Discrete Math.* **1** (2007), 438–445.

[Roy] Royle G. F., The rank of cographs, *Electronic J. Combin.* **10** (2003) Paper N11.

[Run] Runge F., *Beiträge zur Theorie der Spektren von Graphen und Hypergraphen*, Dissertation, TH Ilmenau, 1976.

[Rut] Rutherford D. E., Some continuant determinants arising in physics and chemistry, *Proc. Roy. Soc. Edinburgh* **62A** (1947), 229–236.

[Sac1] Sachs H., Abzählung von Wäldern eines gegebenen Typs in regulären und biregulären Graphen I, *Publ. Math. Debrecen* **11** (1964), 74–84.

[Sac2] Sachs H., Beziehungen zwischen den in einen Graphen enthalten Kreisen und seinem characterischen Polynom, *Publ. Math. Debrecen* **11** (1964), 119–134.

[Sac3] Sachs H., Über Teiler, Faktoren und charakteristische Polynome von Graphen, Teil II, *Wiss. Z., T. H. Ilmenau* **13** (1967), 405–412.

[SaSt] Sachs H., Stiebitz M., Automorphism group and spectrum of a graph, in *Algebraic Methods in Graph Theory* Vol. II (eds. Lovász, Sós V. T), North-Holland (Amsterdam) 1981, pp. 657–670.

[Sch1] Schwenk A.J., Almost all trees are cospectral, in *New Directions in the Theory of Graphs*, Proc. 3rd Ann Arbor Conf. on Graph Theory held at the University of Michigan, Oct. 21-23, 1971 (ed. Harary F.), Academic Press (New York) 1973, pp. 275–307.

[Sch2] Schwenk A. J., Computing the characteristic polynomial of a graph, in *Graphs and Combinatorics* (eds. Bari, R., Harary, F.), Springer-Verlag (New York), 1974.

[Sch3] Schwenk A. J., Spectral reconstruction problems, *Ann. N. Y. Acad. Sci.* **328** (1978), 183–189.

[Sch4] Schwenk A.J., Removal-cospectral sets of vertices in a graph, in *Proc. 10th South Eastern Conference on Combinatorics, Graph Theory and Computing, 1979, Congr. Num.* XXIII-XXIV, Utilitas Mathematica (Winnipeg, Manitoba), 1979, pp. 849–860.

[SchWi] Schwenk A. J., Wilson R. J., On the eigenvalues of a graph, in: *Selected Topics in Graph Theory* (eds. Beineke L. W., Wilson R. J.), Academic Press (New York) 1978, pp. 307–336.

[Sci1] Sciriha I., Polynomial reconstruction and terminal vertices, *Linear Algebra Appl.* **356** (2002), 145–156.

[Sci2] Sciriha I., Polynomial reconstruction: old and new techniques, *Rendiconti Sem. Mat. Messina* **8** (2002), 163–179.

[SciFo] Sciriha I., Formosa M. J., On polynomial reconstruction of disconnected graphs, *Utilitas Math.* **64** (2003), 33–44.

[Sco] Scott L.L., A condition on Higman's parameters, *Notices Amer. Math. Soc.* Abstract A97 (January 1973).

[Sed] Sedláček J., Über Inzidenzmatrizen gerichteter Graphen (Czech, Russian and German summaries), *Časopis Pěst. Mat.* **84** (1959), 303–316.

[Sei1] Seidel J. J., On two-graphs and Shult's characterization of symplectic and orthogonal geometries over $GF(2)$, T. H. Report 73-WSK-02, University of Eindhoven, 1973.

[Sei2] Seidel J. J., Graphs and two-graphs, in *Proc. 5th South Eastern Conf. on Combinatorics, Graph Theory and Computing, Boca Raton (Fla), 1974, Congr. Num.* X, Utilitas Mathematica (Winnipeg, Manitoba), 1974, pp. 125–143.

[Sei3] Seidel J. J., A survey of two-graphs, in *Proc. Coll. Theorie Combinatorie*, Acc. Naz. Lincei (Roma) 1976, pp. 481–511.

[Sei4] Seidel J. J., Strongly regular graphs, in *Surveys in Combinatorics* (ed. Bollobás B.), Cambridge University Press (Cambridge) 1979, pp. 157–180.

[She] Shee S. C., A note on the $C$-product of graphs, *Nanta Math.* **7** (1974), 105–108.

[Shi] Shier D., Maximizing the number of spanning trees in a graph with $n$ nodes and $m$ edges, *J. Res. Nat. Bureau Standards* **78B** (1974), 193–196.

[ShrBh] Shrikhande S. S., Bhagwandas, Duals of incomplete block designs, *J. Indian Stat. Assoc.* **3** (1965), 30–37.

[ShuHW] Shu J., Hong Y., Wen K. R., A sharp upper bound on the largest eigenvalue of the Laplacian matrix of a graph, *Linear Algebra Appl.* **347** (2002), 123–129.

[Sim1] Simić S. K., On the largest eigenvalue of unicyclic graphs, *Publ. Math. Inst. (Beograd)* **42**(56) (1987), 13–19.

[Sim2] Simić S. K., Some results on the largest eigenvalue of a graph, *Ars Combinatoria* **24A** (1987), 211–219.

[Sim3] Simić S. K., On the largest eigenvalue of bicyclic graphs, *Publ. Math. Inst. (Beograd)* **46**(60) (1989), 1–6.

[Sim4] Simić S. K., An algorithm to recognize a generalized line graph and its root graph, *Publ. Inst. Math. (Beograd)* **49**(63) (1990), 21–26.

[Sim5] Simić S. K., A note on reconstructing the characteristic polynomial of a graph, in: Proc. 4th Czechoslovakian Symposium on Combinatorics, Graphs and Complexity, (eds. Nešetřil J., Fiedler M.), North-Holland (Amsterdam) 1992, pp. 315–319.

[Sim6] Simić S. K., Some notes on graphs whose second largest eigenvalue is less than $(\sqrt{5} - 1)/2$, *Linear Multilin. Algebra* **39** (1995), 59–71.

[Sim7] Simić S. K, Complementary pairs of graphs with the second largest eigenvalue not exceeding $(\sqrt{5} - 1)/2$, *Publ. Inst. Math. (Beograd)* **57**(71) (1995), 179–188.

[Sim8] Simić S., Arbitrarily large graphs whose second largest eigenvalue is less than $(\sqrt{5} - 1)/2$, *Rendiconti Sem. Mat. Messina* **8** (2002), 181–205.

[SimKo] Simić S., Kocić V., On the largest eigenvalue of some homeomorphic graphs, *Publ. Inst. Math. (Beograd)* **40**(54) (1986), 3–9.

[SimMB] Simić S. K., Li Marzi E. M., Belardo F., Connected graphs of fixed order and size with maximal index: structural considerations, *Le Matematiche* **59** (2004), 349–365.

[SimRa] Simić S. K., Radosavljević Z., The nonregular, nonbipartite, integral graphs with maximum degree four, *J. Combin. Inform. System Sci.* **20** (1995), 9–26.

[SimSta1] Simić S. K., Stanić Z., The polynomial reconstruction of unicyclic graphs is unique, *Linear Multilin. Algebra* **55** (2007), 35–43.

[SimSta2] Simić S. K., Stanić Z., The polynomial reconstruction is unique for the graphs whose deck-spectra are bounded from below by $-2$, *Linear Algebra Appl.* **428** (2008), 1865–1873.

[SimSta3] Simić S. K., Stanić Z., $Q$-integral graphs with edge-degrees at most five, *Discrete Math.* **308** (2008), 4625–4634.

[SimSte] Simić S., Stevanović D., Two shorter proofs in spectral graph theory, *Univ. Beograd, Publ. Elektrotehn. Fak. Ser. Mat.* **14** (2003), 94–98.

[SimmMe] Simmons H. E., Merrifield R. E., Paraspectral molecular pairs, *Chem. Phys. Letters* **62** (1979), 235–237.

[SinJe] Sinclair A., Jerrum M., Approximate counting, uniform generation and rapidly mixing Markov chains, *Inform. Computing* **82** (1989), 93–133.

[Smi] Smith J. H., Some properties of the spectrum of a graph, in *Combinatorial Structures and Their Applications* (eds. Guy R., Hanani H., Sauer N., Schönheim J.), Gordon and Breach (New York) 1970, pp. 403–406.

[So] So W., Rank one perturbations and its application to the Laplacian spectrum of a graph, *Linear Multilin. Algebra* **46** (1999), 193–198.

[Spia] Spialter, L., The atom connectivity matrix characteristic polynomial (ACMCP) and its physico-geometric (topological) significance, *J. Chem. Doc.* **4** (1964), 269–274.

[Sta] Stanić Z., There are exactly 172 connected integral graphs up to 10 vertices, *Novi Sad J. Math.* **37** (2007), 193–205.

[Stan] Stanley, R. P., A bound on the spectral radius of graphs with $e$ edges, *Linear Algebra Appl.* **67** (1987), 267–269.

[Ste1] Stevanović D., Nonexistence of some 4-regular integral graphs, *Univ. Beograd, Publ. Elektrotehn. Fak., Ser. Mat.* **10** (1999), 81–86.

[Ste2] Stevanović D., 4-regular integral graphs avoiding ±3 in the spectrum *Univ. Beograd, Publ. Elektrotehn. Fak., Ser. Mat.* **14** (2003), 99–110.

[Ste3] Stevanović D., The largest eigenvalue of nonregular graphs, *J. Comb. Theory Ser. B* **91** (2004), 143–146.

[SteAFD] Stevanović D., de Abreu N. M. M, de Freitas M. A. A., Del-Vecchio R., Walks and regular integral graphs, *Linear Algebra Appl.* **423** (2007), 119–135.

[StewMa] Stewartson K., Maechter R. T., On hearing the shape of a drum – further results, *Proc. Cambridge Phil. Soc.* **69** (1971), 353–363.

[Str] Strang G., *Linear Algebra and its Applications*, 4th edn., Thomson (Belmont), 2006.

[Tan] Tanner R. M., Explicit concentrators from generalized $n$-gons, *SIAM J. Alg. Disc. Meth.* **5** (1984), 287–293.

[Tre] Trent H. M., A note on the enumeration and listing of all possible trees in a connected linear graph, *Proc. Nat. Acad. Sci. USA* **40** (1954), 1004–1007.

[Tri] Trinajstić, N. *Chemical Graph Theory*, CRC Press (Boca Raton), 1983; 2nd revised edn., 1993.

[Tur1] Turner J., Point-symmetric graphs with a prime number of vertices, *J. Combin. Theory* **3** (1967), 136–145.

[Tur2] Turner J., Generalized matrix functions and the graph isomorphism problem, *SIAM J. Appl. Math.* **16** (1968), 520–526.

[Tut1] Tutte W. T., Lectures on matroids, *J. Res. Nat. Bur. Standards Sect. B* **69** (1965), 1–47.

[Tut2] Tutte W. T., All the king's horses: a guide to reconstruction, in *Graph Theory and Related Topics*, Proc. Conf. University of Waterloo, Waterloo, Ont. 1977, (eds. Bondy J. A., Murty U. S. R.), Academic Press (New York) 1979, pp. 15–33.

[Vah] Vahovskiĭ E. V., On the eigenvalues of the neighbourhood matrix of simple graphs (Russian), *Sibir. Mat. J.* **6** (1965), 44–49.

[VanBo] Vandenberghe L., Boyd S., Semidefinite programming, *SIAM Review* **38** (1996), 49–95.

[Wang] Wang J. F., A proof of Boesch's conjecture, *Networks* **24** (1994), 277–284.

[Wan1] Wang L. G., A survey of results on integral trees and integral graphs, Memorandum No. 1763, Twente (2005), 1–22, ISSN 0169-2690.

[Wan2] Wang L. G., Integral trees and integral graphs, Wöhrmann Print Service, Enschede (2005), ISBN 90-365-2177-7.

[WaXu] Wang W., Xu C-X., The $T$-shape tree is determined by its Laplacian spectrum, *Linear Algebra Appl.* **419** (2006), 78–81.

[Wat] Watanabe M., Note on integral trees, *Math. Rep. Toyama Univ.* **2** (1979), 95–100

[Wei] Wei T. H., *The Algebraic Foundations of Ranking Theory*, Ph.D. thesis, University of Cambridge, 1952.

[Wes] West D. B., *Introduction to Graph Theory*, Prentice Hall, 1996.

[Whi] Whitney H., Congruent graphs and the connectivity of graphs, *Amer. J. Math.* **54** (1932), 150–168.

[Wilf] Wilf H. S., Spectral bounds for the clique and independence numbers of graphs, *J. Combin. Theory Ser. B* **40** (1986), 113–117.

[Wils] Wilson R. M., Decomposition of complete graphs into subgraphs isomorphic to a given graph, in *Proc. 5th British Combinatorial Conf., 1975* (ed. Nash-Williams C. St J. A., Sheehan J.), *Congr. Num. XV*, Utilitas Mathematica (Winnipeg, Manitoba), 1976, pp. 647–659.

[Witt] Witt E., Über Steinersche Systeme, *Abh. Math. Sem. Hamburg* **12** (1938), 265–275.

[WoNe1] Woo R., Neumaier A., On graphs whose smallest eigenvalue is at least $-1 - \sqrt{2}$, *Linear Algebra Appl.* **226–228** (1995), 577–591.

[WoNe2] Woo R., Neumaier A., On graphs whose spectral radius is bounded by $\frac{3}{2}\sqrt{2}$, *Graph. Combinator.* **23** (2007), 1–14.

[Yan] Yan W., On the minimal energy of trees with a given diameter, *Appl. Math. Lett.* **18** (2005), 1046–1052.

[Yong] Yong X., On the distribution of eigenvalues of a simple undirected graph, *Linear Algebra Appl.* **295** (1999), 73–80.

[ZhaZZ] Zhang F. J., Zhang Z. N., Zhang Y. H., Some theorems about the largest eigenvalue of graphs (Chinese, English Summary), *J. Xinjiang Univ. Nat. Sci.* **3** (1984), 84–90.

[Zha] Zhang X.-D., Two sharp upper bounds for the Laplacian eigenvalues, *Linear Algebra Appl.* **376** (2004), 207–213.

[ZhWi] Zhu P., Wilson R. C., A study of graph spectra for comparing graphs *Pattern Recognition* **41** (2008) No. 9, 2833–2841.

[ZiTR] Živković T., Trinajstić N., Randić M., On conjugated molecules with identical topological spectra, *Mol. Phys.* **30** (1975), 517–533.

# Index of symbols

# Index of terms

absolute bound, 75
adjacency algebra, 69
algebraic connectivity, 197
algorithm
  reconstruction, 130
amenable to switching, 160
angle, 14
  main, 15
automorphism, 7

$B$-graph, 8
Bannai-Ito conjecture, 80
binary number, 124
Binet-Cauchy Theorem, 20
bipartition width, 200
blossom, 8
boundary, 199
branch, 122

carbon skeleton, 267
Cauchy's inequalities, 18
Cayley's formula, 191
characteristic polynomial, 1
  conjugate, 254
Chebyshev polynomial
  first kind, 46
  second kind, 47
chessboard, 181
chromatic number, 90
circuit, 264
claw, 6
clique, 6
clique number, 88
coalescence, 30

coclique, 6
cograph, 253
colouration, 83
colouring, 90
complexity of a graph, 189
conductance, 205
cone, 7
conjugated hydrocarbon, 266
connectivity
  edge, 198
  vertex, 198
corona, 32
cospectral graphs, 4
cospectral mates, 118
Coulomb integral, 267
Courant-Weyl inequalities, 19
cut, 199
cycle
  Hamiltonian, 7

deck, 250
  polynomial, 250
decomposition, 162
degree sequence, 131
deletion-contraction, 36
design, 7
  line graph of, 116
  symmetric, 7
diameter, 59
dimer, 263
distance matrix, 78
divisor, 83
dominating set, 7
double cone, 7